DEGRADATION AND STABILISATION OF PVC

DEGRADATION AND STABILISATION OF PVC

Edited by

E. D. OWEN
Department of Chemistry, University College, Cardiff, Wales, UK

ELSEVIER APPLIED SCIENCE PUBLISHERS
LONDON and NEW YORK

ELSEVIER APPLIED SCIENCE PUBLISHERS LTD
Ripple Road, Barking, Essex, England

Sole Distributor in the USA and Canada
ELSEVIER SCIENCE PUBLISHING CO., INC.
52 Vanderbilt Avenue, New York, NY 10017, USA

British Library Cataloguing in Publication Data

Degradation and stabilisation of PVC.
1. Polyvinyl chloride
I. Owen, E.D.
668.4′36 TP1180.V48

ISBN-13: 978-94-010-8978-4 e-ISBN-13: 978-94-009-5618-6
DOI: 10.1007/978-94-009-5618-6

WITH 34 TABLES AND 96 ILLUSTRATIONS

Softcover reprint of the hardcover 1st edition 1984

PREFACE

Polyvinyl chloride has played a key role in the development of the plastics industry over the past 40 years and continues to be a polymer of major importance. The reasons for its enormous versatility and range of application derive from a combination of the basic structure which gives rise to a relatively tough and rigid material and its ability to accept a range of plasticisers and other additives which can modify its physical characteristics to produce a range of flexible products. Two major problems, however, have tested the skill and ingenuity of PVC technologists since earliest times. One is the thermal instability of the material at the temperatures required for melt processing and fabrication, and the second is the photochemical instability which until recently has limited the potentially large range of outdoor applications. Both problems have been handled in a commercially satisfactory way by the gradual development of a range of stabilisers, lubricants and other processing aids and the high quality material which has resulted has led to massive utilisation of PVC by industry.

Totally adequate stabilisation requires a detailed understanding of the mechanisms by which degradation processes are initiated and propagated. Although great advances have been made in this respect in recent years the problem remains incompletely understood. This book presents an account of the present position and the problems which remain to be solved. Modern spectroscopic techniques which have played a vital role in the recent progress are described and stabilisation of the polymer by the use of processing aids is included. The book should be of interest to those concerned with any aspect of PVC structure, processing or development whether at the production or research level.

E. D. Owen

CONTENTS

LIST OF CONTRIBUTORS

CHRISTIAN DECKER

Laboratoire de Photochimie Générale associé au CNRS, Ecole Nationale Supérieure de Chimie, 68093 Mulhouse, France

T. HJERTBERG

The Polymer Group, Department of Polymer Technology, Chalmers University of Technology, S-412 96 Gothenburg, Sweden

JOHN T. LUTZ JR

Rohm and Haas Company, Research Laboratories, PO Box 219, Bristol, Pennsylvania 19007, USA

W. F. MADDAMS

The British Petroleum Company plc, BP Research Centre, Chertsey Road, Sunbury-on-Thames, Middlesex TW16 7LN, UK

E. D. OWEN

Department of Chemistry, University College, PO Box 78, Cardiff CF1 1XL, Wales, UK

E. M. SÖRVIK

The Polymer Group, Department of Polymer Technology, Chalmers University of Technology, S-412 96 Gothenburg, Sweden

D. A. TESTER

Imperial Chemical Industries plc, Petrochemicals and Plastics Division, Bessemer Road, Welwyn Garden City, Herts AL7 1HD, UK. Present address: *3 Valley Green, Welwyn Garden City, Herts AL8 7DQ, UK*

Chapter 1

THE ROLE OF PVC IN THE MARKET PLACE

D. A. TESTER†

Imperial Chemical Industries plc,
Petrochemicals and Plastics Division, Welwyn Garden City, UK

1.1 INTRODUCTION

PVC found its first industrial applications before the Second World War and has played an important role in the plastics industry for the last forty years. The fact that we are still striving to increase our understanding of the degradation and stabilisation of PVC reflects both the intrinsic scientific interest in this subject, and the continuing industrial importance of the polymer. It is the intention in this introductory chapter to indicate, albeit briefly, the historic and continuing role of PVC in the plastics industry, and the vital importance that the control of degradation has played in developing the major applications.

The almost bewildering variety of PVC applications derives from a combination of properties unique among the major polymers. The structure of the polymer, with a chlorine atom situated on each alternate carbon of the polymer chain, leads to a very rigid and relatively tough plastic, in the unplasticised state. In the presence of a plasticiser, however, the dipole bonding between polymer chains is much reduced, leading to increased freedom of chain movement and thus to a flexible material. By suitable choice of plasticiser level and type, a whole range of flexible products can be produced.

The chain structure which provides the remarkable versatility of PVC as a commercial plastic is also responsible for processing problems which have presented a considerable challenge to the industry. In particular, the

† Present address: 3 Valley Green, Welwyn Garden City, Herts AL8 7DQ, UK.

polymer is thermally unstable at temperatures required for melt processing and fabrication, especially in unplasticised compositions, so that heat stabilisers must be incorporated. The development, over the years, of a wide range of effective heat stabilisers, melt lubricants and processing aids, has been a vital ingredient in the exploitation of PVC. Whilst the earliest advances were highly empirical, they prompted intensive studies of the thermal and melt behaviour of PVC, and of the role of additives. These studies, in turn, led the way to improved processes and formulations, and to the massive utilisation of PVC by industry today. Advances in PVC technology continue to be made and the complexities of the polymer's behaviour, particularly in the area of degradation and stabilisation, continue to invite fundamental study, as witnessed by this volume.

1.2 HISTORICAL PERSPECTIVE

The preparation of vinyl chloride was first recorded in 1835 by Regnault and in 1872 Baumann described the preparation of polyvinyl chloride from the monomer (though it is not clear whether he understood the nature of the process). The industrial origins of PVC have been traced in an excellent study by Morris Kaufmann,[1] who considered that the foundations of the PVC industry were provided by the work of Fritz Klatte, of the German company Griesheim Elektron. In 1912, Klatte patented a process for the production of vinyl chloride by the reaction of hydrogen chloride and acetylene, the route to the monomer which was later to be exploited in the early years of industrial manufacture. Commercial production of PVC did not really begin, however, until 1931 in Germany and 1933 in the USA.

The earliest commercial applications of PVC utilised the flexible, plasticised form, partly because of the greater ease of processing plasticised compositions, but also because a major objective was to find synthetic substitutes for rubber. This was particularly so in Germany, where it was the policy, in the years leading up to the Second World War, to become less dependent on imported natural rubber. Plasticised PVC was seized upon as an alternative cable insulation, and the cable industry is still a major consumer of PVC world-wide. The cornerstone of the flexible PVC industry may be fairly said to have been the work of Waldo Semen, at B. F. Goodrich, in 1932–1935. During this period a variety of plasticising additives were evaluated and patented in the USA and in Germany, including the phthalate esters, which remain the most popular general purpose plasticisers. The Second World War saw a greatly increased interest in PVC manufacture, not least in the UK, once the Japanese invasion of Malaya had cut off a major source of natural rubber. By the end of the war, flexible PVC was firmly established in the cable industry and finding use in a variety of other applications, albeit in compositions whose performance would be considered very inferior by the more demanding standards of today.

Rigid applications developed more slowly, largely because of the greater difficulty in processing unplasticised compositions, which were found to degrade readily at the melt temperature required. In this context, Morris Kaufmann[1] relates the alarming experience of German workers in the early 1930s. It appears that Deutsche Celluloid Fabrik tried to adapt celluloid processing equipment to PVC, but the polymer decomposed so vigorously that the workers had to wear gas masks to cope with the hydrochloric acid evolved! The key to safer processing of PVC, and especially unplasticised PVC, lay in the development of heat stabilisers. German technologists recognised, quite early, the possibilities for prestabilisation of the polymer by additions at the polymerisation stage. It was in the USA, however, that the most crucial developments occurred, in finding effective heat stabilisers for addition to compositions at the processing/fabricating stage.

1.2.1 The role of heat stabilisers

Since the degradation of PVC is always associated with evolution of hydrogen chloride, the earliest attempts to stabilise compositions (as with prestabilisation) involved the addition of mild alkalis or organic bases to react with the acid formed. However, it was soon realised that all basic additives were not equally effective (some actually appeared to accelerate degradation) and that stabilisation was more complex than simply reacting with hydrogen chloride. By 1934, US patent literature was claiming the effectiveness of lead salts and metal soaps, groups of additives which still fill a dominant role in PVC stabilisation. In 1936, there was mention of tin alkyls, to be followed by more complex organotin compounds, such as dialkyltin laurates and maleates, which were to be the basis of the very important class of tin stabilisers in growing use today.

The primary function of these additives was to prevent the colour development, usually attributed to formation of conjugated unsaturation in polymer chains, which occurred so readily during high temperature processing. Technologists came increasingly to appreciate, however, that 'stabilisation' covers a range of differing, though interrelated, phenomena. In addition to the suppression of initial colour development, there is a need to minimise the effects of partial stagnation of melt within processing equipment, to allow equipment to be run as long as possible between cleaning sessions. There is also, in many operations, a need to accept 're-work' of scrap material, without undue colour formation from this second processing of some of the polymer. In terms of laboratory tests of colour formation, these processing requirements have been found to correlate more with longer term heat stability, rather than the initial appearance of colour. There is also concern, in some applications, for the long term stability of appearance and mechanical properties, particularly during external weathering. Whilst photodegradation has been clearly distinguished from thermal decomposition, the interrelation of these phenomena is demonstrated by the common observation that the sensitivity of PVC artefacts to discoloration during outside exposure is dependent on the previous heat history during processing and fabrication. Combinations of additives have therefore been developed to satisfy a range of stability requirements, whilst numerous extraneous considerations also dictate the actual choice of stabiliser. The most obvious differentiation is between opaque and transparent products. For example, for the former, solid lead stabilisers may be chosen, whilst the latter may require organotin or liquid barium/cadmium stabiliser systems. The application of PVC in both rigid and flexible food packaging demands the choice of non-toxic stabiliser systems, commonly based on the fatty acids of calcium and zinc. Physical form, compatibility with other formulation ingredients, toxicity status, and, of course, cost effectiveness, can all play a part in determining the particular stabilisers selected by the PVC processor from the wide range provided by the additive manufacturers.

1.3 THE PVC MARKET

1.3.1 Current consumption

The success of the plastics industry in exploiting the inherent advantages of PVC, and in overcoming problems in processing and stabilisation, is amply demonstrated by the very considerable tonnages consumed throughout the world. Consumption in the USA has averaged between two and three million tonnes per annum in recent years, whilst the consumption in Western Europe in 1981, a year of recession, was estimated to be 3·3 million tonnes.[2] The consumption of PVC in Western Europe in 1980 (including the UK) split into broadly defined application areas, is shown in Table 1.1. Consumption in 1981 and 1982 is expected to be very different in the pattern of usage.

Table 1.1 West European consumption of PVC (1980)[2]

Application	*Tonnes × 10^3*
Bottles	258
Rigid foil and sheet	369
Rigid injection moulding	84
Pipe and conduit	908
Rigid profiles	560
Records	75
Miscellaneous	69
Total rigids	2 323
Coated fabrics and paper	189
Flexible film and sheet	358
Flooring	199
Flexibles profiles and tubing	145
Cables	352
Miscellaneous	43
Total flexibles	1 286
Other unspecified	141
Total consumption	3 750

It is evident from Table 1.1 that rigid applications account for over 60% of the total usage, having grown considerably in the last two decades, whilst some of the flexible applications which were the earliest to be exploited have shown only modest growth rates. The biggest single application area is clearly that of pipe and conduit extrusion, primarily made up of domestic drainage and sewage pipe, land drainage, waste pipe, potable water pipe, rainwater goods and conduit for electric cables. The second largest application area is 'rigid profiles', which covers a wide variety of products, but is dominated by building applications, and in Germany, by the large volume of PVC window frame manufacture. At the time of writing, economic recession has affected the building industry in Europe particularly harshly, so that demand for PVC pipes and profiles, depending heavily on building activity, is at best stagnant. Revival of the Western European economies should see a resumption of significant growth for these dominant application areas, though not generally to the very high growth rates of past years.

The PVC market in the UK shows many similarities but certain significant differences from that of Western Europe as a whole. It must also be noted that the per capita consumption of PVC in Britain is lower than that of other major European communities, notably France and Germany, and lower than in the USA. In 1981 the figures were 10·6 kg/head in the USA, as compared with 9·7 kg/head for Western Europe as a whole, and 6·5 kg/head in the UK. To some extent this reflects a more conservative attitude here to the use of plastics, and to a lower rate of development of new products and industries. The figures are also influenced, however, by specific national differences, such as the large PVC water bottle market in France and Benelux, and the highly developed PVC window frame industry in Western Germany. These points will be considered further under the heading of particular applications. UK consumption, by application area, is shown in Table 1.2, in comparison with the figures for 1976.

It is seen from Table 1.2 that total consumption in 1981, at 385 000 tonnes, was lower, as might be expected, than the 421 000 tonnes for 1976. In spite of the recessionary influences on the economy, and building construction in particular, the consumption in rigid applications actually increased slightly, from 176 000 tonnes to 180 000 tonnes, and from 42 % to 47 % of total consumption. Among flexible applications, the largest percentage decreases in consumption occurred in fabric coating,

Table 1.2 UK consumption of PVC for 1981 compared with 1976[3]

Application	*Consumption in tonnes × 10³*	
	1981	*1976*
Rigid pipe (and fittings)	100	95
Rigid profile extrusion	23	20
Rigid sheet and foil	19	24
Bottles	22	15
Records	16	22
Total rigid	180	176
Cables and other flexible extrusions	62	70
Flexible film and sheet	37	47
Fabric coating	7	14
Paper coating	18	19
Flooring (solid and foamed)	23	31
Footwear	9	14
Miscellaneous flexibles (including conveyor belting, cap sealing, toys, and surgical gloves)	49	50
Total flexibles	205	245
Grand total	385	421

footwear and in traditional solid (as distinct from foamed) PVC flooring. This indicates the continuing trend towards a higher proportion of consumption in rigids, due to the faster growth of newer rigid applications, as against slower growths or actual reduction in some of the oldest flexible markets.

Although the UK application areas have been defined rather differently, comparison of Table 1.2 with the Western Europe table (Table 1.1) immediately indicates salient similarities and differences. Firstly, the proportion of consumption in rigid applications, though evidently growing, is still appreciably lower in the UK than in Western Europe as a whole. In both Tables 1.1 and 1.2, rigid pipe and conduit constitute the single biggest application area. The largest difference is in the proportion of consumption in the rigid profile extrusion area, which amounts to over 15 % for Western Europe, but only 6 % for the UK. In both Tables 1.1 and 1.2, cables and flexible film and sheet will be seen to be the largest flexible application areas. Gramophone records at 4 % of total consumption constitute quite an important outlet for PVC in the UK, indicating the relatively high level of record manufacture and demand, compared with 2 % of total consumption for Western Europe. Whether this level of demand will survive the increasing popularity of tape cassettes, and the introduction of video systems, remains to be seen.

1.3.2 The position of PVC among the bulk thermoplastics

The relative importance of PVC among the bulk thermoplastics, in tonnage terms, is illustrated for Western European consumption in 1981 in Table 1.3.

From Table 1.3, the highest tonnage plastic is seen to be low density polyethylene, followed very closely by PVC, with considerably lower

Table 1.3 Consumption of major thermoplastics in Western Europe in 1981[4]

Polymer	*Tonnes* × 10^3
Low density polyethylene	3 783
High density polyethylene	1 394
Polypropylene	1 327
Polystyrene	1 640
PVC	3 607
ABS	336
Total	12 087

consumptions for the other materials. Whilst there is, of course, competition between commodity plastics, each tends to dominate the particular application areas to which it is specially suited. Thus, a high proportion of LDPE is utilised in film products for packaging, horticulture, etc., whilst a dominant application for PVC is the extrusion of rigid pipe, sections, and profiles. The newest entry to the list of high tonnage thermoplastics is polypropylene, which is expected to continue a higher than average growth of consumption, especially in packaging and for rigid injection moulded parts, particularly in the automotive industry. As will be seen later, current estimates do not indicate any considerable change in the relative consumption of these commodity plastics during this decade.

1.4 SOME MAJOR APPLICATIONS

1.4.1 Rigid pipe and conduit

As already indicated, this is the single largest application in the UK and in Western Europe as a whole. Penetration of the market has generally been at the expense of established traditional materials, but in the case of rainwater goods (guttering and down pipes) the traditional material, cast iron, has been virtually displaced, at least in the UK. The earliest utilisation of PVC in pressure pipe for potable water was in small diameters, but it is now commonplace to extrude pipe for this exacting duty in diameters up to 600 mm. In drainage, PVC is generally in competition with clay, a basically cheap raw material, and growth of market penetration in any particular environment depends critically on the relative (installed) cost of the two systems.

In the UK, and in the majority of European countries, solid lead compounds are most usually employed as heat stabilisers for pipe compositions, though in the USA organotin stabilisers are more usual. The prime requirement is to provide protection during the extrusion process, since the precise control of product colour or subsequent weathering performance are not generally of crucial importance. An adjunct to the extrusion of pipe is the injection moulding of the various ancillary fittings, which can constitute quite large and intricate mouldings, as illustrated in Fig. 1.1. The rigid injection moulding process is one of the most demanding, in terms of requiring correct lubrication and stabilisation of the composition, but with the appropriate choice of equipment, polymer grade, and additives, the thermal and viscous properties of the melt need not present any serious problem.

Fig. 1.1 Injection moulded fittings for PVC pipes. (Courtesy ICI Petrochemicals and Plastics Division.)

In some pipe applications, there is a limited degree of competition between PVC and polyethylene or polypropylene, but the superior stiffness of PVC, together with toughness and good chemical resistance, guarantee a major role in plastic pipe manufacture.

1.4.2 Rigid profiles and sections

The extrusion of rigid sections, other than pipes, covers a multitude of products, but the most important, in tonnage terms, are all concerned with building construction. These applications include external cladding, window frames and sills, patio doors, architraves, skirting boards, and ranch fencing. A very evident application in the UK is the cladding of houses with hollow or foamed sections, resembling painted weather boards, but maintaining their attractive appearance without the need for decorating or costly maintenance. An extensive use of tongued and grooved PVC cladding sections is illustrated in Fig. 1.2. These houses were exhibited by Guildway Limited at an Ideal Home Exhibition, but are typical, in respect of their use of PVC cladding, of homes to be seen throughout this country.

Fig. 1.2 An example of PVC cladding in domestic building. (Courtesy ICI Petrochemicals and Plastics Division.)

It has already been pointed out that rigid profile extrusion represents a smaller proportion of PVC consumption in the UK than in Western Europe generally, certainly less than in West Germany. The single largest application for PVC profiles in West Germany is the construction of window frames, which has consumed up to 100 000 tonnes/annum in recent years. The long history of PVC in West Germany, a high rate of building construction, the building techniques employed, and more recently, financial encouragement for the installation of double-glazing, all combined to provide a rapid growth of the PVC window frame market. The continuing good performance of PVC window frames installed 15–20 years ago has reinforced confidence in their durability. In the UK, by contrast, the market for PVC window frames is very small, but is now beginning to grow quite rapidly. In the main, PVC window frames have so far been installed as replacements for softwood windows that had deteriorated badly, particularly where there was also a desire to install double-glazing. Initially, the favoured material in this market was aluminium, but the superior thermal insulation properties and good appearance of the PVC systems have won PVC a growing share of the market. PVC window frames

Fig. 1.3 Window frames and patio door constructed in rigid PVC. (Courtesy ICI Petrochemicals and Plastics Division.)

are also beginning to be chosen in the construction of new homes, where the requirement is for high quality double-glazed windows, requiring the minimum of future maintenance and decoration. It is confidently predicted that this PVC application will continue to grow in the UK, in other European countries and in the USA. Attractive designs, tailored to British requirements, are now appearing in this country, as illustrated by the PVC window frame and patio door shown in Fig. 1.3, manufactured by Marley Extrusions.

The requirements of window frame profiles, in terms of stability, are of course very demanding. Precise control of the colour and surface quality of the extrudates is essential, and any subsequent discoloration during long term weathering must be minimal, and consistent. Whilst successful performance in some environments is claimed for coloured window frames, there is greater experience and assurance of good long term stability for white frames, which are also the most popular choice, aesthetically, in the UK market. A further requirement is for high impact strength, generally conferred by incorporation of a rubbery impact modifier, either at the polymerisation stage, or as a separate additive in the window frame formulation. To avoid oxidative degradation of the impact modifier during service, with subsequent loss of impact strength, it is necessary to choose good weathering-modifier systems, and particularly to avoid the residual unsaturation of the butadiene copolymer rubbers.

The heat stabiliser systems employed in European window frame compositions, on the basis of proven experience, are generally based on solid barium/cadmium stabilisers, or lead salts, or combinations of these. In the USA, however, following the practice with cladding profiles, the use of organotin stabilisers, with high levels of titanium dioxide to protect against UV degradation, is likely to dominate the development of window profile compositions.

1.4.3 Packaging applications

PVC packaging embraces both rigid and flexible compositions and a variety of fabricating techniques, so that packaging applications fall within several of the general categories already listed, as well as 'bottles'. The combined packaging market ranks in tonnage terms among the major PVC uses, and the packaging of foodstuffs alone has been conservatively estimated to consume about 300 000 tonnes/annum of PVC in Western Europe.[5]

(1) Bottles

The use of PVC bottles will be familiar to most readers, since PVC is employed so extensively for bottling fruit squashes, edible oils and vinegar, as well as a wide range of toiletries and domestic products. The UK usage pales into insignificance, however, compared with the 150 000 tonnes/annum of PVC consumed by bottle manufacture in France, predominantly for the packaging of mineral water. The market for bottled water in the UK is miniscule, by comparison, though it appears to be growing here, and mineral water in PVC bottles is to be found on our supermarket shelves. The uses of PVC in packaging, and particularly the wide variety of PVC bottles, are illustrated in Fig. 1.4.

The manufacture of PVC bottles makes particular demands on the fundamental heat stability of the polymer, and on the heat stabilisers employed in formulations. The longer term heat stability must be sufficient to allow the blow moulding operation to proceed in long production runs without excessive down time for cleaning. At the same time, there must be minimal development of colour in the bottle, especially when it is designed to give a glass-clear 'water white' appearance. Further, the chosen stabiliser

Fig. 1.4 A range of bottles blow moulded in PVC. (Courtesy ICI Petrochemicals and Plastics Division.)

system must be toxicologically acceptable when foodstuffs are to be packaged, whilst the organoleptic properties must also be very good, particularly if the bottle is designed to hold mineral water, in which the slightest suggestion of taint is wholly unacceptable. The two types of heat stabiliser system used for bottle manufacture are based either on organotin compounds or on calcium/zinc salts. Some organotin compounds have food grade toxicity status, among which the octyl tin stabilisers are very widely used. Calcium/zinc systems are now available, however, which are capable of approaching the excellent sparkle, clarity, and colour obtained with octyl tin stabilisers.

Since 1976, PVC has shown a progressive price advantage over glass in bottle manufacture, and it is confidently expected that the use of PVC bottles will continue to grow, at the expense of glass. In some bottle applications, PVC is in competition with other polymers, and in this context, particular mention must be made of polyethylene terephthalate (PET). Through orientation during bottle manufacture, PET can give a superior stiffness/weight ratio, with excellent clarity and good barrier properties. Feedstock costs are still higher than with PVC, however, and likely to remain so, at least in the short to medium term. At present, the two polymers may be considered to a large extent to be complementary, with PET bottles being used for applications such as highly carbonated drinks, where PVC is not really suitable in any case, but PVC generally holding the many applications where it is completely acceptable in performance.

(*2*) *Rigid thermoformed foil*

The PVC employed for foil manufacture may be a homopolymer of relatively low molecular weight or a similar polymer containing around 10% of copolymerised vinyl acetate, to assist flow in the production of deeper thermoformings. The foil for thermoformings, typically in the thickness range of 125–375 μm, can be produced by either calendering or direct foil extrusion. Applications include the manufacture of blister packs, where the toughness and clarity of the PVC foil provides an attractive but pilfer-proof presentation at the point of sale. In food packaging, thermoformed foils are widely used for nestings in cake, biscuit and confectionery packaging, and for some dairy products. For food contact applications, only non-toxic heat stabilisers and other formulation additives are employed, as with bottles. The UK consumption of PVC in thermoformed foil applications was estimated to be around 9000 tonnes in 1981, whilst West German consumption reached the much higher level of 40 000–50 000 tonnes. It must be noted, however, that a considerable tonnage of thermoforming foil is imported into the UK from German and Italian sources.

(*3*) *Cling film*

This thin film (typically 10–20 μm) finds major application in the packaging of foodstuffs, especially red meat, poultry, fruit, vegetables and cheese. The film is plasticised, and the level and balance of plasticisers are carefully chosen to achieve the appropriate transmission of oxygen, carbon dioxide and water vapour, to allow the packaged foodstuff to 'breathe' but to be adequately protected. In addition to providing controlled barrier properties, the PVC film enhances the presentation of the product whilst enabling it to be readily inspected. With the emphasis on food contact, calcium/zinc stabilisers are used, boosted by epoxidised soya bean oil, which acts as both a heat stabiliser and a secondary plasticiser.

The consumption of PVC in cling film is currently around 10 000 tonnes/annum in the UK, a modest tonnage, but corresponding to a very large area of thin film! In Tables 1.1 and 1.2 this application is listed under the general heading of 'flexible film and sheet' which also includes thicker film and calendered sheet. Whilst constituting a major application area overall, this heading covers a considerable range of diverse applications, including shower curtains, groundsheets, rainwear, sacks, wallets and record covers, to name but a few. Some of these uses were established very early in the commercial exploitation of PVC and no longer have the growth potential of newer applications.

1.4.4 Cables

The insulation and sheathing of cables constituted the first important use for plasticised PVC, and is still a major application in every developed community. Though not suitable for very high voltage cables, PVC is used successfully for power transmission at intermediate voltages, and is generally the natural choice for domestic power and lighting, electrical appliance leads, car harness wiring, switch-boards, and telephone leads. The stabilisation of cable compositions does not, in general, present special problems, and the formulation technology, both to optimise the wire coating extrusion, and to tailor compositions for 'high temperature' or 'low temperature' performance, etc., has become highly developed over the years.

Although the evolution of hydrogen chloride is not a practical problem in cable manufacture, it has given rise to concern regarding the behaviour of PVC cables in fires. PVC compositions, properly formulated, can be of very low flammability, providing an advantage in cable and other applications. However, once involved in fires caused by other materials, the heated PVC will degrade and the evolution of hydrogen chloride, suppressed at processing temperatures, becomes very rapid. Since hydrogen chloride is a toxic and irritant gas it has been claimed that the presence of PVC in fire situations can add significantly to the hazard of toxic fumes to the fire victims or fire fighters. Although there is no 'stabiliser' that can possibly prevent the dehydrochlorination of PVC at the temperatures encountered in fires, it is possible to trap the HCl as calcium chloride, by reaction with mineral fillers such as calcium or magnesium carbonates. Cable compounds have been developed with sufficiently high levels of fine particle calcium carbonate to absorb 80–90 % of the potential HCl, but it is doubtful whether the complete prevention of HCl evolution into the atmosphere is possible in practical fire situations. There may, therefore, be justification for limiting the use of PVC cables in certain sensitive and enclosed environments, but more generally, the paramount need would seem to be the prevention of fires in the first place, and the limitation of fire spread. Here, PVC has a positive role, in providing both rigid and flexible materials of inherently low flammability.

1.4.5 Paste applications

The so-called 'paste grades' of PVC are polymers produced in the form of very fine but dense particles, which have the property of interacting with a plasticiser at ambient temperature just sufficiently to form a stable dispersion, of much higher viscosity than that of the plasticiser alone. These plastisols or 'pastes' can be processed by spreading, dipping, or rotational casting techniques, into coatings of appropriate thickness, to be subsequently gelled by heat to give a flexible product.

Since the processing techniques and requirements are so different from those of melt processing in extrusion, moulding, or calendering, it is convenient to consider paste applications as a separate class, although the diverse products fall under various of the headings in Tables 1.1 and 1.2, namely 'fabric coating', 'paper coating', 'foamed flooring', 'cap sealing', 'conveyor belting', and 'miscellaneous flexible'. Paste techniques are of value whenever it is required to spread fairly thin layers of flexible PVC onto a substrate, as in the manufacture of vinyl-coated wallpaper or vinyl-coated fabrics. Whilst such products can often be produced by lamination or direct spreading from a melt, the processing of thin layers of viscous heat-sensitive melt creates problems which are obviated by spreading plastisols at ambient temperature, and only applying heat at the subsequent gelling stage. Similarly, PVC gloves are produced by the dipping of formers into a paste, with gelling and stripping of the retained coating, whilst by casting pastes, a variety of hollow articles—play balls, flexible toys, etc.—are conveniently fabricated.

Some of the paste applications are among the earliest PVC products to be exploited and now have limited growth potential, or are of declining importance. An example is fabric coating, generally as 'leathercloth', which now competes with a variety of newer upholstery fabrics and techniques, especially in car seating. On the other hand, some newer paste applications have been among the faster growing PVC areas in recent years. In particular, it will be noted that whereas solid PVC flooring has shown a decline in tonnage in the UK, from 1976 to 1981, foamed flooring held its consumption (and would have shown continuing growth, but for the prevailing recession). Foamed PVC flooring is manufactured from several layers of spread paste, with a chemically foamed layer sandwiched below the top wear layer, to give a cushioned effect. Foamed flooring thus combines the hard wearing and easy to clean qualities of PVC sheeting, with something of the cushioned softness of carpets.

Since pastes are processed at ambient temperature, the demands on heat stabilisers are less than with melt processing, although some protection at the gelation stage and in the finished article is generally required. The choice of heat stabilisers and other formulation ingredients is very much dominated, however, by the rheological requirements of the fluid plastisol. The presence of heat stabilisers should not affect the tendency of the plastisol to increase in viscosity during ageing, or adversely affect dilatant or thixotropic characteristics, and should enhance rather than retard deaeration. In practice, liquid heat stabiliser systems based on barium/cadmium/zinc, or calcium/zinc systems, are most frequently employed.

Paste applications, though representing only a minor proportion of total PVC usage (around 15 % currently in the UK), include some applications of greater than average growth potential, and paste polymers are expected to maintain their share of PVC manufacture.

1.5 THE FUTURE OF PVC

At the time of writing, the world economies are still suffering from the recessionary forces set in motion by the enormous rise in oil prices of the mid-1970s. The European PVC manufacturers are feeling the effects of over capacity and extremely weak prices. It is pertinent, therefore, to consider whether the scope and pattern of PVC consumption outlined here will carry forward into future years.

In trying to answer that question, it must be remembered that the scope of PVC usage is very wide, with a particular emphasis on building construction. This means that PVC cannot 'buck the trend' of the general economy, and consumption is bound to fall during times of economic recession or stagnation. By the same token, an increase in housing starts, public building, home improvements and leisure spending must see a marked increase in PVC consumption. In terms of competitiveness with other polymers, PVC has the advantage of being more than half inorganic (i.e. 57 % chlorine) and therefore potentially less vulnerable to oil feedstock fluctuations. In terms of energy content, expressed as the oil equivalent required for manufacturing the raw material and converting to the end-product, PVC products compare favourably with those made from most competitive materials.

On the question of the future of the plastics industry in general, it is accepted that the era of 10–15% per annum growth rates in Western European plastics consumption is over. Nevertheless, the plastics industry is expected to grow at a rate 1–2% in excess of the general economy during the next decade, according to Dr W. Madden of ICI.[6] His forecast for the consumption of both commodity and speciality plastics in Western Europe in 1990, compared with actual performance in 1980, is shown in Table 1.4.

It is seen from Table 1.4 that PVC is expected to show a growth rate over the decade of 3% per annum, bringing annual consumption to around 5×10^6 tonnes in 1990. The relative consumptions of the commodity plastics are not very much changed from a 1980 position, except for the expected higher growth for polypropylene, the youngest of this family of bulk polymers.

Growth in European PVC consumption is expected to be largely from growth and extension of the existing applications emphasised in this chapter, rather than from completely novel application areas. Likewise, dramatic improvements in the fundamental properties of the polymer are not anticipated, although manufacturers will undoubtedly continue to improve their technology and the quality and consistency of their products, not least in heat stability performance. Further improvements in the range

Table 1.4 Plastics consumption in Western Europe: the next decade[6]

Material	*Consumption (tonnes × 10^3)*		
	1980	*1990*[a]	*Growth rate (%/yr)*
Commodity thermoplastics			
LDPE/LLDPE	3 810	5 300	3
HDPE	1 320	2 070	4
PP	1 230	2 550	7·5
PS/EPS	1 535	1 900	2
PVC	3 710	5 140	3
Speciality thermoplastics			
ABS	280	390	3·5
Nylons	180	335	6
Acrylics	130	190	3·5
Polyesters	104	420	15
Polycarbonate	60	130	8
Polyacetal	48	80	5
Polyphenylene oxide (modified)	35	90	10
Fluoroplastics	7	11	5

[a] Estimated.

of heat stabilisers and other additives available will also assist in the formulation of compounds to meet the exacting demands of product performance, appearance and environmental acceptance.

Theoretical considerations and studies of model compounds have suggested that the idealised regular structure of polyvinyl chloride should be inherently stable—certainly less sensitive to heat than the commercially available polymer. It has therefore been postulated that if the polymer could actually be manufactured without the known sites of instability, such as internal or terminal allylic chlorides, and tertiary chlorides, the product would show a remarkable improvement in heat stability. This would seem to suggest a most valuable future target, but unfortunately the practical considerations of commercial manufacture make its achievement most unlikely. The future technologist must therefore expect to deal with products probably showing only modest improvements over the invaluable but imperfect polymer discussed in this chapter.

REFERENCES

1. Kaufmann, M. *The History of Polyvinyl Chloride*, Maclaren & Sons Ltd, London, 1969.
2. *Mod. Plast. Int.*, **12**(1) (January 1982).
3. ICI plc. Unpublished data.
4. Kridl, A. G. *Standford Research Institute Presentation to Plastics Industry Symposium*, Zurich, 1982.
5. Pearson, R. B. *Food Chem.*, **8**(3) (1982), 85.
6. Madden, W. *Address to the Finnish Plastics Association, Mod. Plast. Int.*, **12**(2) (February 1982), 6.

Chapter 2

THERMAL DEGRADATION OF PVC

T. HJERTBERG and E. M. SÖRVIK
Department of Polymer Technology,
Chalmers University of Technology, Gothenburg, Sweden

2.1 INTRODUCTION

This chapter summarizes the research related to thermal degradation of unstabilised polyvinyl chloride (PVC) at conditions characteristic of processing, i.e. low oxygen levels and temperatures up to 250 °C. The presentation concentrates on recent work and covers the recent literature. In the relevant sections, special emphasis is given to our own findings.

At temperatures necessary for its processing, PVC undergoes degradation by dehydrochlorination. The double bonds formed are conjugated in polyene sequences which causes unacceptable levels of discoloration even at 0·1 % of dehydrochlorination. The polyenes are active in secondary reactions leading to crosslinking and the formation of aromatic pyrolysates. Furthermore, the dehydrochlorination is autocatalytic and would rapidly cause 'burning' in processing equipment. Addition of thermal stabilisers is therefore necessary. This solution however is not ideal for technical–economical as well as environmental reasons.

When exposed to the high temperatures in fires the degradation pattern of PVC also causes problems in that large amounts of black smoke and corrosive fumes are generated. Thermo-oxidative degradation is to some extent important for the ageing of PVC as well. The sensitivity towards UV-irradiation is, however, much more important in this respect. Photodegradation is also characterised by dehydrochlorination and polyene formation. Several of these aspects are discussed in other chapters in this book.

A huge amount of research and development work has been carried out in order to improve the thermal stability of PVC. These efforts have been summarised in several literature surveys.[1-10] During the last seven to eight years there has been considerable progress in the understanding of the degradation of PVC. This is mainly due to the improvement in NMR and other characterisation techniques. A research cooperation, started in 1977 under the auspices of IUPAC ('IUPAC Working Party on PVC Defects'), has promoted fruitful discussions about the new findings.

The thermal dehydrochlorination of PVC starts at about 100 °C,[11] which is much lower than the degradation temperatures for low molecular weight compounds taken as models for the 'ideal' PVC structure, e.g. 2,4,6-trichloroheptane.[12] It has therefore been assumed that the low stability of PVC is caused by irregular structures in the polymer, which activate the carbon–chlorine bond, creating 'labile chlorine'. The following structures have, for example, been discussed:

1. Initiator residues.
2. Unsaturated chain ends.
3. Internal unsaturation.
4. Branches.
5. Head-to-head structures.
6. Oxidation structures.

Based on experiments with low molecular weight model compounds it has been considered that chlorine adjacent to internal double bonds (internal allylic chlorine) and chlorine on branch carbons (tertiary chlorine) should be the most labile structures in PVC, and the main reasons for the low thermal stability (see, for example, refs. 1–6).

Work with model polymers has supported these ideas. Copolymers of vinyl chloride and pentyne,[13] 2-chloropropene[14-17] or 2,4-dichloropentene[18] have lower stabilities than the homopolymer. Based on degradation experiments with copolymers, Berens[16] concluded that the presence of 1–2 tertiary chlorines per 1000 monomer units (1000 VC) would account for the thermal lability observed in ordinary PVC. According to general opinion,[2-4,6] internal allylic chlorine is expected to be still more reactive.

Earlier attempts to reveal the nature of the labile structures actually present in ordinary PVC were inconclusive. A major reason for this was the lack of adequate characterisation methods. With the advent of modern NMR equipment it has *per se* become possible to determine odd structures, even at the low levels indicated above. In order to determine the relation between structural defects and stability, it is necessary to work with a set of

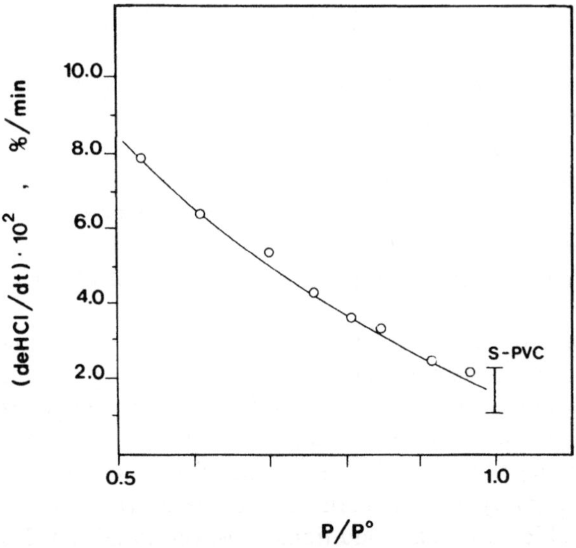

Fig. 2.1 The rate of dehydrochlorination (deHCl)/dt) at 190 °C in nitrogen for u-PVC obtained at 55 °C. P = monomer pressure; P^0 = saturation pressure of vinyl chloride at 55 °C. Values for suspension PVC (s-PVC) from ref. 19.

polymers with adequate differences in properties. However, most commercial PVC samples are rather similar in stability, which is measured as the rate of dehydrochlorination at elevated temperatures in an inert atmosphere (see, for example, ref. 19). More or less successful attempts have therefore been made to enhance the formation of labile structures by modifications in the normal polymerisation procedure. Our approach was to prepare polymers at subsaturation conditions, i.e. without the presence of a liquid monomer phase.[20–30] By changing the monomer pressure it is possible to simulate the conditions prevailing in ordinary polymerisations at and after the 'pressure drop'. As shown in Fig. 2.1, this resulted in polymers with a thermal stability ranging from normal to very low values. Using these kinds of polymers (u-PVC), we could establish definite relations between certain irregular structures and stability. The nature of the irregular structures and their influence on the initiation of the dehydrochlorination are discussed in the two following sections of this chapter. In subsequent sections, the propagation, secondary reactions of the polyenes and the influence of the HCl formed are discussed. The degradation behaviour during processing is treated separately. Finally, the possibility of obtaining PVC with improved heat stability is reviewed.

2.2 THE STRUCTURE OF PVC

As mentioned in the Introduction, the improvement in knowledge of the PVC structure is mainly due to the development of NMR and other characterisation methods. A comprehensive survey of the different analytical techniques is given in Chapter 4. Several recent reviews concerning the structure of PVC and analytical techniques have appeared in the literature.[27,31–33] The spectroscopic[34,35] and chemical[36] methods used in the Working Party for the characterisation of PVC will be reported on soon.

Production of PVC is usually performed by batchwise radical polymerisation.[37] The overwhelming part of the propagation occurs in a swollen polymer gel in contact with a liquid monomer phase.[38–43] The tendency for chain transfer to monomer is extremely high and strongly temperature dependent.[41,44,45] It is therefore possible to control the molecular weight by the polymerisation temperature alone. These features result in stable reaction conditions. Conventional PVC is therefore characterised by a uniform molecular structure, including a narrow Gaussian molecular weight distribution with M_w/M_n 2·0–2·2.[41,44,45] The constant conditions prevail until the liquid monomer phase is used up after about 70% conversion.[46] This point is referred to as the 'pressure drop'. After that, the monomer concentration in the gel gradually decreases. This results in decreasing molecular weight[22,24,47] and increasing probability for side reactions (see below). In conventional PVC production the polymerisation is usually stopped after 85–90% conversion.

The propagation proceeds predominantly in a head-to-tail fashion and with some preference for syndiotactic conformations. The stereoregular sequences are, however, rather short (see Chapter 4). With decreasing polymerisation temperature the degree of syndiotacticity increases,[48] but within the rather narrow temperature range normally used in commercial production (40–75 °C), the changes in stereoregularity are small. Within each class of steric structures there are naturally several conformations (see Chapter 4). Some characteristic values of molecular weight and tacticity are given in Table 2.1.

The growth of a PVC macroradical is normally terminated by chain transfer to monomer.[41,44,45] Kinetic measurements have shown that 4–5 polymer molecules are obtained per initiator residue.[50] The majority of the end groups should therefore be a result of chain transfer. Until recently, the transfer step was believed to involve a direct reaction between the ordinary macroradical and a monomer molecule.[51] It has now been shown that the

Table 2.1 Typical values of molecular weight averages and tacticity for ordinary suspension PVC[a]

Polymerisation temperature (°C)	$\bar{M}_n \times 10^{-3,b}$	$\bar{M}_w \times 10^{-3,b}$	$M_w/M_n{}^b$	α^c
43	58·1	141	2·44	0·56
55	43·6	90·6	2·08	0·54
75	26·2	52·7	2·01	0·52

[a] From ref. 49.
[b] Determined by GPC.
[c] Amount of syndiotactic dyads determined by ^{1}H-NMR.

mechanism is more complicated.[9,23,26–28,52–54] It involves an occasional head-to-head addition followed by 1,2-chlorine migration and loss of a chlorine atom, which effects the actual chain transfer to monomer:

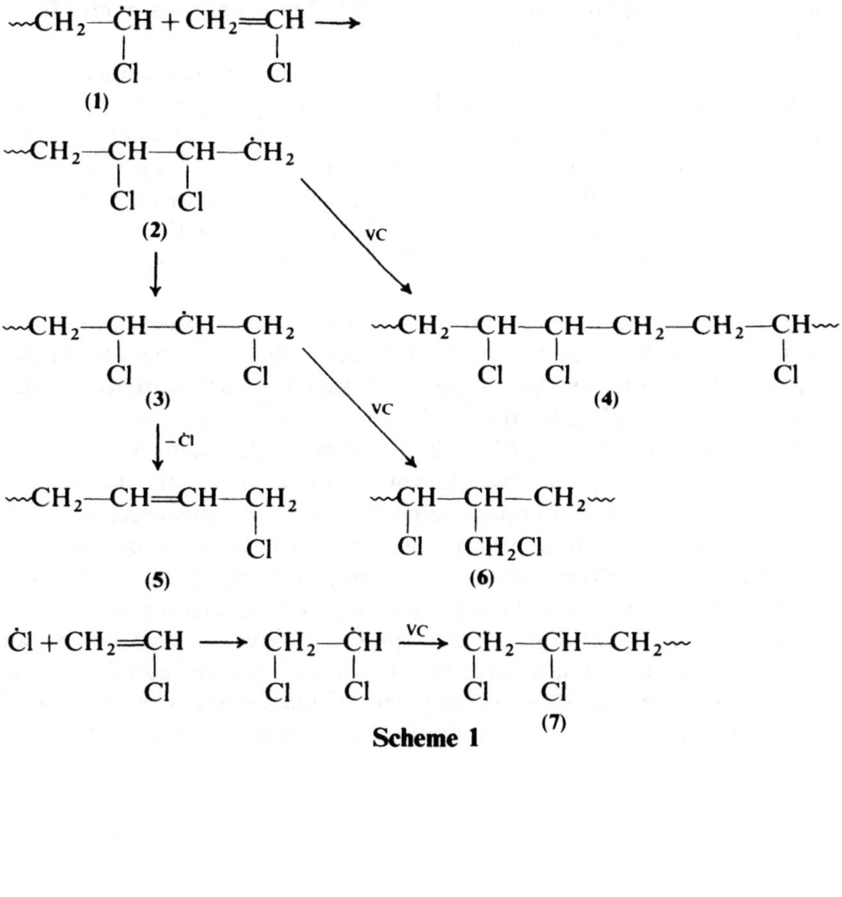

Scheme 1

The 1-chloro-2-alkene (**5**) and 1,2-dichloro alkane (**7**) structures should thus be the most frequent end-groups in PVC. Addition of vinyl chloride to the intermediate radicals (**2**) and (**3**) would result in internal head-to-head structures (**4**) and chloromethyl short chain branching (**6**), respectively. The head-to-head addition was originally proposed by Rigo *et al.*[55] to explain the formation of chloromethyl branches. Abbås,[56] Bovey[57] and their co-workers proved the presence of this structure by ^{13}C-NMR measurements on reduced PVC. The content of chloromethyl branches is 4–5 per 1000 VC.[23,25,27,28,52,56,57] Using poly(vinyl-α-D-chloride), Starnes *et al.*[52] could prove the proposed mechanism.

The expulsion of chlorine (**3**)–(**5**) was suggested by Bezdadea *et al.*,[58] while Starnes *et al.*[52] suggested the subsequent chlorine attack on the monomer. The presence of the end-groups (**5**) and (**7**) has been shown by ^{1}H-NMR.[23,26,27,34,59,60] Structure (**7**) was also verified with ^{13}C-NMR measurements on PVC reduced with Bu_3SnD.[23,27,28,53,61,62] According to our measurements, the content of these structures is somewhat less than one per macromolecule in ordinary PVC.[26,27] The proportion of *cis*-(**5**) has tentatively been given as 10–25%.[34]

Using an iodometric method, Mitani *et al.*[63] claimed that ordinary PVC contains 6–7 internal head-to-head structures (**4**) per 1000 VC. Measurements with ^{13}C-NMR[52] however have shown that the content of structure (**4**) is less than 2 per 1000 VC (detection limit). Recent investigations of the iodometric method in our laboratory have shown that this method gives mainly a measure of the saturated 1,2-dichloro end-group (**7**).[30] It was also found that the content of structure (**4**) must be very low—at most 0·2 per 1000 VC.

At increasing monomer starvation, the balance between the two reactions from (**3**) is expected to change towards chlorine expulsion (**3**)–(**5**). This would result in decreasing molecular weight, which is in accordance with experimental observations.[27,28]

The content of end-group (**7**) is about one per molecule in subsaturation PVC, demonstrating that the chlorine atom is the chain-transferring species even at reduced monomer concentration.[26,27] However, although the total content of double bonds is constant (one per molecule), the relative amount of (**5**) decreases at decreasing monomer pressure. There is also an increasing amount of $\sim$CHCl—CH_2—CH_2Cl chain ends.[23,26–28] The change in the end-group structures in u-PVC is evidence of an increased frequency of side reactions at monomer-starved conditions, i.e. after the pressure drop in a conventional polymerisation. The presence of internal double bonds was demonstrated by oxidative cleavage quite a long

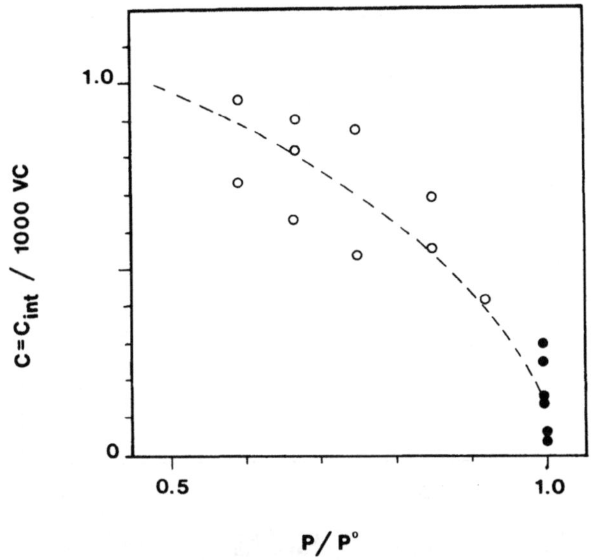

Fig. 2.2 Relation between the amount of internal double bonds ($C{=}C_{int}$) and the relative monomer pressure (P/P^0). ○, u-PVC; ●, fractions of a commercial suspension PVC (s-PVC). After ref. 29.

time ago.[64–71] Our investigations on subsaturation PVC showed that the content of internal unsaturation increases with decreasing monomer concentration (Fig. 2.2).[23,27,29] These and other[19,72–74] measurements indicate that the amount of internal unsaturation in conventional PVC is 0·1–0·2 per 1000 VC.

With the technique used to determine internal unsaturation, it is not possible to discriminate between isolated and conjugated double bonds. Measurements with UV and resonance Raman spectroscopy (see Chapter 4) have, however, shown that the content of polyene sequences must be low. Values in the range $1{\cdot}0 \times 10^{-4}$–5×10^{-4} per 1000 VC have been reported.[75–77] Besides increased amounts of internal unsaturation, we have shown that increased monomer starvation leads to increased amounts of branching. The presence of long chain branching in u-PVC was first demonstrated by combining GPC and viscometry measurements.[20–22,24] This was confirmed by our ^{13}C-NMR studies of PVC reduced with Bu_3SnH.[23,25,27,28] It was also shown that butyl branches are present. The amount of these two structures increases with increasing monomer starvation (Fig. 2.3). The ratio of butyl and long-chain branches is approximately 2:1.

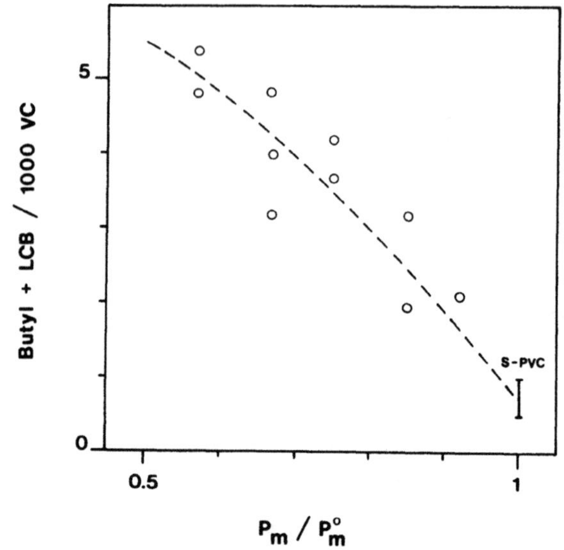

Fig. 2.3 Relation between the total content of butyl and long-chain branches and the relative monomer pressure at polymerisation (P/P^0). ○, u-PVC. After ref. 28.

^{13}C-NMR measurements on these polymers after reduction with Bu_3SnD showed that all butyl branches and the major part of the long-chain branches were associated with tertiary chlorine.[23,27,28] It was estimated that up to one-third of the long-chain branch points could contain tertiary hydrogen:

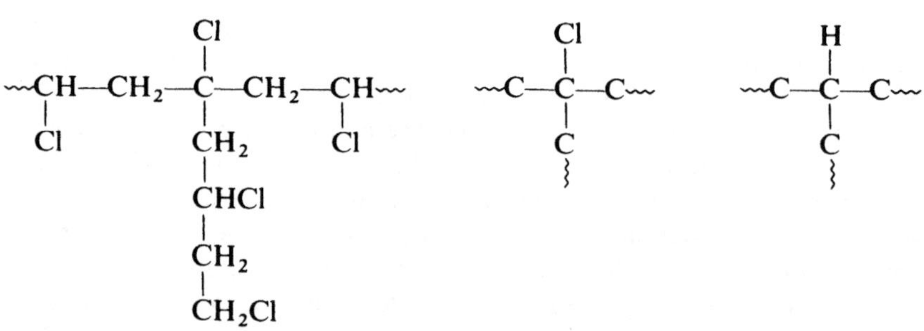

The amounts of butyl and long-chain branches in fractions of a commercial suspension PVC were found to be 0·5–1 per 1000 VC, decreasing with increasing molecular weight.[23,27,28] Park *et al.*[78] have examined several commercial polymers (with ^{13}C-NMR after reduction) and found about one butyl branch per 1000 VC.

Increased amounts of butyl and long-chain branches have been reported by Starnes *et al.*[62] in a polymer prepared at high temperature (100 °C). They confirmed the microstructure of the butyl branches. They further reported that the 100 °C polymer contained 2-chloroethyl branches[53,61,62] and claimed the presence of this structure in conventional PVC. We have not been able to find this branch type in either u-PVC or conventional PVC.[23,25,27,28] Considering all available data, it can be stated that the amount of tertiary chlorine in ordinary PVC is 0·5–1·5 per 1000 VC. The tertiary chlorines are to a large extent associated with butyl branches. This structure is the one expected from the back-biting mechanism suggested by Bovey and Tiers:[79]

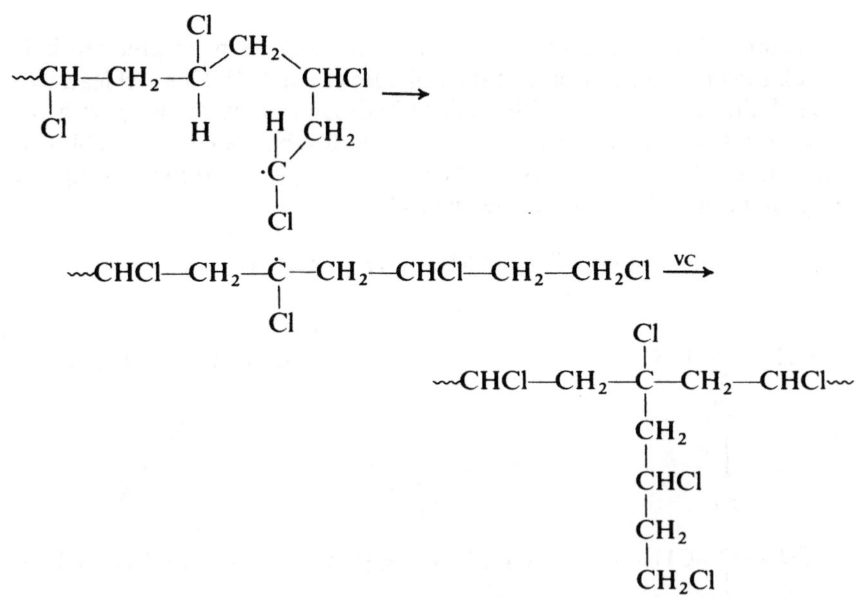

(8)

Scheme 2

The other internal defects—long-chain branching and internal unsaturation—should be the result of intermolecular chain transfer to polymer. It is likely that the macroradical selectively attacks the chloromethylene groups, leading to long-chain branch points with tertiary chlorine:[23,27,28,53]

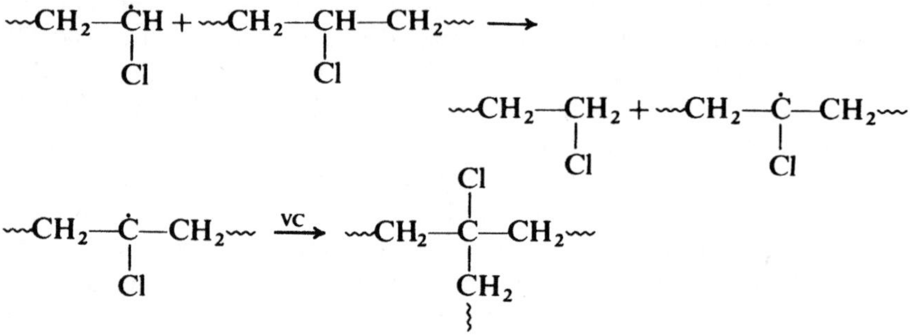

(9)

Scheme 3

In spite of its high reactivity, the chlorine atom seems to preferentially attack the methylene groups in the polymer chain.[80,81] This will lead to a long-chain branch point with tertiary hydrogen or, in analogy with the reactions from (3) to (5), to an internal double bond after chlorine expulsion.[23,27–29] An attack on the chloromethylene group will lead to a long-chain branch point with tertiary chlorine:

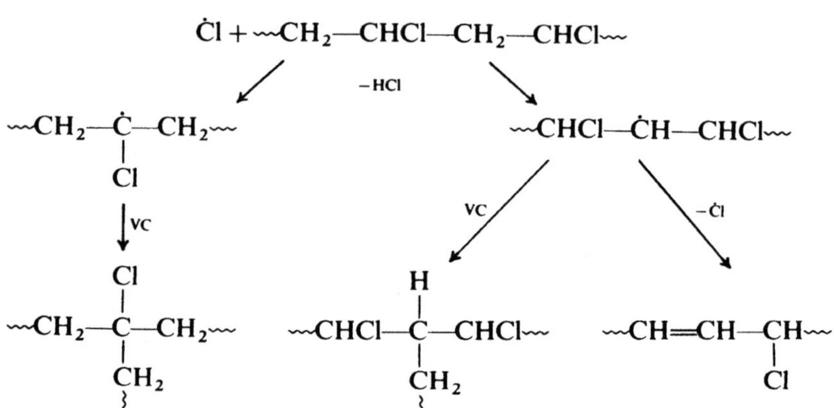

(10) **(11)**

Scheme 4

Minsker *et al.*[82] have suggested that structure (**11**) is oxidised to ketoallylic groups during polymerisation or storage:

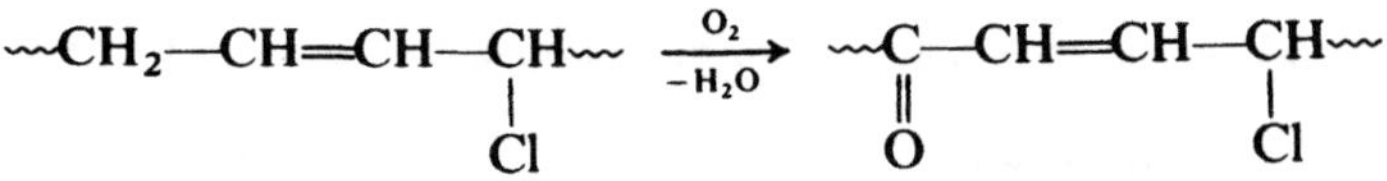

(11) **(12)**

This research team considers the ketoallylic group to be the most important labile structure in PVC.[83–85] The assumed high probability of this reaction must be questioned, as structure (**5**) is found to be the most frequent unsaturated structure in ordinary PVC. Using ^{1}H-NMR, we have not been able to detect structure (**12**) in u-PVC with a high content of internal double bonds.[27,29]

A number of other oxygen-containing structures have been considered as potential labile structures. Apart from those originating from initiator residues, few oxygenated groups are found in ordinary PVC.[34] Polymerisations carried out in the presence of large amounts of oxygen ($>0{\cdot}16$ mol %) give low molecular weight polyperoxides with low thermal stability.[86,87] When the oxygen content is lower, acrylic acid units are the major oxygen containing structures found.[87] This structure is formed by copolymerisation between vinyl chloride and carbon monoxide[87,88] which is formed by decomposition of polyperoxides:[89]

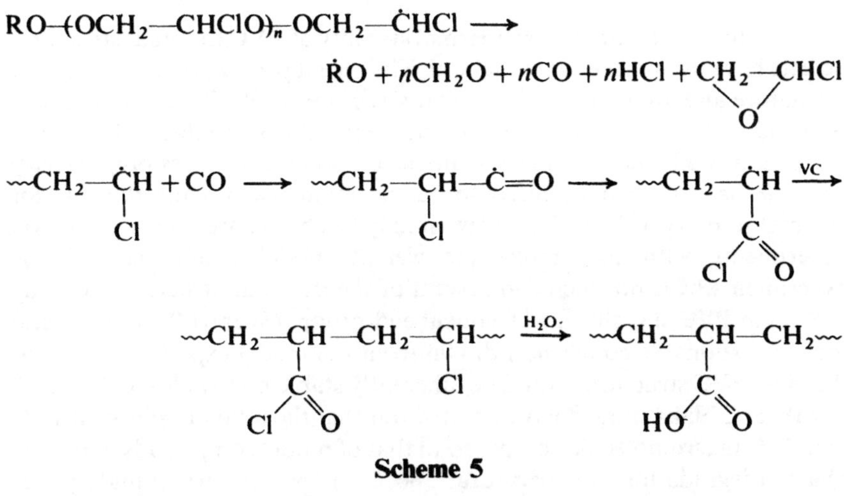

Scheme 5

In conclusion it can be stated that knowledge of the nature and content of the irregular structures in PVC is now rather well established. The content of such structures in ordinary PVC is summarised in Table 2.2.

Table 2.2 Irregular structures determined in ordinary PVC[a]

Structure	*Content*
$\sim CH_2—CHCl—CH_2Cl$	0·8–0·9/molecule
$\sim CH_2—CH{=}CH—CH_2Cl$	*c.* 0·7/molecule
Initiator residues	*c.* 0·2/molecule
Total unsaturation	1/molecule
Internal double bonds	0·1–0·2/1 000 VC
Polyene sequences	$<5 \times 10^{-4}$/1 000 VC
Branches	
chloromethyl	4/1 000 VC
2-chloroethyl	<0·5/1 000 VC
2,4-dichlorobutyl	1/1 000 VC
long branches	<0·5/1 000 VC
Tertiary chlorine	0·5–1·5/1 000 VC
Head-to-head groups	<0·2/1 000 VC

[a] For references see text.

2.3 RELATION BETWEEN STRUCTURE AND THERMAL STABILITY

Earlier efforts to relate the low thermal stability of PVC to actual structural defects have been rather unsuccessful. The recent progress in knowledge of the nature and abundance of irregular structures in PVC gives a far better possibility of elucidating the relation between defects and thermal stability.

As indicated above, internal allylic chlorine and, possibly, tertiary chlorine have been considered to be the main reasons for the inferior thermal stability of PVC. This view mainly has been based on degradation experiments with low molecular weight model compounds. These experiments have not included several of the irregular structures actually present in PVC, e.g. the most frequent end-groups (**5**) and (**7**). Another still more questionable conclusion drawn from the model experiments is that the ideal PVC structure should be inherently stable up to at least 200 °C.[2,4] Suzuki and Nakamura[90] have pointed out that the different physical state of a PVC macromolecule compared to that of model compounds in the gas phase (degradation temperature 360 °C[12]) or in the liquid phase (degradation temperature 257 °C[90]), may lead to an overestimation of the limiting stability of the PVC structure.

The following discussion is restricted to structures actually determined in PVC. These are normally the result of reactions occurring during the radical polymerisation of vinyl chloride, but some of them could in principle originate from copolymerisation of impurities in the monomer. Internal double bonds can, for example, be formed by copolymerisation with butadiene or acetylenic impurities. Nowadays, the levels of such impurities are very low in polymerisation-grade vinyl chloride, e.g. butadiene < 10 ppm, acetylenic impurities < 1 ppm.[49] This source of irregular structures can therefore be ignored. The level of oxygen present in commercial polymerisation is also low. As discussed above, oxygen at low concentrations leads to acrylic acid units in the polymer chain. Braun and Sonderhof[87] have stated that this structure does not influence the thermal stability to any major extent. Extraneous impurities have also been considered as potential reasons for the low thermal stability of PVC.[91] Examples of materials of this kind are initiators or their decomposition products, suspension stabilisers, emulgators and anti-scaling agents. Compounds of these types with prodegradant effects are known[92] and are carefully avoided in commercial production. Normally, reprecipitation of commercial PVC qualities does not improve the thermal stability to any significant extent.[19,66,93,94] Such impurities cannot therefore explain the inherent thermal lability of PVC. In earlier investigations, chain ends were suggested as important initiation sites for the dehydrochlorination. Most of the interest was concentrated on the unsaturated end-groups but initiator residues were also considered.[91,95–98] Due to the frequent chain transfer to monomer, the content of this type of end-groups is rather low, *c.* 0·2 per molecule (see above). Investigations on model substances have further shown that such residues should be rather stable.[2–4,65] From these statements, it can be assumed that the initiator residues are of minor importance for the instability of PVC. For some initiator types, e.g. azobisisobutyronitrile, reduced stability has been reported.[95–97] This may be caused by attack on the polymer by initiator radicals, rather than being an effect of the built-in initiator fragments.

The 1,2-dichloro alkane end-group (**7**) is the most frequent saturated chain-end structure. No appropriate model substance has been investigated. Experiments with chlorinated 1,4-polybutadiene[64,99–101] and (vinyl chloride)–*trans*-1,2-dichloroethylene copolymers[102] have shown that internal 1,2-dichloro structures are rather stable. The most frequent unsaturated chain-end is the 1-chloro-2-alkene end-group (**5**). This structure contains an allylic halogen which is easily substituted with, for example, phenol.[23,26,27] Furthermore, it has been suggested that (**5**) may undergo 1,4-dehydrochlorination via an ionic mechanism during thermal degradation:[9,26]

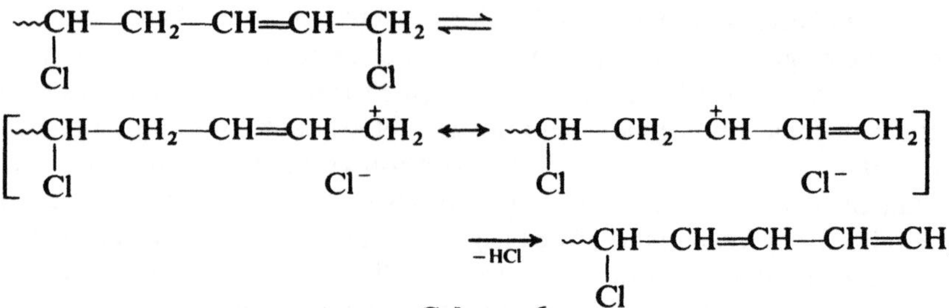

Scheme 6

Amer and Shapiro[103] have suggested, that *cis*-**(5)** would give 1,4 elimination but via a cyclic intermediate. These reactions would initiate the formation of a polyene sequence. Furthermore, **(5)** could undergo an HCl-catalysed rearrangement:[104,105]

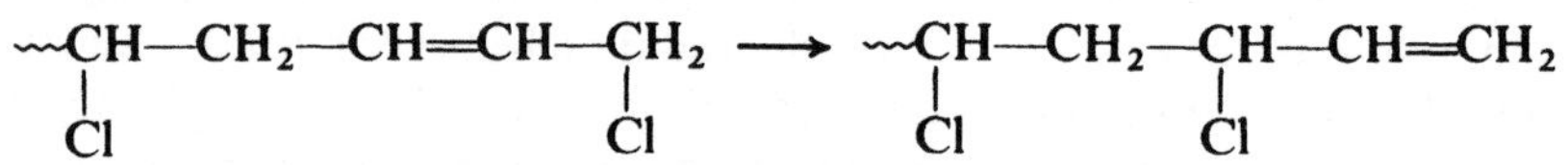

Scheme 7 **(13)**

From studies on low molecular weight compounds, **(13)** is known to dehydrochlorinate faster than the ordinary monomer unit, although not as fast as an internal β-chloroallylic structure.[12] If these reactions are operative, a pronounced dependence of the dehydrochlorination rate on the molecular weight would be expected.

Much work has been devoted to the influence of the end-groups. In most cases fractions have been used. Some workers have reported definite relationships between the thermal instability and the content of end-groups,[106,107] while others did not find any conclusive results.[108–110] Often the lowest molecular weight fractions showed low stability, while more high molecular weight material showed no dependence on the molecular weight.[2,73,109] In none of these investigations was the content of other structures measured. In our work, we have studied the influence of the molecular weight as well as that of tertiary and internal allylic chlorine. From these measurements we could show that the end-groups in ordinary PVC have a negligible influence on the dehydrochlorination rate.[23,26–29] This conclusion has also been drawn by Starnes.[9] Very recently, van den Heuvel and Weber[111] made a direct investigation of the behaviour of the end-groups **(5)** and **(7)** during thermal degradation. They degraded a low molecular weight fraction to about 5% dehydrochlorination. Proton NMR measurements revealed that the content of **(5)** and **(7)** did not change. The most frequent end-groups thus seem to be without noticeable influence on the thermal instability of PVC.

Earlier, tertiary chlorine was considered to be an important labile site and many attempts have been made to find evidence for its negative effect in the polymer. One way has been to relate the thermal stability to the total content of branches. The number of branches was measured by IR-spectroscopy after reduction with $LiAlH_4$ (see Chapter 4). It was found that the rate of dehydrochlorination was independent of this measure.[19,112] This is in good agreement with the present knowledge, i.e. in ordinary PVC the chloromethyl branches with tertiary hydrogen account for 70–80 % of the total branching. This structure should be rather stable, and Suzuki and Nakamura[90] have indeed reported a relatively high decomposition temperature for a model substance with a similar structure. In polymers obtained by copolymerisation of vinyl chloride and allyl chloride, Lewis *et al.*[113] even claimed a small stabilising effect of the chloromethyl branches. They considered this to be due to the blocking action of the CH_2Cl branches on the growth of polyene sequences.

As it was shown that mild chlorination increased the thermal stability,[64] it was assumed that double bonds were deleterious. Suzuki and Nakamura[90] and Burille *et al.*[73,114] obtained linear relations between the rate of dehydrochlorination and the total unsaturation determined by bromination. Boissel[115] did not find any correlation. Referring to Table 2.2 it can be concluded that the amount of the total unsaturation reflects mainly the content of the stable 1-chloro-2-alkene end-group. Total unsaturation is therefore not an adequate measure for labile structures.

Yet another approach to quantify the total amount of thermolabile structures is to measure the content of chlorine which is exchangeable by chemical agents, e.g. cadmium acetate,[116,117] potassium acetate,[118] trialkylaluminium,[119] phenol,[120,121] thiophenol[73,114] and thiophenate.[122,123] These methods are based on the assumption that chlorines vulnerable to exchange reactions are responsible for the initiation of the thermal degradation. In the light of present knowledge this is doubtful. As already mentioned, phenol will, for example, react with the 1-chloro-2-alkene end-groups. A quite inconsistent result was therefore obtained in our comparative measurements of s- and u-PVC (Fig. 2.4). The same content of 'labile' chlorine resulted in a different thermostability.[23,26,27] Furthermore, the exchange reaction may not be quantitative or may involve secondary chlorine in ordinary monomer units.[118,122,123]

The degradation experiments with high molecular weight models showed that, if present, tertiary and internal allylic chlorine should be very active as initiation sites for the dehydrochlorination. The inconclusive results of the attempts to relate the total branch content with thermal stability created

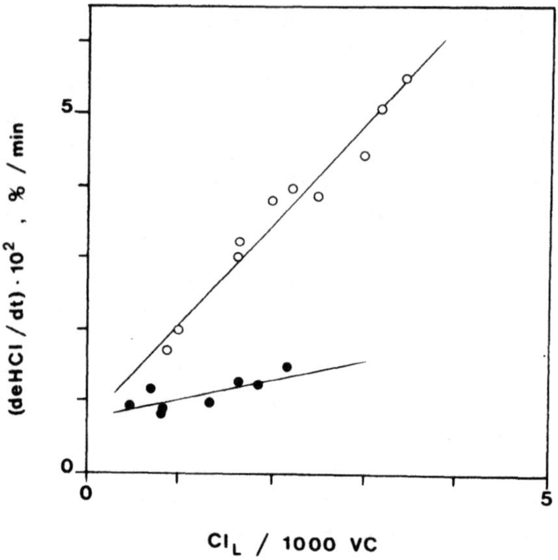

Fig. 2.4 Relation between rate of dehydrochlorination (deHCl/d*t*) at 190 °C in nitrogen and the amount of labile chlorine determined by phenolysis. ○, u-PVC; ●, s-PVC. After ref. 26.

the impression that no tertiary chlorine exists in PVC. Earlier work with phenolysis was interpreted in the same way.[120,121] On the other hand, the presence of internal unsaturation was confirmed and adequate characterisation methods based on oxidative cleavage with ozone were developed.[64–68,70–72] Using this technique, Abbås and Sörvik,[19] Minsker *et al.*,[67] Burille[73] and Braun *et al.*[74] have reported linear relations between the rate of dehydrochlorination and the content of internal double bonds. As a consequence, internal allylic chlorine has been considered to be the most important labile structure in PVC.

Recently, the development of ^{13}C-NMR has allowed a more direct determination of tertiary chlorine. Bovey *et al.*[61] found 2-chloroethyl groups with tertiary chlorine in a polymer prepared at 100 °C. We found 2,4-dichlorobutyl and long-chain branches with tertiary chlorine in polymers prepared in monomer-starved conditions.[23,27,28] The content of butyl and long-chain branches decreased with increasing monomer pressure, but was still quite considerable in a series of fractions of an ordinary PVC, 0·5–1 per 1000 VC. These polymers were prepared at 55 °C, i.e. within the normal commercial temperature range. No ethyl branches could be observed. Starnes *et al.*[62] have confirmed the presence of butyl and

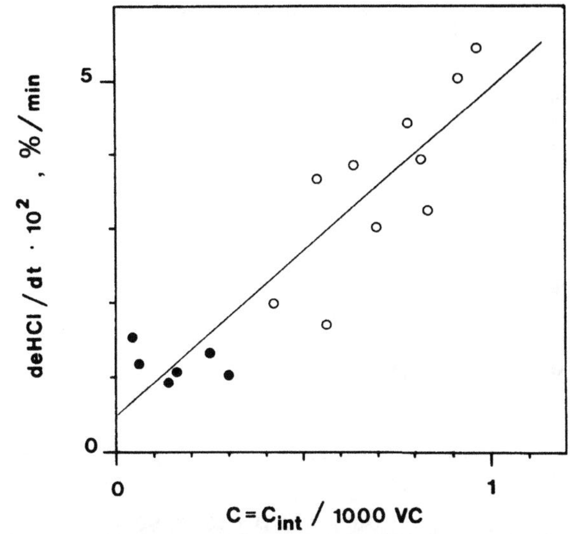

Fig. 2.5 Relation between the rate of dehydrochlorination (deHCl/d*t*) at 190 °C in nitrogen and the amount of internal double bonds ($C{=}C_{int}$). Correlation coefficient = 0·88; ○, u-PVC; ●, s-PVC. After ref. 29.

long-chain branches with tertiary chlorine. The available data[23,27,28,62,78] point to a content of about one tertiary chlorine per 1000 VC in conventional PVC. As the amount of tertiary and internal allylic chlorine is low and rather constant, the possibility of determining their relative importance for the thermal stability by investigations of conventional PVC is small. However, with the use of u-PVC this problem can be overcome. As shown above, the content of these labile structures increases considerably with increasing monomer starvation (Figs 2.2 and 2.3). The relation between dehydrochlorination rate and the amount of defect structure can thus be established with adequate accuracy. As shown in Figs 2.5 and 2.6, both structures give linear relations, although the correlation obtained for tertiary chlorine is much better. Furthermore, the content of this structure is about five times higher than that of internal allylic chlorine. Contrary to earlier statements, we therefore consider tertiary chlorine to be the most important labile structure in PVC.[23,27–29] This conclusion is in line with Berens'[16] statement that the presence of one to two tertiary chlorines per 1000 VC could *per se* explain the instability of PVC. With reference to Berens and their own measurements of tertiary chlorine, Starnes *et al.*[9,62] also consider tertiary chlorine to be the main reason for the low thermal stability of PVC.

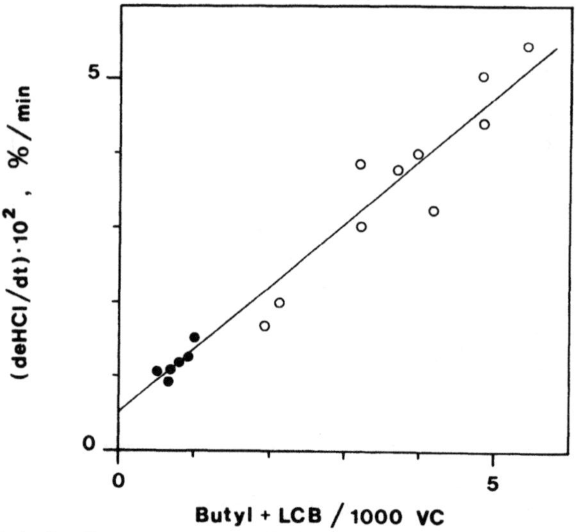

Fig. 2.6 Relation between the rate of dehydrochlorination (deHCl/d*t*) at 190 °C in nitrogen and the amount of tertiary chlorine given as the total content of butyl and long-chain branches. Correlation coefficient = 0·97; ○, u-PVC; ●, s-PVC. After ref. 28.

This conclusion implies that the reactivity of internal allylic chlorine is not considerably higher than that of tertiary chlorine. This is in contrast to the opinion expressed in several reviews (see, for example, refs. 3, 4 and 6), in which references are made to experiments with low molecular weight models. The literature data on the rate constants of dehydrochlorination for different model compounds are, however, confusing. Depending on the reaction conditions, values between 0·04 and 5×10^6 can be found for the ratio between the rate constant for internal allylic and tertiary chlorine.[27,29] On the other hand, there is conclusive evidence from model compounds that an enhanced reactivity is to be expected from conjugated unsaturation.[4] However, as mentioned above, the content of polyene sequences is negligible, $< 5 \times 10^{-4}$ per 1000 VC. It has not been possible to find direct evidence for the presence of ketoallylic groups. Minsker *et al.*[83–85] consider this structure to be very reactive, but this is disputed by Starnes.[9] Svetly *et al.*[124–126] have suggested that ketoallylic groups in the PVC chain might act as catalysts for dehydrochlorination from ordinary monomer units. This idea is supported by degradation experiments in which small quantities of the cisoidal enone 2-methylene-5α-cholestane-3-one were added.[125,126] According to this view, the ketoallylic group acts as

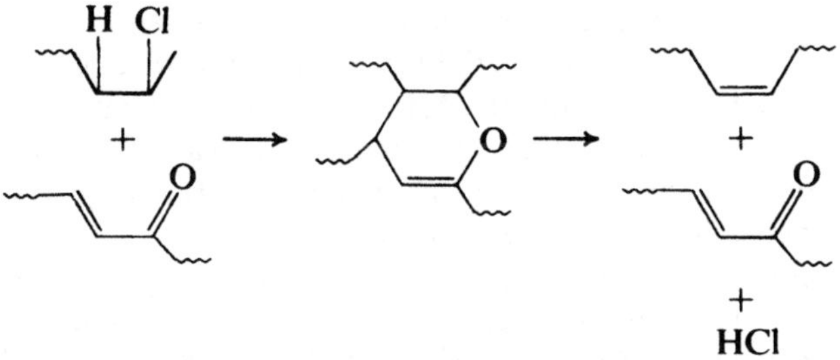

Scheme 8

a 'living defect', creating internal allylic chlorine. Our measurement with a sample prepared at highly starved conditions showed that the content of ketoallylic groups was at most 5% of the internal unsaturation,[27,29] and it is likely that this relation holds for ordinary PVC as well. This implies that the content of ketoallylic groups is less than 0·01 per 1000 VC.

In an attempt to quantify the relative reactivity of tertiary and internal allylic chlorine we have performed a multiple regression analysis of the data given in Figs. 2.5 and 2.6.[27,29] Assuming that one-third of the long-chain branch points contain tertiary hydrogen, the following relations were obtained:

$$\text{deHCl} = k_T\text{Cl}_T + k_A\text{Cl}_A + \text{const.}$$

s-PVC $$\text{deHCl} = 1{\cdot}05\,\text{Cl}_T + 0{\cdot}67\text{Cl}_A + 0{\cdot}30 \qquad (1)$$

u-PVC $$\text{deHCl} = 0{\cdot}91\text{Cl}_T + 0{\cdot}56\text{Cl}_A + 0{\cdot}15 \qquad (2)$$

where deHCl is the rate of dehydrochlorination at 190 °C in nitrogen (%/min) × 10^2; k_T and k_A are the relative rate constants of dehydrochlorination initiated by tertiary and internal allylic chlorine, respectively; and the concentrations Cl_T and Cl_A are given per 1000 VC. These relations indicate that the reactivity of tertiary and internal allylic chlorine is of the same order. As already stated, the more frequently occurring tertiary chlorine must therefore be considered as the most important labile structure in PVC. It can also be concluded that the effect of eventual ketoallylic groups is of minor importance.

The non-zero rate of dehydrochlorination at zero content of tertiary and internal allylic chlorine includes, for example, the influences of unspecified labile groups. Another important factor might be random dehydrochlorination from ordinary monomer units. This possibility was neglected for a long time but has received increased attention in recent years.[83-85,103,127-130] Kinetic models have been proposed in which the effects of random dehydrochlorination as well as labile structures are considered.[83-85,127-129] In the light of present knowledge, these models cannot be considered appropriate. Minsker *et al.*[83-85] have, for example, excluded the presence of tertiary chlorine, while Tvaroska[127] has used far too high values for the content of this structure. Other objections against these approaches have also been raised.[7]

The existence of random dehydrochlorination is supported by the fact that a residual instability is observed after treatments intended to convert the defects into stable structures.[64,93,94,119,131-4] The reactions have been performed with a variety of reactants (see below). The achieved values of dehydrochlorination rate are reported to be about 10–30 % of the original ones. This low level of instability is also observed with PVC prepared with butyllithium as initiator.[90,135] It is not likely that this polymerisation system generates tertiary or internal allylic chlorine via transfer to polymer.

The suggestion that random dehydrochlorination occurs at a relatively low temperature is not in accordance with the conclusions drawn from the experiments with low molecular weight models for the ideal PVC structure. As mentioned earlier, these experiments might not be representative for the degradation of ideal PVC macromolecules due to the differences in aggregation state. Millán *et al.*[122,123] have considered initiation by dehydrochlorination from ordinary monomer units. They came to the conclusion that GTTG isotactic and TTTG heterotactic triads should be most susceptible to thermal dehydrochlorination. Based on reductions of PVC with thiols,[94] Starnes *et al.*[9,136] suggest that dehydrochlorination from ordinary monomer units occurs via a cyclic chloronium cation:

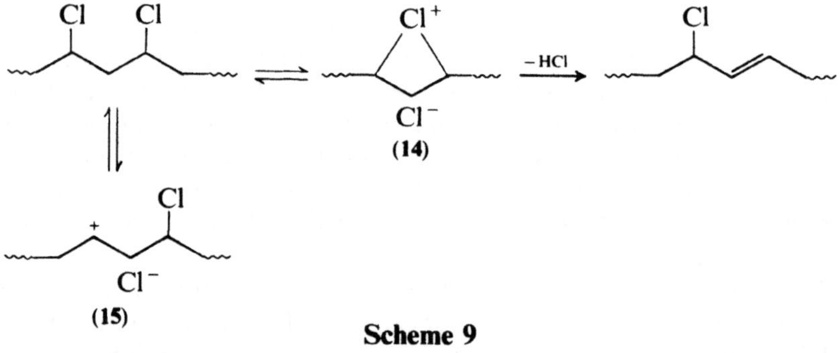

Scheme 9

The possibility of dehydrochlorination via the cyclic cation is supported by molecular orbital theory calculations.[136] These authors showed that the cyclic cation (**14**) is energetically more favoured than the linear ion

(**15**). For Scheme 9, the following reactivity order is predicted:[136] isotactic > heterotactic > syndiotactic.

There is thus strong support for the existence of dehydrochlorination from ordinary monomer units. Apart from random dehydrochlorination, the residual thermal lability after chemical pretreatments may be caused by the reactivity of the introduced groups or by unreacted labile structures. The lowest level of remaining thermolability obtained so far for ordinary PVC, is in the order of 10–15 % of that in the original sample.[93,94,133,134] It may be assumed that this instability is caused by random dehydrochlorination. Combined with our results obtained with s- and u-PVC, summarised in eqns. (1) and (2), this assumption will give the following tentative estimate of the contribution to the initiation of the dehydrochlorination from different structures:

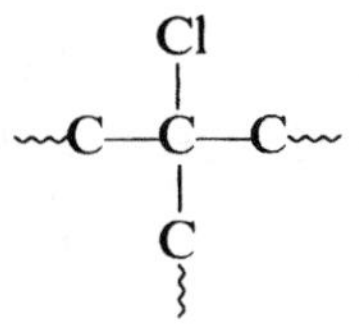
70–80 %

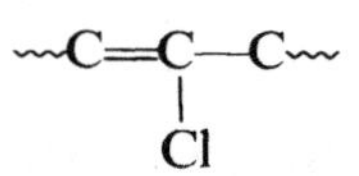
10–15 %

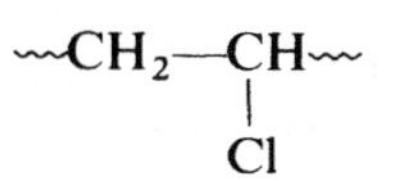
10–15 %

This illustrates the new view of the initiation of the dehydrochlorination attained during the last few years, namely that internal allylic chlorine is not the dominating labile structure but accounts for only a minor part of the instability of PVC. Random dehydrochlorination must also be considered, but branch points containing tertiary chlorine are the dominating labile structures in PVC.

2.4 POLYENE PROPAGATION

Irrespective of the nature of the initiation site, the generation of HCl is mainly a matter of polyene propagation. One important feature of this propagation is the restricted length of the resulting polyene sequences. The maximum number of conjugated double bonds is about 30,[137] while mean values in the range 6–14 have been reported.[72,138–140] The sequence distribution does not gradually change towards a higher mean value. Instead, 'long' sequences are observed already at low levels of dehydrochlorination[141] and the distribution is shifted towards shorter sequences at higher conversions.[72,139,141] The fast development of long sequences has been ascribed to a rapid 'zipper' process for the growth of the polyenes.

Over the years, several different mechanisms for polyene propagation have been discussed. The suggestions include radical, ionic and unimolecular (concerted) paths. In recent years few new contributions have been made. In the 1950s much attention was given to radical mechanisms. Winkler[142] and Stromberg *et al.*[97] suggested similar schemes in which chlorine atoms are the chain carriers:

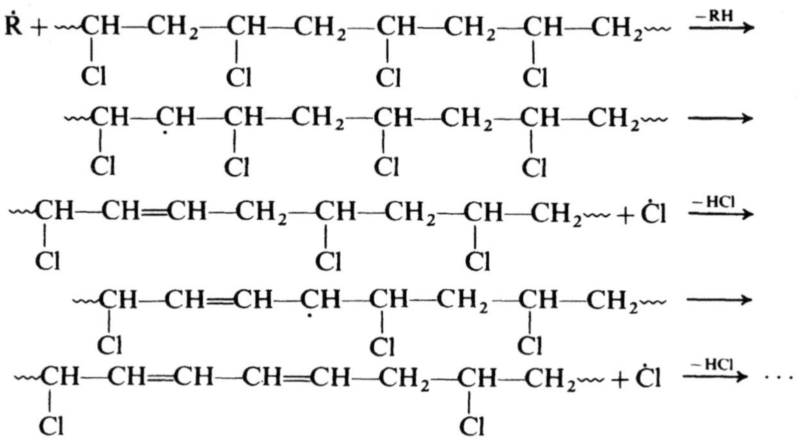

Scheme 10

As support for a radical mechanism, the following observations have been advanced; namely (a) increased rate of degradation in the presence of radical initiators,[106,143] γ-irradiation[11,144–147] and UV irradiation,[148–150] (b) the presence of radicals as detected by ESR-spectroscopy,[65,138,151–154] (c) grafting of methylmethacrylate to PVC during degradation,[155] (d) the retarding effect of radical traps,[156,157] (e) increased rate of degradation of polymethylmethacrylate[158–160] and polypropylene[156] in the presence of PVC, (f) decreased rate of degradation of polystyrene and PVC in PS–PVC blends,[161] and (g) exchange reaction with toluene labelled with tritium or ^{14}C.[155] However, using the same techniques, other workers have reported contradictory results, i.e. the presence of radical initiators[162,163] or inhibitors[5,164] have no effect on the rate of dehydrochlorination. It has also been claimed that the observed ESR-signals originate from species which are not active in the polyene propagation but are due to thermoactivated electrons in the polyene systems[165] or to radicals connected with aromatic structures formed in secondary reactions[166] (see below). Although it has been argued that HCl catalysis is consistent with a radical mechanism,[146] most workers consider the influence of HCl as strong evidence against a radical mechanism for the polyene propagation. Recently, Starnes and Edelson[167,168] and Amer and Shapiro[103] have raised another objection, namely that free chlorine atoms are very mobile and rather unselective[169] in their abstraction of hydrogen. This would lead to isolated double bonds rather than polyene sequences.

The main support for ionic processes arises from observations that (a) the rate of dehydrochlorination in solution is influenced by the nature and the dielectric constant of the solvent,[163,170,171] (b) the degradation is catalysed by HCl,[172,173] other acids,[174] strong bases[164,175] and some metal salts,[162,170,176,177] (c) the degradation behaviour of low molecular weight model compounds[178] and (d) that the degradation of PVC follows first order kinetics.[4] Some workers find support for a concerted mechanism from the elimination behaviour of low molecular weight compounds.[2,179] The observation of first order kinetics is also in accordance with a concerted mechanism.[163,164]

Based on calculations of bond energies, Imoto and Nakaya[180] proposed an elimination mechanism with a cyclic, polarised transition state:

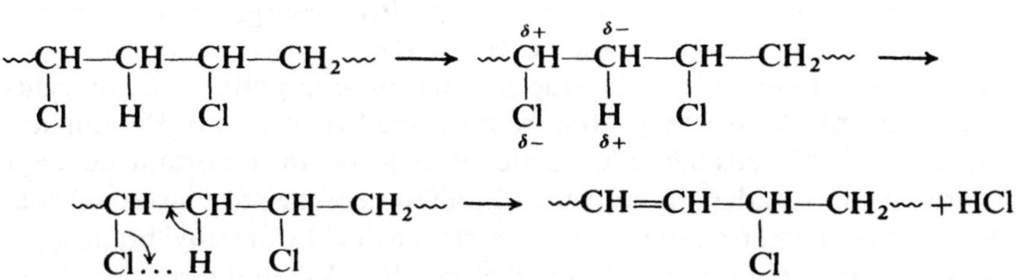

Scheme 11

This mechanism can *per se* explain random initiation. It has also been assumed that such an elimination would be facilitated by allyl activation, thus explaining the zipper reaction.[64,162,164,181] Mukherjee and Gupta[10] have pointed out that there is no evidence that a concerted elimination is activated by an adjacent double bond. On the other hand, it is known that ionic elimination is strongly allyl activated. An ion-pair mechanism according to Scheme 12 is considered to account for the data relating to the growth of the polyenes:[9,167,168]

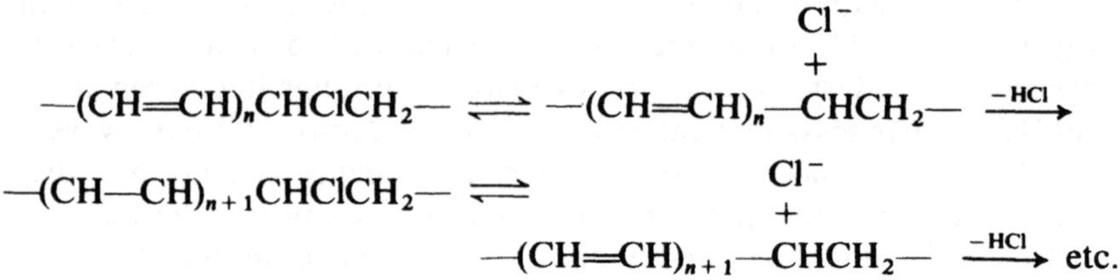

Scheme 12

By molecular orbital calculations and analogies with the formation of tropenylium chloride ion pairs in solution, Haddon and Starnes[183,184] have concluded that low equilibrium concentrations of polyenyl chloride ion pairs are likely to be present during degradation.

Resonance stabilisation has been assumed to be an important factor in explaining the restricted length of the polyene sequences.[4,140,142,185] At increasing length an additional HCl split-off will be less and less favourable from a thermodynamical point of view. The work with molecular orbital calculations referred to above showed that this assumption is not valid for neutral polyenes and that the incremental conjugation energy is constant,[183,184] which should disprove molecular mechanisms. On the other hand, the conjugation energy of the polyenyl cations formed in an ionic elimination reaches a limiting value for long sequences.[184] The calculations further revealed that the positive charge shows a strong tendency to concentrate in the centre of the polyenyl cation. This will decrease the ability of the chloride counterion to migrate to the sequence end in order to abstract a proton from the methylene group. Haddon and Starnes[9,183,184] consider this kinetic effect to be an important factor in restricting the length of the polyenes. Limitation of the polyene growth may also be due to other reasons, e.g. the presence of chloromethyl branches.[72] Assuming four chloromethyl branches per 1000 VC and a mean polyene sequence length of 10, about 10 % of the sequences should be terminated by these branch points. Another reason may be secondary reactions, e.g. inter- and intramolecular cyclisations (see below).

The stereoregularity is another important structural feature which influences the propagation and/or the termination of the polyene sequences. It has been reported that the polyene length increases with increasing syndiotacticity, both in thermal[123,186–189] and UV degradation.[190] The increased length may be due to faster propagation along favoured (*trans–trans*) conformations.[186–190] Assuming an ion pair mechanism according to Scheme 12[9] (or a concerted mechanism), the stereochemistry of the HCl elimination would be *cis*. An elimination from a syndiotactic dyad would thus lead to a *trans* alkene, which will not be liable to intramolecular cyclisation. An intramolecular cyclisation demands the presence of a *cis* alkene, which to some extent should be formed from isotactic dyads. The relation between cyclisation and polyene sequence length is illustrated by the finding that increasing syndiotacticity also leads to decreasing formation of benzene.[191,192] Amer and Shapiro[103] have suggested a mechanism in which the propagation is restricted to syndiotactic sequences, at least at low levels of dehydrochlorination. It is based on the observation that allylic chlorides with *cis* configuration undergo faster HCl elimination than the corresponding *trans* isomers.[193] This faster reaction has been ascribed to 1,4 elimination via a six-membered cyclic transition state. Dehydrochlorination of PVC is therefore presumed to proceed via 1,4-eliminations from *cis* double bonds. The resulting diene (or polyene) undergoes double bond shift by HCl-catalysed isomerisation. A syndiotactic sequence allows the regeneration of a *cis* double bond:

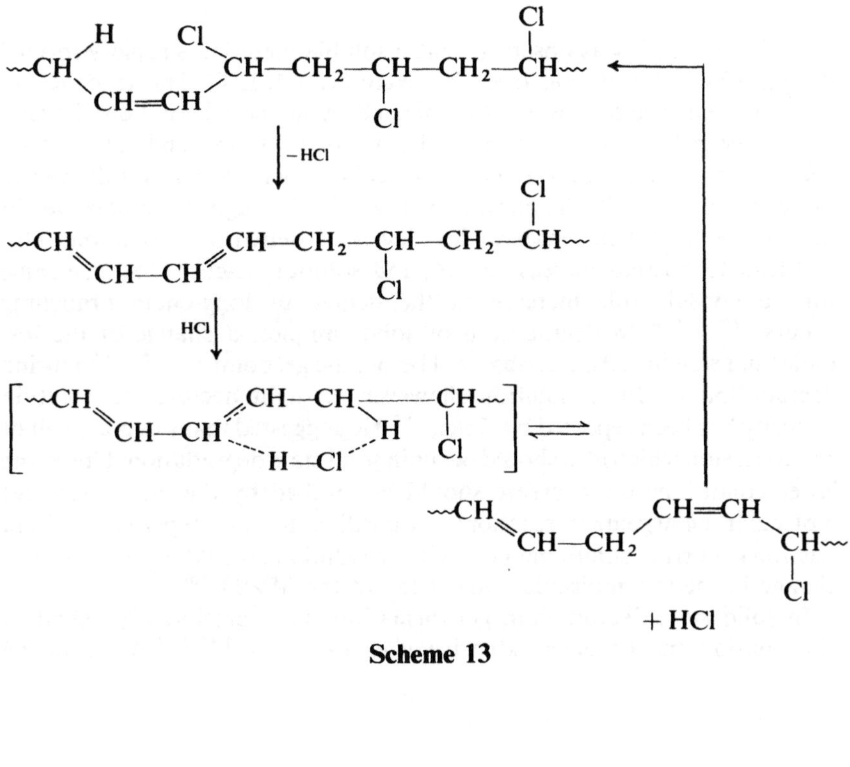

Scheme 13

The generation of isolated double bonds is supposed to proceed either by random concerted 1,2 elimination or by a free radical process initiated by tertiary or allylic chlorine. Like Starnes and Edelson,[167,168] Amer and Shapiro[103] consider that the chlorine atoms are not participating in a zipper mechanism but attack the polymer chain randomly. If a methylene proton is abstracted a new double bond is formed by chlorine expulsion, c.f. Scheme 4. The suggested mechanism accounts for the influence of tacticity, HCl catalysis and the presence of radicals. As will be discussed below, radicals may also be formed by secondary reactions.

As shown by this section on polyene propagation there are few new experimental data elucidating the nature of the reaction mechanism. Although it is likely that radicals are present, an ion-pair mechanism for the polyene propagation is more plausible. It accounts for the influence of solvent, HCl and polar additives, as well as the influence of tacticity. It also gives a reasonable explanation for the restricted length of the polyene sequences.

2.5 SECONDARY REACTIONS

In earlier work, it was observed that insoluble material is rapidly formed during PVC degradation (see, for example, ref. 194). The tendency to molecular enlargement was also indicated by increased solution viscosity before the gel point.[106,162] With the use of GPC in combination with viscosity measurements the molecular enlargement could be followed in more detail.[72,195,196] As shown by Fig. 2.7, the high molecular weight tail of the MWD increases at low levels of dehydrochlorination. This will lead to a rapid increase in $\bar{M}_w$ and solution viscosity. At the same time a considerable increase in the degree of long-chain branching occurs.[72,195,196] In degradation of solid samples, a change of the low molecular weight tail is not observed before the gel point.[72,195-199] During degradation in dilute solution, however, a slight decrease in solution viscosity has been reported by Kelen.[197] He suggested that this is a result of chain scission which also should occur in solid state degradation. Under the latter conditions the decrease should be masked by the more extensive molecular enlargement reactions. According to our experience, chain scission occurring simultaneously with crosslinking would be revealed by a change in the low molecular weight tail of the MWD.[200]

In solid state degradation, gel formation starts around 1% dehydrochlorination and becomes extensive after about 2%.[195,199] After the gel point, the molecular weight and the viscosity of the soluble material decrease rapidly.[196,197] However, this cannot be taken as an indication of chain scission, but as a result of the higher probability of larger molecules crosslinking.

Earlier, the network formation was assumed to occur by intermolecular HCl elimination between growing polyene sequences and ordinary methylene groups.[194,195] This mechanism would account for the increased crosslinking and decreased mean polyene length observed at increased degradation temperature.[195] It has also been considered to explain the decrease in polyene length at increasing degree of degradation,[195] but this statement is questionable. Diels–Alder cycloaddition between two polyenes (Scheme 14)[198,201–203] is more in accordance with the pertinent facts:

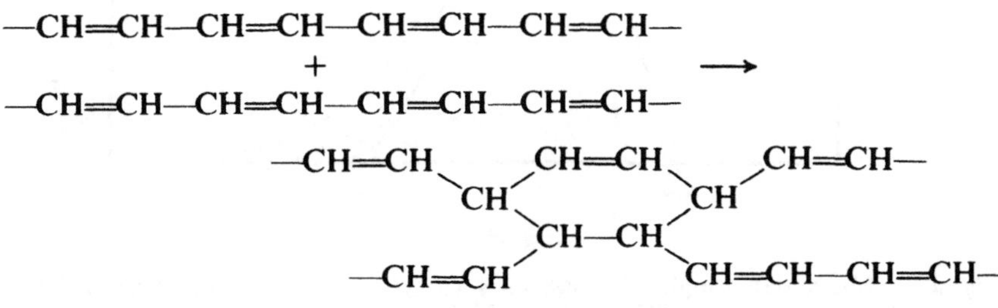

Scheme 14

Kelen *et al.*[198] have shown that this mechanism accounts for a major part of the crosslinking. By treating degraded, but still soluble PVC with maleic anhydride (MA), they observed a decrease in the high molecular weight material formed during degradation. For samples degraded beyond the gel point, a considerable decrease in the gel content was observed. These observations were explained by a displacement in the Diels–Alder/retro-Diels–Alder equilibrium:

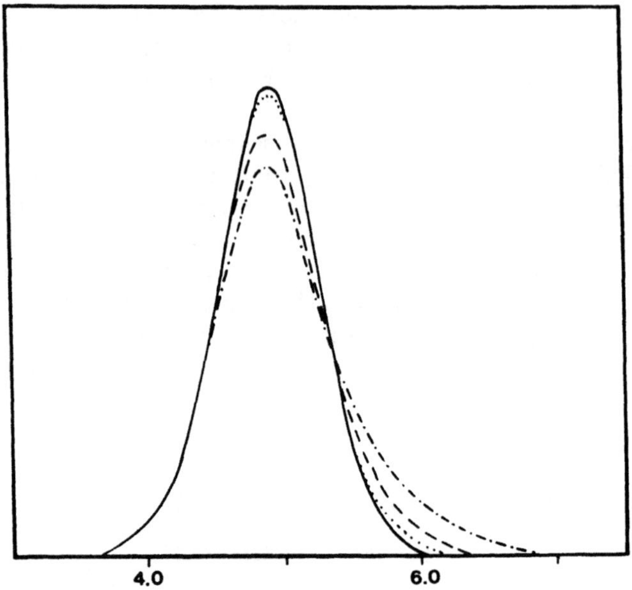

Fig. 2.7 Molecular weight distributions for a commercial suspension PVC degraded in nitrogen at 190 °C. ——, Original polymer; · · · ·, at 0·3 % conversion; ---, at 0·54 % conversion; -·-·-, at 1·0 % conversion. After ref. 72.

The existence of other crosslinking mechanisms was not excluded, but it was stated that at least 60–70 % of the crosslinking during thermal degradation of PVC is due to Diels–Alder reactions.

The formation of volatile aromatic pyrolysates, mainly benzene, is another result of secondary reactions.[191] At the conditions characteristic of PVC processing and stability testing, about 0·01 mole benzene per mole evolved HCl is formed.[197] O'Mara[204] pointed out that the cyclic product in Scheme 14 could be a precursor of benzene. However, further investigations by O'Mara and other workers with deuterium-labelled PVC have shown that benzene must be formed via an intramolecular route. Degradation of mixtures of PVC and PVC-d_3 results in benzene-h_6 and benzene-d_6 but only traces of other labelled benzenes.[204–206] Four different intramolecular routes have been suggested to explain the formation of benzene. One of them involves cyclisation of polyenes formed at the end of the polymer chain. Bellenger *et al.*[207] found that the evolution of benzene was dependent on the molecular weight, i.e. the concentration of end-groups, and that several benzene molecules were evolved per initial chain end after extensive degradation. They therefore suggested a repeated cyclisation and scission process, starting from 1-chlorovinyl ends:

$$\sim\!CHCl{-}CH_2{-}CHCl{-}CH_2{-}CHCl{-}CH{=}CHCl \xrightarrow{-HCl}$$

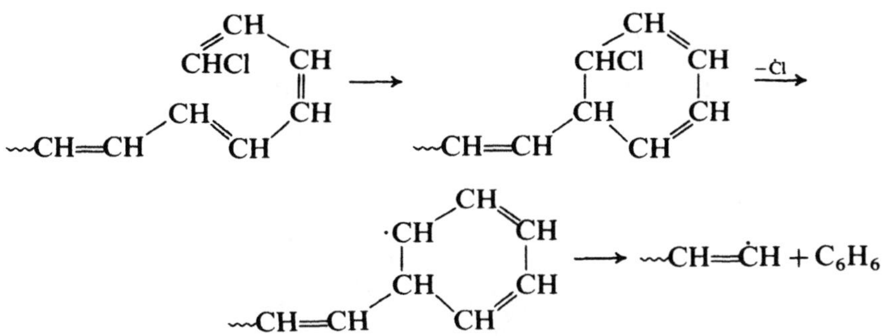

Scheme 15

However several facts speak against this mechanism. The content of 1-chlorovinyl ends is very low (see above) and the formation of vinyl radicals ($\sim\!CH{=}\dot{C}H$) is unlikely.[199] The most frequent unsaturated end-group, structure (**5**), should not react in the suggested way and is, in fact, found to remain unchanged during degradation.[111]

Another suggestion is that the benzene precursor is formed via cyclisation of triene segments in the triplet state.[208] At the same time there is a homolytic chain scission next to the ring. Benzene is then formed by another homolytic scission. This mechanism has been disputed, as the two first steps have no experimental support and no analogous reactions are known.[167,168] Intramolecular cyclisation of polyenes into cyclohexadiene structures by an electrocyclic reaction is more likely. Benzene is then formed by two successive homolytic cleavages:[166]

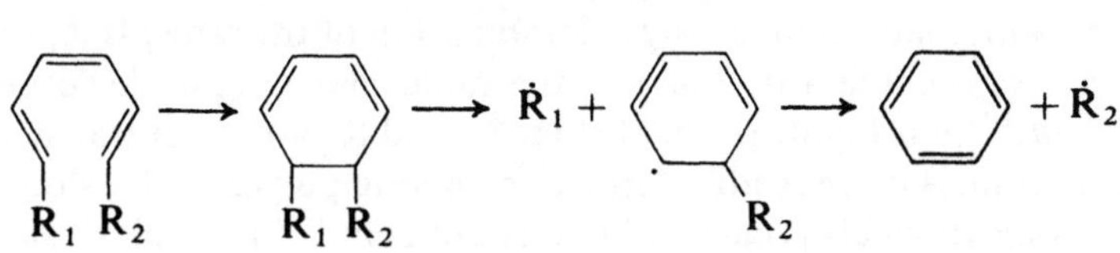

$$(R_1, R_2 = {-}CH_2{-}CHCl\!\sim,\ {-}CHCl{-}CH_2\!\sim,\ {-}CH{=}CH\!\sim)$$

Scheme 16

As mentioned above, pyrolysis with labelled polymers resulted in only a minor extent of benzenes with mixed labelling. Lattimer and Kroenke[206] have therefore suggested an extension of the mechanism above to include intermolecular abstraction of a proton:

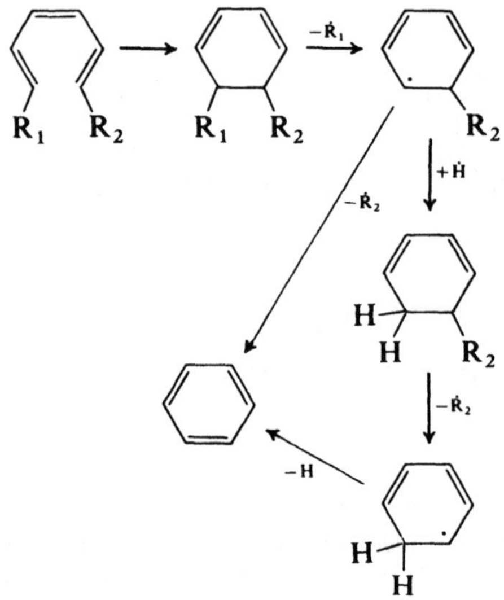

(R_1, R_2 = —CH_2—CHCl〰, —CHCl—CH_2〰, —CH═CH〰)

Scheme 17

These mechanisms result in chain scission. Starnes and Edelson[167,168] have pointed out that Kelen[197] found a decrease in the viscosity during degradation in dilute solution. They take this as support for Scheme 16 (and 17) including no coupling of $\dot{R}_1$ and $\dot{R}_2$. Nagy *et al.*[199] have, however, emphasised that the observed decrease in viscosity is small enough to affirm that the scission process is negligible compared to the formation of benzene. A mechanism involving a combination of the radicals $\dot{R}_1$ and $\dot{R}_2$ would *per se* exclude a decrease in the molecular weight. According to Tüdos *et al.*[141] it is highly probable that $\dot{R}_1$ and $\dot{R}_2$ are unsaturated, as the cyclic structure is more readily formed from long polyenes. Scission of the rather strong vinyl–alkyl carbon bond is unlikely.[199] Instead, cleavage of the C—H bond of the tertiary carbons in the ring is plausible.[167,168] This would lead to aromatic structures within the polymer chains:

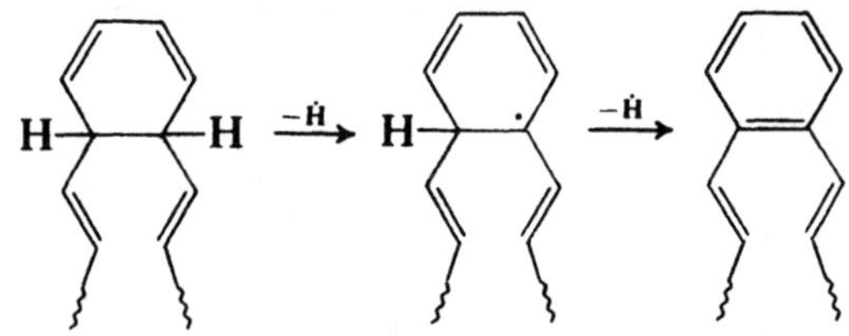

In order to account for the formation of benzene without accompanying chain scission, Nagy *et al.*[199] prefer another electrocyclic reaction scheme suggested by Starnes and Edelson.[167,168] It involves a bicyclo intermediate

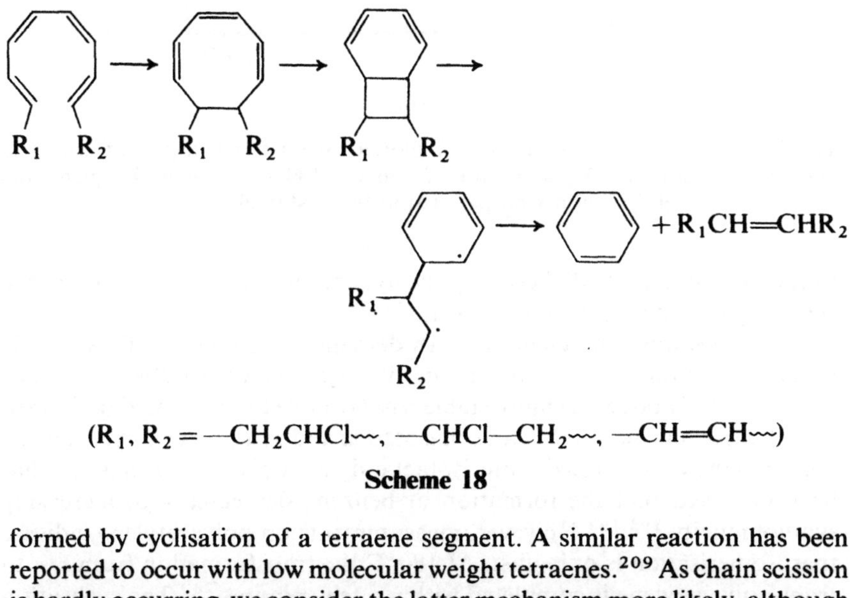

(R_1, R_2 = —CH_2CHCl〰, —$CHCl$—CH_2〰, —CH=CH〰)

Scheme 18

formed by cyclisation of a tetraene segment. A similar reaction has been reported to occur with low molecular weight tetraenes.[209] As chain scission is hardly occurring, we consider the latter mechanism more likely, although the former one (Schemes 16 and 17) cannot be totally excluded.

Few attempts have been made to directly prove the existence of the cyclic structures. Owen and Read[210] have investigated thermally degraded PVC with fluorescence spectroscopy. The spectra indicated the presence of a diene contained in a rigid skeleton. Furthermore, the excitation spectrum corresponded to the absorption spectrum of cyclohexadiene. Brauman and Chen[211] studied the soluble part of thermally degraded PVC with ^1H-NMR. They found resonances between 6·9 and 8·1 ppm, which they ascribed to benzenoid structures and assumed the presence of such groups

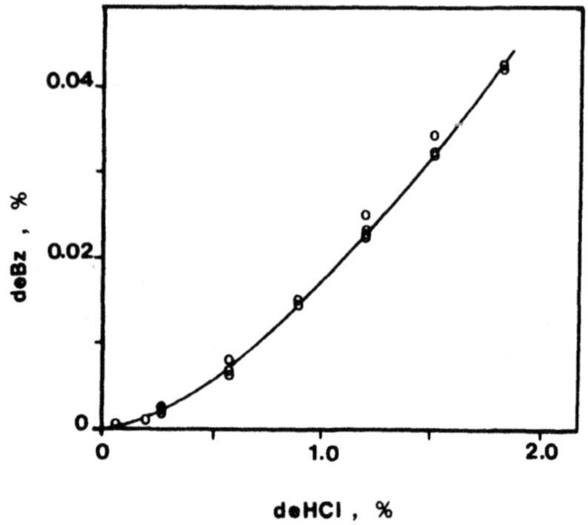

Fig. 2.8 Amount of benzene (deBz) plotted against the degree of dehydrochlorination (deHCl). Degradation performed at 180 °C in argon. Adapted from ref. 197 with kind permission from Marcel Dekker.

in the insoluble part. If present, polyenyl cations show resonances within this range of chemical shifts as well.[212]

The cyclisations in Schemes 16–18 demand the presence of *cis* double bonds; in the former case one and in the latter two consecutive *cis* bonds. As mentioned above, the more stable *trans* double bonds should be formed by dehydrochlorination from both isotactic and syndiotactic dyads, while the *cis* bonds are formed from isotactic dyads only. Accordingly, it has been observed that the formation of benzene decreases with increasing syndiotacticity.[191,192] However, measurements on polyacetylene indicate that isomerisation is quite fast. At 180 °C the transformation from *cis*- to *trans*-polyacetylene is completed within a few minutes.[263] The cyclisation reactions during degradation of PVC should thus be relatively fast. It is also observed that the relative yield of benzene increases with the degree of dehydrochlorination (Fig. 2.8). This behaviour might reflect the fact that the cleavage reactions are relatively slow compared to the HCl elimination. A fast cyclisation would contribute to the termination of the polyene sequence growth.

It may be concluded that the secondary reactions proceed by cyclisation of polyenes. Intermolecular combinations occur by Diels–Alder addition and result in rapid network formation. Intramolecular electrocyclic reactions lead to built-in cyclohexadiene or benzene groups and the formation of benzene and other aromatic compounds. All these reactions occur without any significant cleavage of the polymer chains.

2.6 INFLUENCE OF HCl

Once started, the 'burning' of PVC in processing equipment is very rapid. This is generally ascribed to a catalytic influence of the HCl evolved. Nowadays, there is overwhelming experimental evidence that HCl catalyses the thermal degradation,[103,130,170,196,213–221] although a few workers have reported that HCl has no catalytic effect under inert conditions.[222,223] This important feature of the degradation process was earlier a matter of much discussion (see, for example, refs. 1, 2, 6 and 8), although rather few experimental investigations have been directly devoted to the problem. This may be due in part to the difficulty of following the dehydrochlorination in the presence of deliberately added HCl. Various means of measuring the degree of dehydrochlorination in such systems have been applied, e.g. by following the changes in pressure,[209] volume[91] or HCl concentration by IR spectroscopy[103] in closed systems, or measuring the weight loss of the sample in a flow system.[196] In some investigations a closed degradation system has been used without removal of HCl, thus allowing the HCl concentration to increase gradually.[213,216–218] The extent of dehydrochlorination was determined by stopping the degradation at different time intervals and measuring the total amount of HCl evolved.

In many of these investigations the strong promotion of the rate of dehydrochlorination by HCl has been demonstrated. In our study of HCl catalysis,[196] we followed the weight changes during degradation in different mixtures of HCl and nitrogen in a flow system. As shown in Fig. 2.9, HCl exerted a strong influence on the rate of dehydrochlorination. It is also important to note that the catalytic effect increased with conversion, i.e. the presence of polyenes is a prerequisite for HCl catalysis. The same conclusion has also been drawn by Razuvaev *et al.*[218] and Morikawa.[224]

The structural changes occurring during degradation in the presence of HCl have been studied in order to reveal further details of the process. By comparing the UV spectra of PVC degraded as a film or in bulk, Thallmaier and Braun[225] showed that higher levels of HCl resulted in longer polyene sequences, while Carenza *et al.*[219] found the opposite effect in similar experiments. Entrapped oxygen was discussed as an alternative reason to the shortening in the latter case. In the experiment referred to in Fig. 2.9 we

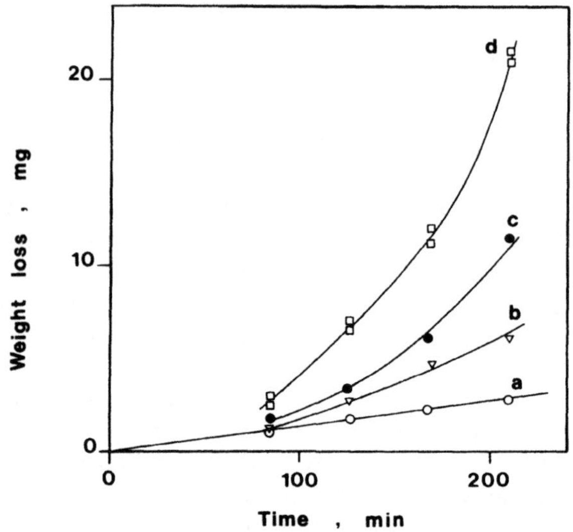

Fig. 2.9 Relation between weight loss and time during degradations at 190 °C in mixtures of nitrogen and HCl. (a) Pure nitrogen; (b) 10 % HCl; (c) 20 % HCl; (d) 40 % HCl. After ref. 196.

observed an increase in polyene sequence length at the lowest level of HCl (10 %). At higher HCl content no further increase in the sequence length could be noticed. Working with an HCl pressure of only 1 cm Hg, Amer and Shapiro[103] did not notice any change in the sequence distribution compared to degradation in vacuum. However, Nagy *et al.*[221] observed an almost momentary increase in the content of long polyene sequences when the removal of HCl by freezing-out was stopped during degradation in sealed ampoules. In these experiments the amount of short sequences also increased.

In some cases the course of the secondary reactions has been followed. By measuring the gel content Marks *et al.*[170] found an increased rate of molecular enlargement in the presence of HCl. After the above-mentioned rapid increase in the content of long polyenes, Nagy *et al.*[221] noticed that the distribution was shifted faster towards shorter sequences in the presence of HCl than in vacuum. These findings were taken as evidence for HCl catalysis of the secondary reactions. However, in these investigations the results are given in relation to the time of degradation instead of the extent. Such a representation is liable to be misinterpreted, as the rate of dehydrochlorination increases rapidly with the HCl content. Neiman *et al.*[213] found that the formation of benzene increased in the presence of HCl. They concluded, however, that HCl has no direct influence on the benzene formation and that the increased rate was an indirect effect of the increased rate of dehydrochlorination. As a measure of the secondary reactions we followed the molecular enlargement using GPC and viscometry.[196] If the molecular weight or the degree of long-chain branching is plotted against the conversion, no influence of HCl can be observed (Fig. 2.10). The two latter investigations show that HCl does not catalyse the secondary reactions leading to crosslinking and benzene formation. If the data presented in the two former investigations are rearranged in a similar way, it is easy to observe that they point to the same conclusion. Readdition of HCl to polyene sequences has been suggested to occur.[220] Using tritium-labelled HCl, Tüdos and Kelen[139] showed that a fast proton exchange takes place during degradation conditions. In some investigations a decelerating effect has been observed at relatively high levels of HCl in the atmosphere.[103,171] This has been ascribed to readdition of HCl.

As mentioned earlier, the existence of HCl catalysis has been taken as evidence for a non-radical mechanism in the (uncatalysed) thermal dehydrochlorination. Accordingly, most of the mechanisms suggested to explain the catalytic effect of HCl are ionic[173,215,218] or molecular,[103,196,218,224] although radical mechanisms have been suggested as well.[146,218] The majority of these reaction schemes involve catalysis of the dehydrochlorination from allyl-activated sites, i.e. catalysis of the propagation of polyene sequences. Minsker *et al.*[215] have argued that HCl exclusively catalyses the propagation. Faster propagation is *per se* in accordance with the increased polyene sequence length obtained in the presence of HCl.

Referring to the discussion concerning polyene propagation, it can be concluded that the dehydrochlorination most likely occurs via an ionic mechanism. The restricted length of the polyenes has been suggested mainly to be due to a kinetic effect for the ion pair in Scheme 12.[9,167,168] The increased length in the presence of HCl might be a result of an increased possibility of polarisation of the allylic carbon–chlorine bond, or a result of a change in the kinetic equilibrium. This in turn could be due to a less firm association between the chloride ion and the positive charge concentrated to the middle of the polyene. The rapid increase in mean polyene length observed if HCl is allowed to accumulate after a certain time of degradation in vacuum[221] is, however, important in this context. Obviously, already existing sequences become longer as well as those newly formed. Nagy *et al.*[221] have suggested that proton exchange may lead to migration of the polyenes. In this way a secondary chlorine may be allyl-activated:

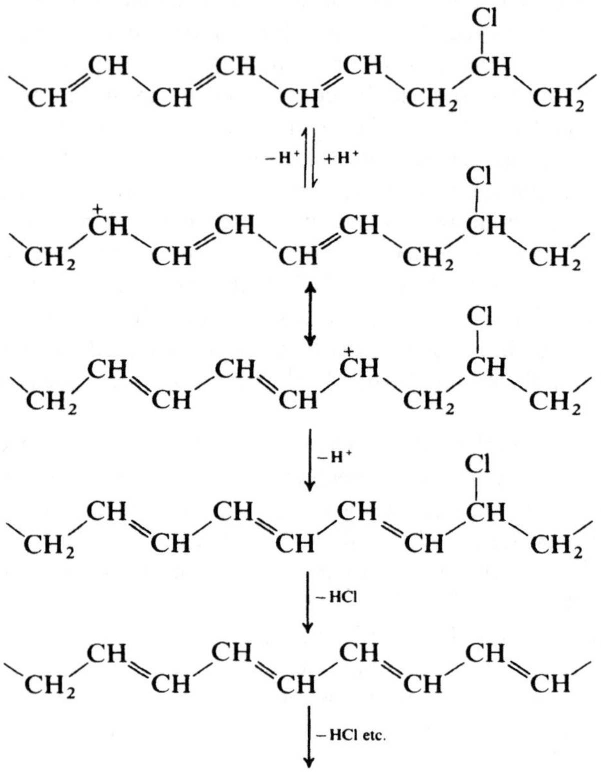

Scheme 19

This reinitiation of sequences terminated earlier would give longer polyenes followed by a corresponding dehydrochlorination. However, the data published by Nagy *et al.*[221] indicate that the rapid increase in polyene sequence length is not accompanied by a similar increase in the dehydrochlorination. Owen *et al.*[226] have shown that the sequence distribution of chemically degraded PVC shifts towards longer sequences by treatment with trifluoroacetic acid. They suggested that this may be due to protonation of polyene sequences followed by migration along the polymer chain until two sequences are unified. It was also argued that HCl would have the same effect during thermal degradation of PVC. In this

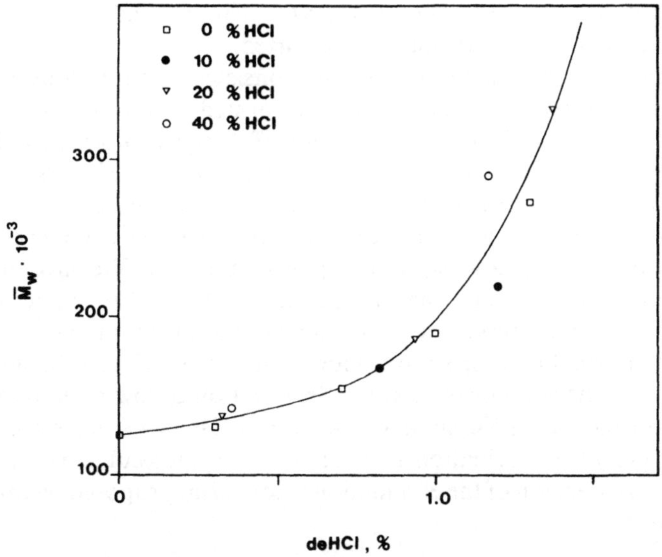

Fig. 2.10 Relation between $\bar{M}_w$ and the degree of dehydrochlorination after degradation at 190 °C in atmospheres containing different amounts of HCl. After ref. 196.

way, a sudden addition of HCl should lead to an increased polyene sequence length without any additional dehydrochlorination. The protonation and migration were described by the following reactions. If migration occurs by this mechanism, the polyenes would partly contain chlorine. Furthermore, it should be possible to observe sequences of CH_2 groups. So far, there are few experimental data reported about the detailed

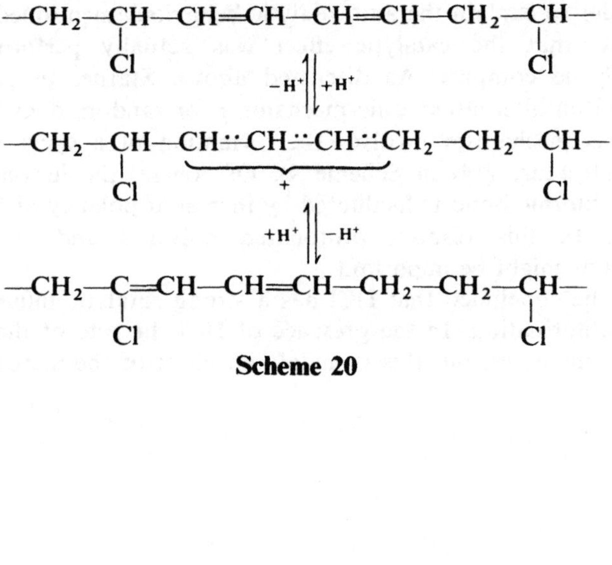

Scheme 20

molecular structure of degraded PVC. The presence of the above-mentioned structures has not been reported.

The mechanisms discussed above are consistent with the longer polyene sequences observed in degradations performed in the presence of HCl. They could explain the difference in rate of dehydrochlorination between the degradations in nitrogen and 10% HCl shown in Fig. 2.9. However, according to our measurements the polyene sequence distribution is unaffected at higher levels of HCl. The strong catalytic influence cannot thus be explained by catalysis of the zipper elimination. We have concluded that random dehydrochlorination is catalysed too.[196] At higher levels of HCl this is the most important effect and it is likely that catalysis of random dehydrochlorination is operative at low contents of HCl as well. Suzuki and Nakamura[90] have reported that the degradation of low molecular weight models for the ideal PVC chain is catalysed by HCl. We suggested that the random dehydrochlorination is catalysed by a molecular mechanism[196] which is an extension of the elimination mechanism proposed by Imoto and Nakaya:[180]

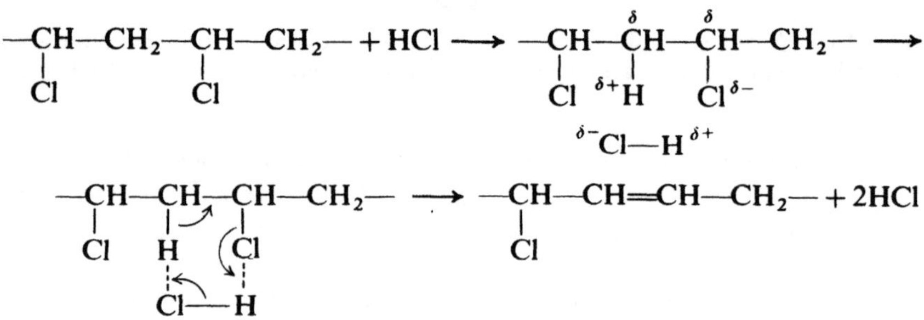

Scheme 21

In order to explain the autocatalytic behaviour mentioned earlier, we suggested that the catalytic effect was actually performed by an HCl–polyene complex. As discussed above, Starnes *et al.*[9,136] have suggested an alternative ionic mechanism for random dehydrochlorination. This mechanism involves the formation of a cyclic chloronium cation (structure (**14**) in Scheme 9). Of course, the ionisation of the carbon–chlorine bond is facilitated by increased polarity of the reaction medium. In this respect, protonated polyenes and their chloride counterions might be important.

It can be concluded that HCl has a strong catalytic influence on the dehydrochlorination. In the presence of HCl the rate of the secondary reactions increases, but this is mainly an effect of the increased rate of polyene sequence formation. It is likely that the most important effect of HCl is an increase in the random dehydrochlorination. The polyenes are involved in the reaction mechanism but the exact nature of the HCl catalysis is not known.

2.7 DEGRADATION DURING PROCESSING

During processing, the polymer is exposed to temperatures which cause thermal dehydrochlorination. In addition, the presence of oxygen will lead to thermooxidative degradation and the mechanical stresses may cause chain scission. Normally, processing is performed with a number of additives, e.g. thermal stabilisers and lubricants, which influence the degradation reactions. The effect of stabilisers is discussed in Chapter 5. In this section a short account of degradation during processing of unstabilised PVC will be given.

Thermooxidative degradation of PVC has not been studied as extensively as thermal degradation. For a recent review the reader is referred to ref. 227. Chapter 3 in this book deals with photooxidative degradation. Like thermal degradation, the main feature in thermo-oxidative degradation is dehydrochlorination. The presence of oxygen causes acceleration and it has been suggested that the rate of dehydrochlorination should be proportional to the square root of the oxygen partial pressure.[107,228] Compared to thermal degradation, the discoloration is not as severe during thermooxidative degradation. Accordingly, the UV–visible spectra of thermooxidatively degraded PVC show that the polyene sequences are shorter.[64,107,162,163,195,229,230] This is obviously a result of reactions between oxygen and the polyenes.[231,232] If the degradation is performed in oxygen, no molecular enlargement can be observed at low conversions.[195] In air, on the other hand, molecular enlargement as well as scission reactions are observed.[106,194,195,233] The effect of the oxygen content of the atmosphere on the changes of molecular weight during degradation is illustrated in Fig. 2.11. At higher conversions, molecular enlargement leads to gel formation, even in pure oxygen.[164,233] By treating gel, formed in thermooxidative degradation, with maleic anhydride, Ivan *et al.*[233] have shown that the majority of the crosslinks are the result of intermolecular Diels–Alder cycloadditions in agreement with the behaviour during thermal degradation.

The thermooxidative degradation of PVC is mostly considered to proceed via a radical mechanism (see, for example, refs. 1, 5, 6, 8).

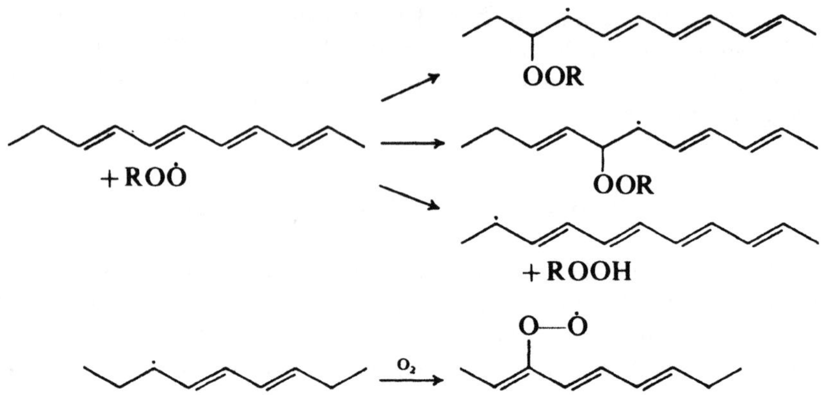

Scheme 22

However, Tüdos *et al.*[227] have pointed out that the overall activation energy of HCl elimination is practically the same for thermal and thermooxidative degradation.[195,232] They therefore suggest that the initial dehydrochlorination proceeds by the same mechanism in thermal and thermooxidative degradation. No specific mechanism was suggested, but referring to the discussion above, an ionic mechanism seems most likely. Oxygen and peroxy radicals will react with the polyenes, forming peroxides.[227] In this way, the polyenes become shorter. The peroxy radicals may also abstract hydrogen from intact monomer units and initiate further HCl loss by thermal dehydrochlorination:

$$\mathrm{RO\dot{O}} + \sim\mathrm{CH_2{-}CHCl{-}CH_2{-}CHCl}\sim \longrightarrow$$

$$\mathrm{ROOH} + \sim\mathrm{CH_2{-}CHCl{-}\dot{C}H{-}CHCl}\sim \xrightarrow{-\dot{\mathrm{C}}\mathrm{l}}$$

$$\sim\mathrm{CH_2{-}CH{=}CH{-}CHCl}\sim \xrightarrow{-n\mathrm{HCl}} \sim\mathrm{CH_2{-}(CH{=}CH)_{n+1}}\sim$$

Scheme 23

For a more complete discussion of the different reactions possible, see ref. 227 and Chapter 3.

Besides heat and oxygen, the polymer may be influenced by mechanical energy during processing. The shear forces will reach a maximum in the beginning of the processing when the material is only partly melted. Scott *et al.*[234] have suggested that chain scission occurs at this stage. The radicals formed then react with oxygen. The chlorine atoms, as well as the alkyl and alkylperoxy radicals, could initiate zipper dehydrochlorination[234] as in Scheme 22. Instead of molecular diminishing, formation of gel was observed after only about one minute of processing. This was explained by

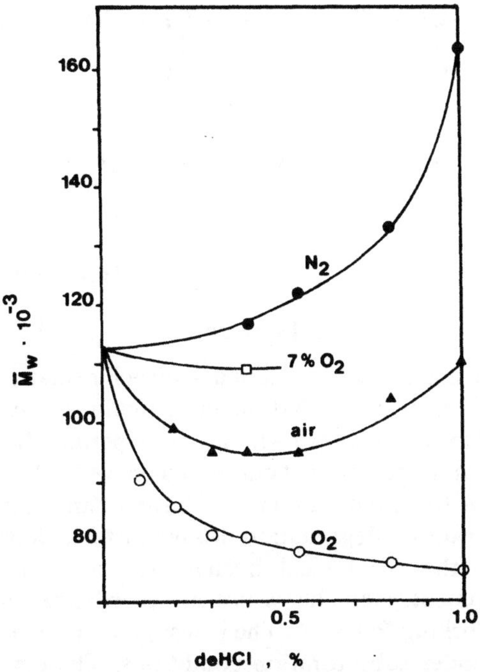

Fig. 2.11 $\bar{M}_w$ as a function of the degree of dehydrochlorination (deHCl). Data for a commercial suspension PVC degraded at 190 °C at different oxygen levels. After ref. 195.

the formation of a thermally unstable gel by addition of alkylperoxy radicals to double bonds.[234] This rapid formation of gel is contrary to results obtained by Nagy *et al.*[199] in similar experiments with unstabilised PVC. They reported that the formation of insoluble material started after about 40 min of processing at 180 °C. Scott *et al.*[234] measured the content of gel as the insoluble part after 10 min refluxing in methylene chloride. It is questionable whether this treatment is enough to completely dissolve the polymer after processing.

Scott and co-workers[234,235] consider that the chain scission process in Scheme 24 is much more important than the ordinary thermal dehydrochlorination for the initiation of the degradation in the first stage of the process. The later stages would be characterised by thermooxidative degradation where the unsaturation formed in the first stage plays an important role. According to this research group, the accelerating effect of HCl is caused by a catalytic effect on the radical decomposition of

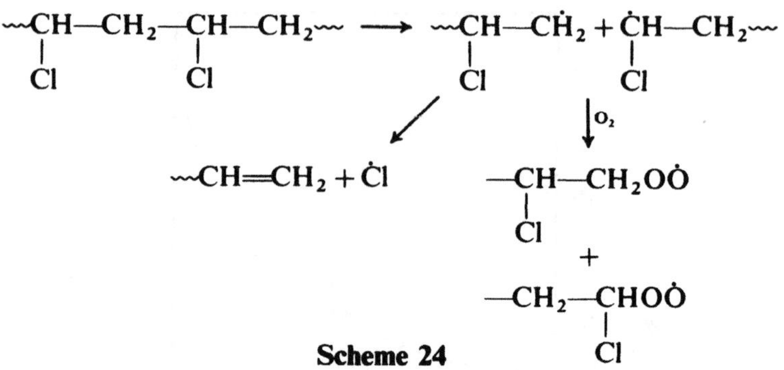

Scheme 24

peroxides, resulting in increased random dehydrochlorination.[236] Accordingly, Scott *et al.*[234] have stated that the most significant damage to PVC during commercial processing occurs in the initial period, due to mechanochemical reactions in the presence of entrapped oxygen. The concentration of thermolabile structures should therefore have a minor influence on the degradation. The rate of degradation would instead depend on other factors related to the rheological behaviour, e.g. molecular weight, morphology, equipment, etc. This is not in accordance with the results obtained by the Working Party.[237] The investigated polymers included a sample prepared under subsaturation conditions. The concentration of tertiary chlorine and internal double bonds is much higher in this polymer than in normal PVC and, consequently, it shows a higher rate of dehydrochlorination in static tests.[238] In spite of the relatively low molecular weight, i.e. low melt viscosity, the subsaturation PVC is also considerably less stable than the reference samples during dynamic tests in the presence of stabilisers.[237] In fact, in some cases the instability did not even allow processing. It must therefore be stated that the concentration of labile structures indeed has a major influence on the degradation of PVC during processing.

In very recent work, Pukánszky *et al.*[239] compared the dynamic and static degradation of PVC. They studied the process by IR and UV spectroscopy and GPC. The $\bar{M}_w$ value increased gradually during

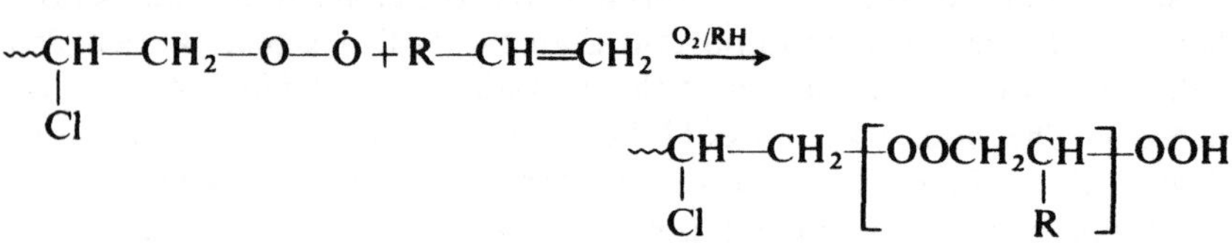

Scheme 25

processing and tended to infinity at the gel point after 40–50 min. The UV spectra obtained for material degraded in processing showed resemblances to those obtained after static degradation in air. However, in the samples obtained with dynamic degradation, the amount of carbonyl groups was much less than in samples degraded in air. Furthermore, static degradation in sealed evacuated tubes in the presence of evolved HCl resulted in similar UV spectra. Pukánszky *et al.*[239] therefore concluded that HCl is an important parameter during processing. Due to the catalytic effect of HCl, the degradation in dynamic tests will be faster than in static dehydrochlorination experiments.

It is obvious that the degradation of PVC during processing is rather complex and that not all the details are known. Different research groups have quite different views of the degradation process and the rôle of, for example, the thermolabile structures present in virgin PVC. Our conclusion is that thermal dehydrochlorination is the dominating reaction. If oxygen is present, a radical process leading to peroxides and initiation of new polyenes also occurs. However, as the oxygen content in the polymer melt during processing is rather low, this effect can be neglected. The HCl evolved seems to be much more important, due to the catalytic influence on the dehydrochlorination. The increase in temperature caused by the mechanical energy will increase the rate of degradation. The shear forces may cause chain scission, but this is not conclusively shown. As in thermal degradation under static conditions, the labile structures in the original polyenes therefore are important for the degradation during processing.

2.8 STABILITY IMPROVEMENT OF PVC

As discussed earlier, present knowledge indicates that the low stability of PVC under processing conditions primarily is due to the content of tertiary chlorine, although internal allylic chlorine and random dehydrochlorination also contribute. The low stability makes the addition of thermostabilisers necessary. However, these additives have several drawbacks, including toxicity, high costs and their influence on the rheological behaviour and end-use properties. It would therefore be of great practical interest to obtain PVC with considerably improved thermal stability. This could be achieved either by changing the polymerisation conditions in order to reduce the content of labile structures, or by eliminating the negative influence of these structures by chemical modifications.

Both tertiary and internal allylic chlorine are the result of chain transfer to polymer. Our work with subsaturation polymerisation has shown that the formation of these structures increases with decreasing monomer concentration. The content will thus increase after the pressure drop in a conventional polymerisation. The availability of monomer at saturated conditions is, however, not enough to prevent these side reactions. In fact, the major part of the tertiary and internal allylic chlorine in commercial PVC is formed before the pressure drop.[27]

An obvious way to reduce the content of labile structures is therefore to increase the monomer concentration in the polymer gel. This can *per se* be achieved by increasing the polymerisation temperature, although the effect is small.[240] However, in practice such a procedure is not possible, as the molecular weight of commercial PVC is controlled by the polymerisation temperature. Furthermore, at higher temperature the tendency to such reactions will increase. As mentioned above, polymers prepared at 100 °C show a relatively high content of tertiary chlorine.[62] Higher polymerisation pressure would raise the monomer concentration at the reaction site. It has been reported that polymerisation rate and molecular weight increases with the pressure.[241–243] The molecular weight of polymers obtained at 2500 atm was doubled compared to that of PVC polymerised in the normal way at the same temperature.[242] No information about the thermal stability of these polymers was reported. Addition of monomer absorbing agents would increase the monomer availability. In most cases, such additions should lead to chain transfer to the additive and a decrease in the molecular weight. The use of tetrahydrofuran, for example, has been found to improve the thermal stability and reduce the molecular weight.[244] The content of labile structures was not determined.

A more adequate method would be to use a polymerisation system in which chain transfer to polymer is unlikely or reduced, e.g. an anionic system. Kolinsky and co-workers,[245–247] as well as other research workers,[90,248] have studied polymerisation of vinyl chloride initiated by butyllithium. The dehydrochlorination rate of such polymers is about one-third of that of ordinary PVC.[90,135] Furthermore, degraded samples were much less discoloured than ordinary PVC dehydrochlorinated to the same extent.[135] The thermal stability of these polymers is, however, not as high as what could be expected from PVC without structural defects. Our NMR measurements have shown that butyllithium-PVC contains short-chain branches with one carbon.[135] This observation indicates that transfer reactions may occur in this system as well. Due to economical reasons, this polymerisation system has not yet been commercialised.

Elimination of labile structures already formed is the other main way of improving the stability. This can be achieved either by substitution of thermolabile chlorine by more strongly bonded groups, or by increasing the stability of weak carbon–chlorine bonds. Removal of allylic chlorine by saturation of the double bond is an example of the latter approach. Mild chlorination [64,67] addition of HF[249] and reaction with di-imide[145] have been used to remove double bonds. Elimination of allylic chlorine should only lead to minor improvements of the thermal stability—about 10–15%. Baum and Wartman,[64] as well as Minsker *et al.*[67] have, however, reported improvements of about 70% by mild chlorination in darkness. This considerable increase in stability implies that tertiary chlorine is deactivated as well. There is no obvious way of explaining such a deactivation.

As originally proposed by Frey and Horst,[116,250] thermal stabilisers may act by substituting labile chlorine, thereby introducing more thermally stable groups. Experiments with low molecular weight models have shown that Ca–Zn carboxylates and tin stabilisers substitute allylic chlorine at processing temperatures.[5,251] For structures containing tertiary chlorine, both substitution[252] and HCl elimination[90] have been reported. Elimination of HCl leads to the formation of allylic chlorine which is subsequently substituted. The thermal stability of the new structures has also been investigated and the following stability order is reported:[90] $—SC_6H_5 > —OH > —OCOR$. It has further been shown that thermal stabilisers are bound to PVC at processing conditions,[251,253] but their selectivity against labile structures has not been elucidated. One important factor for the function of thermal stabilisers is the deleterious effect of the metal chlorides formed, e.g. $ZnCl_2$ or R_2SnCl_2. A further discussion about the stabilisation of PVC can be found in Chapter 5.

Substitution of labile chlorine has also been applied in order to increase the thermal stability of PVC prior to processing. While earlier attempts with thermal stabilisers gave negative results,[254–256] Starnes and Plitz[93,257] obtained good results using di(*n*-butyl)tin bis(*n*-dodecyl mercaptide). The reaction was carried out in *o*-dichlorobenzene solution at high temperature (up to 185 °C), for a long time (up to 240 h) and with a large amount of reactant. Under these conditions the rate of dehydrochlorination was reduced to 10–15% of the original rate. The thermal stability increased with the sulphur content, reaching a maximum level when about 20 mercaptide groups per 1000 VC were incorporated. This is far more than the amount of tertiary and allylic chlorine (including structure (**5**)) in the virgin PVC.

Reaction with secondary chlorine or, more likely, with allylic chlorine formed by random dehydrochlorination during treatment, may explain the high degree of substitution. No attempt was made to check whether the original labile structures were substituted. Starnes and co-workers obtained similar results when they treated PVC in *o*-dichlorobenzene solution with other organotins[258,259] or thiols.[94] If the reaction with thiol was performed in benzophenone solution, they noticed partial reductive dehalogenation.

Several workers have tried to substitute labile chlorine using organo-aluminium compounds. Gaylord and Takahashi[260] treated PVC with diethylaluminiumchloride (Et_2AlCl) followed by addition of methanol. They found an increased thermal stability and assumed that Et_2AlCl attacked labile chlorine, forming a carbocation, which reacted with the alcohol. If butadiene was added instead, grafting occurred and the stability increased.[261] Similar results were obtained by Thame and co-workers[119,262] using butadiene and isobutene. Thame *et al.*[119] investigated non-graft modifications with triethylaluminium (Et_3Al) as well. Improved thermal stability was obtained, especially when toluene was used as the reaction medium. They assumed that toluene was incorporated via a Friedel–Crafts reaction. By treating PVC with triphenylaluminium in CS_2 suspension, Kennedy and Ichikawa[131] obtained improved heat stability.

Mitani *et al.*[133,134] investigated the reaction between PVC and organoaluminium compounds in more detail. They found that Et_3Al was very effective as a stabiliser, but only if the amount was large enough. Ethylaluminium chlorides caused decomposition during reaction. They proposed that tertiary and allylic chlorines in PVC react with trialkyl-aluminium, which is dimeric in solution. The deleterious aluminium chloride is subsequently transformed into a less reactive complex if there is an excess of triethylaluminium:

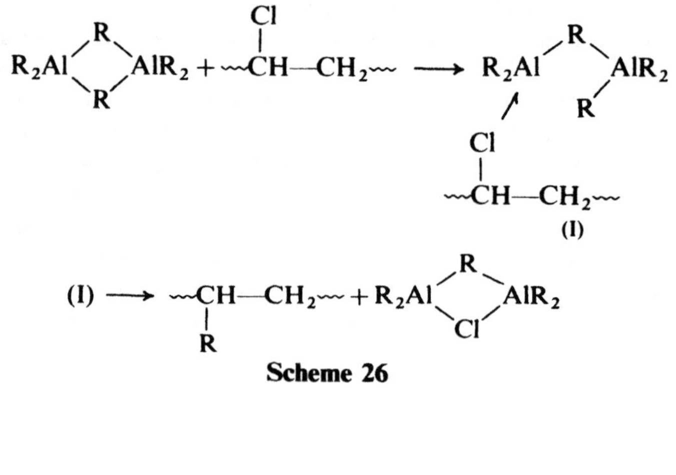

Scheme 26

The solvent or suspending medium has a profound influence on the reaction for several reasons. Carbon tetrachloride reacts with trialkylaluminium forming prodegradant aluminium chloride, while ethers, esters and ketones form strong donor–acceptor complexes with Et_3Al, resulting in negligible alkylation of PVC. Amines not only form complexes with Et_3Al but also induce dehydrochlorination of PVC at high temperatures. The use of 1,2-dichloroethane, benzene and toluene, which are inert and act as solvents or swelling agents for PVC, results in increased thermal stability. The improvement obtained by reaction in benzene slurry was enhanced by subsequent cooling, which resulted in a reduced content of aluminium compounds in the polymer.[134] In spite of the heterogeneous conditions, the reaction in benzene was very fast. It was assumed that 90% of the labile chlorines had reacted after 5 min at 85 °C. Using this method, Mitani *et al.*[134] managed to reduce the rate of dehydrochlorination to 10–15% of the original rate. Furthermore, the mean polyene sequence length was reduced from 14 to 4, as measured by UV spectroscopy.[133] Based on the minimum amount of Et_3Al necessary to avoid decomposition during treatment, they estimated the amount of labile (substitutable) chlorine to be 2·5 per 1000 VC.[134] This value is in good agreement with the content of tertiary and allylic chlorine including structure (**5**) (see above). Treatments with Et_3Al, tin mercaptides or thiols result in approximately the same improvement in heat stability, i.e. by a factor of about 10. As quite different linkages are formed in these reactions, it is plausible to assume that the remaining instability is due to random dehydrochlorination. Although the improvement is impressive, the reaction conditions prohibit the use of these methods in practice.

Substitution of labile chlorine by solvolytic displacement reactions, as discussed by Suzuki,[132] is another way to achieve stabilisation. These reactions are of the S_N1 type and the rate-determining step is the ionisation of the carbon–chlorine bond. The polarity of the reaction medium is therefore very important. As a compromise between this demand and accessibility, Suzuki[132] suggested C_1–C_3 alcohol/water mixtures. Treatment of a low molecular weight model compound of allylic chlorine with refluxing ethanol/water (1:1) gave the following products: hydroxyl substitution 73%, ethoxy substitution 25% and HCl elimination 2%. A model containing tertiary chlorine underwent fast and complete elimination of HCl. Treatment of PVC with the same mixture for 10 h at 150 °C decreased the rate of dehydrochlorination to about 30% but the product was discoloured. By addition of minor quantities of metal carboxylates or epoxy stabilisers this discoloration could be avoided. This method was used commercially a few years ago.

Obviously, there are several ways to improve the thermal stability of PVC by substitution of labile chlorine. By such reactions, improvements of 10–15% have been obtained. Of the reported methods, treatment with triethylaluminium seems to be especially effective, while a method based on solvolytic displacement has advantages from a technical point of view. Much less work has been devoted to the polymerisation procedure in order to depress the formation of labile structures. Polymerisation with butyllithium as initiator is the only system investigated in more detail. With the present knowledge, the development of adequate methods to obtain heat stable PVC would be more effective. However, the improvement will be restricted to the limit set by random dehydrochlorination.

2.9 CONCLUDING REMARKS

As illustrated by this review of the thermal degradation of PVC, a considerable gain in knowledge has been achieved in the last seven to eight years. Much of the progress is due to the development of modern techniques for the characterisation of polymer structure. NMR has played an important role in the elucidation of the nature and content of irregular structures in virgin PVC. It is in this area of degradation studies that the most obvious progress has been made. It is reasonable to believe that the essential features concerning the formation and importance of the dominating labile structures are now known.

The knowledge of polyene propagation and secondary reactions is not established to the same degree. The possibility of following the changes in the molecular weight distribution by GPC increased the understanding of the molecular enlarging and diminishing reactions. In the work now in progress in this field, NMR will certainly be a powerful tool.

The ultimate goal for the research of PVC degradation is to obtain a polymer with greatly enhanced stability. Most of the work concerning this question still remains to be done, although important attempts have been made. Naturally, the increased knowledge of the fundamental processes in degradation will promote the development of PVC with improved heat stability.

ACKNOWLEDGEMENTS

The financial support from the Swedish Board of Technical Development is acknowledged. The authors also want to express their sincere thanks to Anne Wendel for valuable assistance.

APPENDIX

In the papers published during the preparation of this chapter we have noticed the following information.

Starnes *et al.*[264] have discussed in detail their results obtained with ^{13}C-NMR spectroscopy on reduced PVC samples in a very recent paper. The authors' opinion concerning the structure of branches and end-groups in PVC and the reaction mechanisms for their formation is similar to that discussed earlier in this review.

Starnes *et al.*[265] have also demonstrated the possibility of determining internal double bonds by ^{13}C-NMR after reduction. They claim that they have found evidence for the presence of the cyclopentane group formed from the internal double bonds in a reduced PVC sample which was solution polymerised at low monomer concentration. Very recently, we have found an approximate agreement between the content of double bonds determined in this way and that obtained by ozonolysis, in subsaturation PVC obtained at low monomer pressure. Although internal double bonds can be thus determined in a more direct manner, the relatively low sensitivity of NMR restricts the use of this technique to samples with an increased content of double bonds. For ordinary commercial PVC, ozonolysis is still the only available method to determine internal double bonds.

Other investigations concerning the determination of the structure of PVC include H-NMR measurements of end-groups in low molecular weight fractions[266] and nucleophilic substitution of chlorine by phenyl thiocarbonylthio groups.[267] Braun and Holzer have published two papers concerning polymerisation of vinyl chloride at atmospheric pressure.[268,269] They have found that the low monomer concentration results in increased contents of internal double bonds and tertiary chlorine in accordance with our findings discussed earlier. The atmospheric PVC was also found to contain aldehyde groups. Braun and Holzer suggest that such structures can be formed by dehydrochlorination of end-groups formed by coupling with hydroxyl radicals:

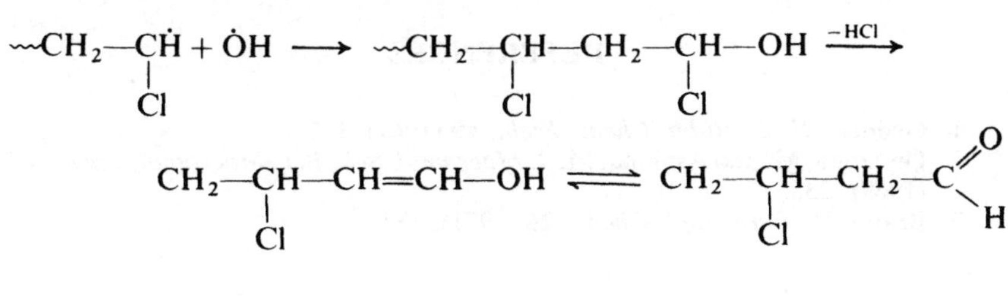

Ivan *et al.*[270] have studied the degradation of PVC containing increased concentrations of internal allylic chlorines obtained by controlled chemical dehydrochlorination. The introduction of this structure decreases the thermal stability and the effect is comparable to that observed by us in u-PVC as discussed earlier. The amount of internal double bonds in the original PVC was 0·1/1000 VC and the results indicate the existence of other unspecified labile structures at a level of about 0·4/1000 VC. The rate constant for initiation from the two defects was found to be of the same order.

These conclusions are totally in accordance with our results obtained with PVC polymerised at subsaturation conditions. The unspecified defect should be tertiary chlorine and its higher content implies that this structure is the most important defect. Besides initiation from defects, Ivan *et al.* also consider random dehydrochlorination. They conclude that ketoallylic groups cannot be the main reason for the instability of PVC. Haynie *et al.*[271] have shown that models containing the latter structure dehydrochlorinate considerably slower than model compounds with allylic or tertiary chlorine. These authors also propose to investigate whether or not the carbonyl-containing models catalyse the thermal degradation of PVC as proposed by Svetly *et al.*[125,126] The latter possibility has also been discussed in a paper concerning the kinetics of the dehydrochlorination.[272]

Millan and his co-workers[273] have presented new data which they claim support their previous view on the influence of the stereostructure on degradation. According to them, initiation occurs through *GTTG* isotactic triads. In this paper they also suggest that the polyene sequence distribution should be strongly influenced by the fraction of this conformation.

An important technological aspect is the initial colouring of virgin PVC. The open literature on this subject is very sparse. However, in a recent paper, Hagii *et al.*[274] have discussed the influence of different organic peroxides on the initial colouring of suspension PVC. They conclude that peroxides susceptible to hydrolysis yielded PVC with less initial colouring.

REFERENCES

1. Geddes, W. C. *Rubb. Chem. Tech.*, **40** (1967), 177.
2. Onozuka, M. and Asahina, M. *J. Macromol. Sci., Rev. Macromol. Chem.*, **C3** (1969), 235.
3. Braun, D. *Pure Appl. Chem.*, **26** (1971), 173.

4. Mayer, Z. *J. Macromol. Sci., Rev. Macromol. Chem.*, **C10** (1974), 263.
5. Ayrey, G., Head, B. C. and Poller, R. C. *J. Polym. Sci., Macromol. Rev.*, **8** (1974), 1.
6. Braun, D. In: *Degradation and Stabilization of Polymers*, G. Geuskens, Ed., John Wiley, New York, 1975, 23.
7. Starnes, W. H. *Adv. Chem. Ser.*, **169** (1978), 309.
8. Braun, D. In: *Developments in Polymer Degradation—3*, N. Grassie, Ed., Applied Science Publishers, London, 1981, 101.
9. Starnes, W. H. In: *Developments in Polymer Degradation—3*, N. Grassie, Ed., Applied Science Publishers, London, 1981, 135.
10. Mukherjee, A. K. and Gupta, A. *J. Macromol. Sci., Rev. Macromol. Chem.*, **C20** (1981), 309.
11. Palma, G. and Carenza, M. *J. Appl. Polym. Sci.*, **14** (1970), 1737.
12. Asahina, M. and Onozuka, M. *J. Polym. Sci.*, **A1, 2** (1964), 3505, 3515.
13. Buruiana, E. C., Robilă, G., Bezdadea, E. C., Barbinta, V. T. and Caraculacu, A. A. *Eur. Polym. J.*, **13** (1977), 159.
14. Caraculacu, A. A. *J. Polym. Sci.*, **A1, 4** (1966), 1839.
15. Braun, D. and Weiss, F. *Angew. Makromol. Chem.*, **13** (1970), 55.
16. Berens, A. R. *Polym. Eng. Sci.*, **14** (1974), 318.
17. Ivan, B., Kennedy, J. P., Kende, I., Kelen, T. and Tüdos, F. *J. Macromol. Sci. Chem.*, **A16** (1981), 1473.
18. Caraculacu, A. A., Bezdadea, E. C. and Istrate, G. *J. Polym. Sci.*, **A1, 8** (1970), 1239.
19. Abbås, K. B. and Sörvik, E. M. *J. Appl. Polym. Sci.*, **20** (1976), 2395.
20. Sörvik, E. M. *J. Polym. Sci., Polym. Lett. Ed.*, **14** (1976), 735.
21. Sörvik, E. M. and Hjertberg, T. *J. Macromol. Sci. Chem.*, **A11** (1977), 1349.
22. Hjertberg, T. and Sörvik, E. M. *J. Polym. Sci., Polym. Chem. Ed.*, **16** (1978), 645.
23. Hjertberg, T. and Sörvik, E. Paper presented at Third International Symposium on Poly(vinyl chloride), Cleveland, Ohio, August 1980, Preprints, 60.
24. Hjertberg, T. and Sörvik, E. Paper presented at Third International Symposium on Poly(vinyl chloride), Cleveland, Ohio, August 1980, Preprints, 41.
25. Hjertberg, T. and Sörvik, E. *J. Polym. Sci., Polym. Lett. Ed.*, **19** (1981), 363.
26. Hjertberg, T. and Sörvik, E. M. *J. Macromol. Sci. Chem.*, **A17** (1982), 983.
27. Hjertberg, T. Dissertation, Chalmers University of Technology, Gothenburg, 1982.
28. Hjertberg, T. and Sörvik, E. M. *Polymer*, **24** (1983), 673.
29. Hjertberg, T. and Sörvik, E. M. *Polymer*, **24** (1983), 685.
30. Hjertberg, T., Sörvik, E. M. and Wendel, A. *Makromol. Chem. Rapid Commun.*, **4** (1983), 175.
31. Carrega, M. E. *Pure Appl. Chem.*, **49** (1977), 569.
32. Caraculacu, A. A. *Pure Appl. Chem.*, **53** (1981), 385.
33. Abbås, K. B. *Pure Appl. Chem.*, **53** (1981), 411.
34. Maddams, W. F. To be published.
35. Park, G. S. To be published.
36. Hjertberg, T. and Sörvik, E. M. To be published.

37. Johnston, C. W. In: *Encyclopedia of PVC*, Vol. 1, L. I. Nass, Ed., Marcel Dekker, New York, 1975, 75.
38. Crosato-Arnaldi, A., Gasparini, P. and Talamini, G. *Makromol. Chem.*, **117** (1968), 140.
39. Ugelstad, J., Lervik, H., Gardinovacki, B. and Sund, E. *Pure Appl. Chem.*, **26** (1971), 121.
40. Ugelstad, J., Flögstad, H., Hertzberg, T. and Sund, E. *Makromol. Chem.*, **164** (1973), 171.
41. Abdel-Alim, A. H. and Hamielec, A. E. *J. Appl. Polym. Sci.*, **16** (1972), 783.
42. Arita, K. and Stannet, V. *J. Polym. Sci., Polym. Chem. Ed.*, **11** (1973), 1565.
43. Olaj, O. F. *Angew. Makromol. Chem.*, **47** (1975), 1.
44. Kuchanov, S. I. and Bort, D. N. *Polym. Sci. USSR*, **15** (1973), 2712.
45. Russo, S. and Stannet, V. *Makromol. Chem.*, **143** (1971), 47.
46. Johnston, C. W. In: *Encyclopedia of PVC*, Vol. 1, L. I. Nass, Ed., Marcel Dekker, New York, 1975, 77.
47. Liegois, J. M. *J. Macromol. Sci. Chem.*, **A11** (1977), 1379.
48. Liu, N.-I., Tong, S. N. and Koenig, J. L. *J. Appl. Polym. Sci.*, **25** (1980), 2205, and references cited therein.
49. Sörvik, E. M. *J. Appl. Polym. Sci.*, **21** (1977), 2769.
50. Hauss, A. F. *J. Polym. Sci.*, **C33** (1971), 1.
51. Frissell, W. J. In *Encyclopedia of PVC*, Vol. 1, L. I. Nass, Ed., Marcel Dekker, New York, 1975, 259.
52. Starnes, W. H., Schilling, F. C., Abbås, K. B., Cais, R. E. and Bovey, F. A. *Macromolecules*, **12** (1979), 556.
53. Starnes, W. H., Schilling, F. C., Plitz, I. M., Cais, R. E., Freed, D. J. and Bovey, F. A. Paper presented at Third International Symposium on Poly(vinyl chloride), Cleveland, Ohio, August 1980, Preprints, 58.
54. Hildebrand, P., Ahrens, W., Brandstetter, F. and Simak, P. *J. Macromol. Sci. Chem.*, **A17** (1982), 1093.
55. Rigo, A., Palma, G. and Talamini, G. *Makromol. Chem.*, **153** (1972), 219.
56. Abbås, K. B., Bovey, F. A. and Schilling, F. C. *Makromol. Chem., Suppl.*, **1** (1975), 227.
57. Bovey, F. A., Abbås, K. B., Schilling, F. C. and Starnes, W. H. *Macromolecules*, **8** (1975), 437.
58. Bezdadea, E. C., Buruiana, E. C., Istrate-Robila, G. and Caraculacu, A. A. *Eur. Polym. J.*, **9** (1973), 445.
59. Petiaud, R. and Pham, Q.-T. *Makromol. Chem.*, **178** (1977), 741.
60. Schwenk, U., König, I., Cavagna, F. and Wrackmeyer, B. *Angew. Makromol. Chem.*, **83** (1979), 183.
61. Bovey, F. A., Schilling, F. C. and Starnes, W. H. *Polym. Prepr., Am. Chem. Soc., Div. Polym. Chem.*, **20** (1979), 160.
62. Starnes, W. H., Schilling, F. C., Plitz, I. M., Cais, R. E. and Bovey, F. A. *Polym. Bull.*, **4** (1981), 555.
63. Mitani, K., Ogata, T., Awaya, H. and Tomari, Y. *J. Polym. Sci., Polym. Chem. Ed.*, **13** (1975), 2813.
64. Baum, B. and Wartman, L. H. *J. Polym. Sci.*, **28** (1958), 537.
65. Landler, Y. and Lebel, P. *J. Polym. Sci.*, **48** (1960), 477.
66. Geddes, W. C. *Eur. Polym. J.*, **3** (1967), 267, 733.

67. Minsker, K. S., Kazachenko, D. V., Abdullina, R. G., Kovler, R. B. and Berlin, A. A. *Vysokomol. Soyed.*, **A15** (1973) 866; *Pol. Sci. USSR*, **15** (1973), 974.
68. Michel, A., Schmidt, G. and Guyot, A. *Polym. Prepr.*, **14** (1973), 655.
69. Braun, D. and Quarg, W. *Angew. Makromol. Chem.*, **29/30** (1973), 163.
70. Michel, A., Schmidt, G., Castaneda, E. and Guyot, A. *Angew. Makromol. Chem.*, **47** (1975), 61.
71. Lindenschmidt, G. *Angew. Makromol. Chem.*, **47** (1975), 79.
72. Abbås, K. B. and Sörvik, E. M. *J. Appl. Polym. Sci.*, **19** (1975), 2991.
73. Burille, P. Dissertation, University Claude Bernard, Lyons, 1979.
74. Braun, D., Michel, A. and Sonderhof, D. *Eur. Polym. J.*, **17** (1981), 49.
75. Sondheimer, F., Ben Efraim, D. A. and Wolovsky, R. *J. Am. Chem. Soc.*, **83** (1961), 1675.
76. Peitscher, G. and Holtrup, W. *Angew. Macromol. Chem.*, **47** (1975), 111.
77. Bassez, C. Dissertation, University of Lille, Lille, 1978.
78. Park, G. S., Maddams, W. F. and Pyszora, H. Unpublished results.
79. Bovey, F. A. and Tiers, G. V. D. *Chem. Ind.* (*London*), (1962), 1826.
80. Colebourne, N. and Stern, E. S. *J. Chem. Soc.* (*London*), (1965), 3599.
81. Lukas, R., Svetly, J. and Kolinsky, M. *J. Polym. Sci., Polym. Chem. Ed.*, **19** (1981), 295.
82. Minsker, K. S., Berlin, A. A., Lisitskii, V. V. and Kolesov, S. V. *Vysokomol. Soyed.*, **A19** (1977), 32; *Pol. Sci. USSR*, **19** (1977), 35.
83. Minsker, K. S., Lisitskii, V. V. and Zaikov, G. E. *J. Vinyl Tech.*, **2** (1980), 77.
84. Minsker, K. S., Lisitskii, V. V. and Zaikov, G. E. *Vysokomol. Soyed.*, **A23** (1981), 483; *Pol. Sci. USSR*, **23** (1981), 535.
85. Minsker, K. S., Lisitskii, V. V., Kolesov, S. V. and Zaikov, G. E. *J. Macromol. Sci., Rev. Macromol. Chem.*, **C20** (1981), 243.
86. George, M. H. and Garton, A. *J. Macromol. Sci. Chem.*, **A11** (1977), 1389.
87. Braun, D. and Sonderhof, D. *Eur. Polym. J.*, **18** (1982), 141.
88. Ratti, L., Visani, F. and Ragazzini, M. *Eur. Polym. J.*, **9** (1973), 429.
89. Mayo, F. R. and Miller, A. A. *J. Am. Chem. Soc.*, **78** (1956), 1017; **80** (1958), 2480, 2493.
90. Suzuki, T. and Nakamura, M. *Jap. Plast.*, **4** (1970), 16.
91. Talamini, G., Cinque, G. and Palma, G. *Mat. Plast.*, **30** (1964), 317.
92. Bjerke, O. KemaNord AB, Sweden, private communication.
93. Starnes, W. H. and Plitz, I. M. *Macromolecules*, **9** (1976), 633.
94. Starnes, W. H., Plitz, I. M., Hische, D. C., Freed, D. J., Schilling, F. C. and Schilling, M. L. *Macromolecules*, **11** (1978), 373.
95. Cittadini, A. and Paolillio, R. *Chim. Ind.* (*Milan*), **41** (1959), 980.
96. Corso, C. *Chim. Ind.* (*Milan*), **43** (1961), 8.
97. Stromberg, R. R., Straus, S. and Achhammer, B. G. *J. Polym. Sci.*, **35** (1959), 355.
98. Park, G. S. and Skene, C. L. *J. Polym. Sci.*, **C33** (1971), 269.
99. Bailey, F. E., Henry, J. P., Lundberg, R. D. and Whelan, J. M. *J. Polym. Sci.*, **B2** (1964), 447.
100. Murayama, N. and Amagi, Y. *J. Polym. Sci.*, **B4** (1966), 115.
101. Crawley, S. and McNeill, I. C. *J. Polym. Sci., Polym. Chem. Ed.*, **16** (1978), 2593.

102. Kitamura, R., Urata, M. and T. Takeuchi, T. *J. Appl. Polym. Sci.*, **9** (1965), 1669.
103. Amer, A. R. and Shapiro, J. S. *J. Macromol. Sci. Chem.*, **A14** (1980), 185.
104. Kharasch, M. S., Kritchevsky, J. and Mayo, F.R. *J. Org. Chem.*, **2** (1937), 489.
105. Suzuki, T., Takakura, I. and Yoda, M. *Eur. Polym. J.*, **7** (1971), 1105.
106. Arlman, E. J. *J. Polym. Sci.*, **12** (1954), 547.
107. Talamini, G. and Pezzin, G. *Makromol. Chem.*, **39** (1960), 26.
108. Bengough, W. I. and Varma, I. K. *Eur. Polym. J.*, **2** (1966), 49, 61.
109. Crosato-Arnaldi, A., Palma, G., Peggion, E. and Talamini, G. *J. Appl. Polym. Sci.*, **8** (1964), 747.
110. Sato and Asahina, M. *Symposium of Chemical Industry of PVC* (*Japan*), (1967), 16; see ref. 50 in ref. 2.
111. van den Heuvel, C. J. M. and Weber, A. J. M. Paper presented at Fifth Annual International Conference on Advances in the Stabilization and Controlled Degradation of Polymers, Zürich, June 1983.
112. Suzuki, T., Nakamura, M., Yasuda, M. and Tatsumi, J. *J. Polym. Sci.*, **C33** (1971), 281.
113. Lewis, J., Okieimen, F. E. and Park, G. S. *J. Macromol. Sci. Chem.*, **A17** (1982), 915.
114. Guyot, A., Bert, M., Burille, P., Llauro, M.-F. and Michel, A. *Pure Appl. Chem.*, **53** (1981), 401.
115. Boissel, J. *J. Appl. Polym. Sci.*, **21** (1977), 855.
116. Frey, A. H. and Horst, R. W. *J. Polym. Sci.*, **40** (1959), 419.
117. Bengough, W. I. and Onozuka, M. *Polymer*, **6** (1965), 625.
118. Lewis, J., Nagvi, M. K. and Park, G. S. *Makromol. Chem., Rapid Commun.*, **1** (1980), 119.
119. Thame, N. G., Lundberg, R. D. and Kennedy, J. P. *J. Polym. Sci.*, **A1, 10** (1972), 2507.
120. Caraculacu, A. A., Bezdadea, E. C. and Istrate, G. *J. Polym. Sci.*, **A1, 8** (1970), 1239.
121. Robila, G., Buruiana, E. C. and Caraculacu, A. A. *Eur. Polym. J.*, **13** (1977), 21.
122. Millan, J., Martinez, G. and Mijangos, C. *Polym. Bull.*, **5** (1981), 407.
123. Martinez, G., Mijangos, C. and Millan, J. *J. Macromol. Chem.*, **A17** (1982), 1129.
124. Svetly, J., Lukas, R. and Kolinsky, M. *Makromol. Chem.*, **180** (1979), 1363.
125. Svetly, J., Lukas, R., Michalocova, J. and Kolinsky, M. *Makromol. Chem., Rapid Commun.*, **1** (1980), 247.
126. Svetly, J., Lukas, R., Pokorny, S. and Kolinsky, M. *Makromol. Chem., Rapid Commun.*, **2** (1981), 149.
127. Tvaroska, I. Paper presented at Second International Symposium on Poly(vinyl chloride), Lyons, July 1976, Preprints, 301.
128. Razuvaev, G. A., Troitskii, B. B. and Troitskaya, L. S. Paper presented at Second International Symposium on Poly(vinyl chloride), Lyons, July 1976, Preprints, 261.
129. Troitskii, B. B., Dozorov, V. A., Minchuk, F. F. and Troitskaya, L. S. *Eur. Polym. J.*, **11** (1975), 277.

130. Nolan, K. P. and Shapiro, J. S. *J. Chem. Soc., Chem. Commun.* (1975), 490.
131. Kennedy, J. P. and Ichikawa, M. *Polym. Eng. Sci.*, **14** (1974), 322.
132. Suzuki, T. *Pure Appl. Chem.*, **49** (1977), 539.
133. Mitani, K., Ogata, T., Nakatsukasa, M. and Eda, Y. *J. Macromol. Sci. Chem.*, **A11** (1977), 2265.
134. Mitani, K., Ogata, T., Nakatsukasa, M. and Mizutani, Y. *Polymer*, **21** (1980), 1463.
135. Hjertberg, T. and Sörvik, E. M. Unpublished results obtained on sample kindly supplied by M. Kolinsky, Prague.
136. Starnes, W. H., Haddon, R. C., Hische, D. C., Plitz, I. M., Schosser, C. L., Schilling, F. C. and Freed, D. J. *Polym. Prepr.*, **21**(2) (1980), 176.
137. Braun, D. and Thallmaier, M. *Makromol. Chem.*, **99** (1966), 59.
138. Geddes, W. C. *Eur. Polym. J.*, **3** (1967), 747.
139. Tüdos, F. and Kelen, T. In: *Macromolecular Chemistry*, Vol. 8, K. Saarela, Ed., Butterworths, London, 1973, 393.
140. Abbås, K. B. and Laurence, R. L. *J. Polym. Sci., Polym. Chem. Ed.*, **13** (1975), 1889.
141. Tüdos, F., Kelen, T., Nagy, T. T. and Turcsanyi, B. *Pure Appl. Chem.*, **38** (1974), 201.
142. Winkler, D. E. *J. Polym. Sci.*, **35** (1959), 3.
143. Stapfer, C. H. and Granick, J. D. *J. Polym. Sci.*, **A1, 9** (1971), 2625.
144. Albarino, R. V., Otocka, E. P. and Luongo, J. P. *J. Polym. Sci.*, **A1, 9** (1971), 15.
145. Nakagawa, T. and Okawara, M. *J. Polym. Sci.*, **A1, 6** (1968), 1975.
146. Palma, G. and Carenza, M. *J. Appl. Polym. Sci.*, **16** (1972), 2485.
147. Salvoey, R. and Bair, W. E. *J. Appl. Polym. Sci.*, **14** (1970), 713.
148. Gibb, W. H. and MacCallum, J. R. *Eur. Polym. J.*, **7** (1971), 1231.
149. Gibb, W. H. and MacCallum, J. R. *Eur. Polym. J.*, **8** (1972), 1223.
150. Owen, E. D. and Bailey, R. J. *J. Polym. Sci.*, **A1, 10** (1972), 113.
151. Gupta, V. P. and St Pierre, L. E. *J. Polym. Sci.*, **A1, 8** (1970), 37.
152. Guyot, A., Bert, M., Michel, A. and Spitz, R. *J. Polym. Sci.*, **A1, 8** (1970), 1596.
153. Liebman, S. A., Reuwer, J. F., Gollatz, K. A. and Nauman, C. D. *J. Polym. Sci.*, **A1, 9** (1971), 1823.
154. Ouchi, I. *J. Polym. Sci.*, **A1, 3** (1965), 2685.
155. Bamford, C. H. and Fenton, D. F. *Polymer*, **10** (1969), 63.
156. Papko, R. A. and Padov, V. S. *Polym. Sci. USSR*, **16** (1974), 1636.
157. Myakov, V. N. and Troitskii, B. B. *Vysokomol. Soyed.*, **4** (1962), 876.
158. McNeill, I. C. and Neil, D. *Makromol. Chem.*, **117** (1968), 265.
159. McNeill, I. C. and Neil, D. *Eur. Polym. J.*, **6** (1970), 143.
160. McNeill, I. C. and Neil, D. *Eur. Polym. J.*, **6** (1970), 569.
161. Dodson, B. and McNeill, I. C. *J. Polym. Sci.*, **A1, 14** (1976), 353.
162. Baum, B. *S.P.E. J.*, **17** (1961), 71.
163. Braun, D. and Bender, R. F. *Eur. Polym. J.* (*Suppl.*), (1969), 269.
164. Bengough, W. I. and Sharpe, H. M. *Makromol. Chem.*, **66** (1963), 31.
165. Hay, J. N. *J. Polym. Sci.*, **A1, 8** (1970), 1201.
166. Troitskii, B. B., Troitskaya, L. S., Myakov, V. N. and Lepaev, A. F. *J. Polym. Sci., Polym. Symp.*, **42** (1973), 1347.

167. Starnes, W. H. and Edelson, D. *Am. Chem. Soc., Div. Org. Coat. Plast. Chem., Pap.*, **41** (1976), 505.
168. Starnes, W. H. and Edelson, D. *Macromolecules*, **12** (1979), 797.
169. Hendry, D. G., Mill, T., Piszkiewicz, L., Howard, J. A. and Eigenmann, H. K. *J. Phys. Chem. Ref. Data*, **3** (1974), 937.
170. Marks, G. C., Benton, J. L. and Thomas, C. M. *S.C.I. Monogr.*, **26** (1967), 204.
171. Zafar, M. M. and Mahmood, R. *Eur. Polym. J.*, **12** (1976), 333.
172. Mayer, Z., Michailov, L., Obereigner, B. and Lim, D. *Chem. Ind. (London)*, (1965), 508.
173. Van der Ven, S. and DeWit, W. F. *Angew. Makromol. Chem.*, **8** (1969), 143.
174. Wolkober, Z. *J. Polym. Sci.*, **58** (1962), 1311.
175. Shindo, Y. and Hirai, T. *Makromol. Chem.*, **155** (1972), 1.
176. Reiche, A., Grimm, A. and Mücke, H. *Kunststoffe*, **52** (1962), 265.
177. Troitskaya, L. S. and Troitskii, B. B. *Plast. Massy*, **7** (1966), 46.
178. Hoang, T. V., Michel, A., Pichot, C. and Guyot, A. *Eur. Polym. J.*, **11** (1975), 469.
179. Varma, I. K. and Grover, S. S. *Makromol. Chem.*, **175** (1974), 2515.
180. Imoto, M. and Nakaya, T. *Kogyo Kagaku Zasshi*, **68** (1965), 2285.
181. Braun, D. and Thallmaier, M. *Kunststoffe*, **56** (1966), 80.
182. Bethell, D. and Gold, V. *Q. Rev. Chem. Soc. (London)*, **12** (1958), 173.
183. Haddon, R. C. and Starnes, W. H. *Polym. Prepr.*, **18** (1977), 505.
184. Haddon, R. C. and Starnes, W. H. *ACS Adv. Chem. Ser.*, **169** (1978), 333.
185. Berlin, A. A., Yanovskii, D. M. and Popova, Z. V. In: *Aging and Stabilization of Polymers*, M. B. Neiman, Ed., Consultants Bureau, New York, 1968, 175.
186. Millan, J. and Guzman, J. *Eur. Polym. J.*, **12** (1976), 299.
187. Millan, J., Madraga, E. L. and Martinez, G. *Angew. Makromol. Chem.*, **45** (1975), 172.
188. Martinez, G., Mijangos, C., Millan, J., Gerrard, D. L. and Maddams, W. F. *Makromol. Chem.*, **180** (1979), 2937.
189. Millan, J., Martinez, G. and Mijangos, C. *J. Polym. Sci., Polym. Chem. Ed.*, **18** (1980), 505.
190. Mitani, K. and Ogata, T. *J. Appl. Polym. Sci.*, **18** (1974), 3205.
191. Liebman, S. A., Ahlström, D. H. and Foltz, C. R. *J. Polym. Sci., Polym. Chem. Ed.*, **16** (1978), 3139.
192. Chang, E. P. and Salovey, R. *J. Polym. Sci., Polym. Chem. Ed.*, **12** (1974), 2927.
193. Robinson, P. J., Skelhorne, G. G. and Wall, M. J. *J. Chem. Soc., Perkin Trans. II* (1978), 349.
194. Druesedow, D. and Gibbs, C. G. *Nat. Bur. Stand., Circular*, **525** (1953), 69.
195. Abbås, K. B. and Sörvik, E. M. *J. Appl. Polym. Sci.*, **17** (1973), 3577.
196. Hjertberg, T. and Sörvik, E. M. *J. Appl. Polym. Sci.*, **22** (1978), 2415.
197. Kelen, T. *J. Macromol. Sci. Chem.*, **A12** (1978), 349.
198. Kelen, T., Ivan, B., Nagy, T. T., Turcsanyi, B., Tüdos, F. and Kennedy, J. P. *Polym. Bull.*, **1** (1978), 79.
199. Nagy, T. T., Ivan, B., Turcsanyi, B., Kelen, T. and Tüdos, F. *Polym. Bull.*, **3** (1980), 613.

200. Holmström, A. and Sörvik, E. M. *J. Polym. Sci., Polym. Chem. Ed.*, **16** (1978), 2555.
201. Razuvaev, G. A., Troitskaya, L. S., Myakov, V. N. and Troitskii, B. B. *Dokl. Akad. Nauk. SSSR. Khim.*, **170** (1966), 1342.
202. Troitskii, B. B. and Myakov, V. N. *Plast. Massy*, (1967), (3), 13.
203. Zilberman, E. N., Perepletchikova, E. M., Getmanenko, E. N. and Pomeranceva, E. G. *Vysokomol. Soyed.*, **B16** (1974), 46.
204. O'Mara, M. M. *Pure Appl. Chem.*, **49** (1977), 649.
205. Lattimer, R. P. and Kroenke, W. J. *J. Appl. Polym. Sci.*, **25** (1980), 101.
206. Lattimer, R. P. and Kroenke, W. J. *J. Appl. Polym. Sci.*, **27** (1982), 1355.
207. Bellenger, V., Carette, L. B., Fontaine, E. and Verdu, J. *Eur. Polym. J.*, **18** (1982), 337.
208. Rasuvayev, G. A., Troitskaya, L. S. and Troitskii, B. B. *J. Polym. Sci.*, **A1, 9** (1971), 2673.
209. Marvell, E. N., Seubert, J., Vogt, G., Zimmer, G., Moy, G. and Siegmann, J. R. *Tetrahedron*, **34** (1978), 1323.
210. Owen, E. D. and Read, R. L. *Eur. Polym. J.*, **15** (1979), 41.
211. Brauman, S. K. and Chen, I. J. *J. Polym. Sci., Polym. Chem. Ed.*, **19** (1981), 495.
212. Sorensen, T. S. *J. Am. Chem. Soc.*, **87** (1965), 5075.
213. Neiman, M. V., Papko, R. A. and Padov, V. S. *Vysokomol. Soyed.*, **A10** (1968), 841; *Polym. Sci. USSR*, **10** (1968), 975.
214. Minsker, K. S., Malinskaya, V. P. and Panasenko, A. A. *Vysokomol. Soyed.*, **A12** (1970), 1151; *Polym. Sci. USSR*, **15** (1970), 1304.
215. Minsker, K. S., Malinskaya, V. D., Artsis, M. I., Razumovskii, S. D. and Zaikov, G. E. *Dokl. Acad. Nauk. SSSR*, **223** (1975), 138.
216. Troitskaya, L. S., Myakov, V. N., Troitskii, B. B. and Razuvayev, G. A. *Vysokomol. Soyed.*, **A9** (1967), 2119; *Polym. Sci. USSR*, **9** (1967), 2392.
217. Troitskii, B. B., Troitskaya, L. S. and Razuvayev, G. A. *Vysokomol. Soyed.*, **A13** (1971), 1183; *Polym. Sci. USSR*, **13** (1971), 1333.
218. Razuvayev, G. A., Troitskaya, L. S. and Troitskii, B. B. *J. Polym. Sci.*, **A1, 9** (1971), 2673.
219. Carenza, M., Moiseev, Y. V. and Palma, G. *J. Appl. Polym. Sci.*, **17** (1973), 2685.
220. Guyot, A. and Bert, M. *J. Appl. Polym. Sci.*, **17** (1973), 753.
221. Nagy, T. T., Kelen, T., Turcsanyi, B. and Tüdos, F. *Polym. Bull.*, **2** (1980), 77.
222. Kelen, T., Balint, G., Galambos, G. and Tüdos, F. *J. Polym. Sci.*, **C33** (1971), 211.
223. Bataille, P. and Van, B. T. *J. Polym. Sci.*, **A1, 10** (1972), 1097.
224. Morikawa, T. *Chem. High Polym. (Jpn)*, **25** (1968), 505.
225. Thallmaier, M. and Braun, D. *Makromol. Chem.*, **108** (1967), 241.
226. Owen, E. D., Pasha, I. and Moayyedi, F. *J. Appl. Polym. Sci.*, **25** (1980), 2331.
227. Tüdos, F., Kelen, T. and Nagy, T. T. In: *Developments in Polymer Degradation—2*, N. Grassie, Ed., Applied Science Publishers, London, 1979, 187.
228. Minsker, K. S., Berlin, A. A. and Abdullin, M. N. *Vysokomol. Soyed.*, **B16** (1974), 439.

229. Berlin, A. A. *Dokl. Akad. Nauk SSSR*, **144** (1962), 1042.
230. Minsker, K. S., Krats, E. O. and Pakhomova, I. *Vysokomol. Soyed.*, **A12** (1970), 483.
231. Nagy, T. T., Kelen, T., Turcsanyi, B. and Tüdos, F. *J. Polym. Sci., Polym. Chem. Ed.*, **15** (1977), 853.
232. Nagy, T. T., Kelen, T., Turcsanyi, B. and Tüdos, F. *Angew. Makromol. Chem.*, **66** (1978), 193.
233. Ivan, B., Nagy, T. T., Kelen, T., Turcsanyi, B. and Tüdos, F. *Polym. Bull.*, **2** (1980), 83.
234. Scott, G., Tahan, M. and Vyvoda, J. *Eur. Polym. J.*, **14** (1978), 377.
235. Cooray, B. B. and Scott, G. In: *Developments in Polymer Stabilization—2*, Applied Science Publishers, London, 1980, 53.
236. Cooray, B. B. and Scott, G. *Eur. Polym. J.*, **16** (1980), 169.
237. Michel, A. To be published.
238. Millan, J. To be published.
239. Pukánszky, B., Nagy, T. T., Kelen, T. and Tüdos, F. *J. Appl. Polym. Sci.*, **27** (1982), 2615.
240. Nilsson, H., Silvergren, C. and Törnell, B., *Eur. Polym. J.*, **14** (1978), 737.
241. Imoto, T., Ogo, Y. and Hashimoto, Y. *Kogyo Kagaku Zasshi*, **70** (1967), 1952; *Int. Chem. Eng.*, **8** (1968), 564.
242. Crosato-Arnaldi, A., Guarsie, G. B. and Talamini, G., *Polymer*, **10** (1969), 385.
243. Ogo, Y., Yoshikawa, M., Ohtani, S. and Imoto, T. *J. Polym. Sci., Polym. Chem. Ed.*, **16** (1978), 1413.
244. Witenhafer, D. E. Private communication.
245. Jisova, V., Kolinsky, M. and Lim, D. *J. Polym. Sci.*, **A1, 8** (1970), 1525.
246. Jisova, V., Kolinsky, M. and Lim, D. *J. Polym. Sci.*, **C42** (1973), 467.
247. Kolinsky, M., Jisova, V. and Lim, D. *J. Polym. Sci.*, **C42** (1973), 657.
248. Wesslen, B. and Wirsen, A. *J. Polym. Sci., Polym. Chem. Ed.*, **13** (1975), 2571.
249. Sicedison Societa per Azioni (L. Scarso, G. Boccato and A. Rigo), French Patent 1 379 227 (1964); *Chem. Abstr.*, **62** (1965), 16460.
250. (a) Frey, A. H. and Horst, R. W. *J. Polym. Sci.*, **45** (1960), 1; (b) Frey, A. H., Horst, R. W. and Paliobagis, M. A. *J. Polym. Sci.*, **A2** (1964), 1765, 1785, 1801.
251. Guyot, A. and Michel, A. In: *Developments in Polymer Stabilization—2*, G. Scott, Ed., Applied Science Publishers, London, 1980, 89.
252. Klemchuk, P. P. *Prog. Chem.*, **85** (1968), 1.
253. Figge, K. and Findeiss, W. *Angew. Makromol. Chem.*, **47** (1975), 141.
254. Morikawa, T. *Kobunshi Kagaku* (1968), 602; *Chem. Abstr.*, **70** (1969), 38417z.
255. Minsker, K. S., Fedoseyeva, G. T., Zavarova, T. B. and Krats, E. O. *Polym. Sci. USSR*, **13** (1971), 2544.
256. Krats, E. O., Zavarova, T. B., Fedoseyeva, G. T. and Minsker, K. S. *Polym. Sci. USSR*, **13** (1971), 1013.
257. Starnes, W. H. and Plitz, I. M. *Polym. Prepr.*, **16**(2) (1975), 500.
258. Plitz, I. M., Starnes, W. H. and Hartless, R. L. *Polym. Prepr.*, **17**(2) (1976), 495.
259. Plitz, I. M., Willingham, R. A. and Starnes, W. H. *Macromolecules*, **10** (1977), 499.

260. Gaylord, N. G. and Takahashi, A. *J. Polym. Sci., Polym. Lett. Ed.*, **8** (1970), 349.
261. Gaylord, N. G. and Takahashi, A. *J. Polym. Sci.*, **8** (1970), 361.
262. Abbås, K. B. and Thame, N. G. *J. Polym. Sci., Polym. Chem. Ed.*, **13** (1975), 59.
263. Bernier, P., Schue, F., Sledz, J., Rolland, M. and Giral, L. *Chem. Scr.*, **17** (1981), 151.
264. Starnes, W. H., Schilling, F. C., Plitz, I. M., Cais, R. E., Freed, D. J., Hartless, R. L. and Bovey, F. A. *Macromolecules*, **16** (1983), 790.
265. Starnes, W. H., Villacorta, G. M., Schilling, F. C., Park, G. S. and Saremi, A. H. *Polym. Prepr., Am. Chem. Soc., Div. Polym. Chem.*, **24**(1) (1983), 253.
266. Barboiu, V., Robila, G., Buruiana, E. and Caraculacu, A. *Makromol. Chem.*, **183** (1983), 2667.
267. Bonnans-Plaisance, C., Gressier, J.-C. and Levesque, G. *Makromol. Chem., Rapid Commun.*, **4** (1983), 387.
268. Braun, D. and Holzer, G. *Angew. Makromol. Chem.*, **113** (1983), 91.
269. Braun, D. and Holzer, G. *ibid.*, **116** (1983), 51.
270. Ivan, B., Kennedy, J. P., Kelen, T., Tüdos, F., Nagy, T. T. and Turcsanyi, B. *J. Polym. Sci., Polym. Chem. Ed.*, **21** (1983), 2177.
271. Haynie, S. L., Villacorta, G. M., Plitz, I. M. and Starnes, W. H. *Polym. Prepr., Am. Chem. Soc., Div. Polym., Chem.*, **24**(2) (1983), 3.
272. Valko, L., Simon, P. and Kellö, V. *Makromol. Chem.*, **183** (1982), 3057.
273. Martinez, G., Mijangos, C. and Millan, J. *J. Appl. Polym. Sci.*, **28** (1983), 33.
274. Hagii, H., Arai, T. and Yamamoto, T. *J. Appl. Polym. Sci.*, **27** (1982), 4879.

Chapter 3

PHOTODEGRADATION OF PVC

CHRISTIAN DECKER

Laboratoire de Photochimie Générale associé au CNRS, Ecole Nationale Supérieure de Chimie, Mulhouse, France

3.1 INTRODUCTION

Most commercial polymers undergo chemical and physical modifications when they are exposed for a long time to the combined action of sunlight, atmospheric oxygen, moisture and pollutants. The extent of this degradation depends primarily on the presence within the polymer of light-absorbing structures. Polymers containing aromatic or carbonyl groups in their monomer unit are likely to absorb the terrestrial sunlight (wavelength $\lambda > 290$ nm) and usually become photosensitive materials. On the other hand, some polymers that do not possess in their structural unit any chromophore absorbing above 250 nm, like polyolefins, still appear to be degraded by sunlight during outdoor exposure. Since a basic law in photochemistry states that a photochemical reaction can only occur if light has been absorbed, it implies that those polymers must contain chromophore groups absorbing the solar radiations, either as external impurities or as functional groups incorporated into the polymer backbone.

Polyvinyl chloride (PVC) belongs to the second class of polymers since its monomer unit contains only saturated chemical bonds (C—C, C—H and C—Cl) which absorb in the far ultraviolet region, at wavelengths below 200 nm. However, although PVC is an important commercial polymer, it is the one which exhibits the highest sensitivity toward UV radiation. When exposed to natural weathering, PVC sheets deteriorate within a few weeks, the irradiated material becoming increasingly coloured and brittle, with a steady drop in the tensile strength, elasticity and impact resistance. Moreover, as photodegradation is primarily a superficial process due to the limited penetration of UV–visible radiation, the surface properties of the degraded polymer are substantially modified, as shown by the observed loss of abrasion resistance, gloss and interfacial free energy. Owing to its large outdoor application, PVC has therefore to be protected effectively against photodegradation by the addition of adequate stabilisers. A chapter (Chapter 5) of this book is devoted to the photostabilisation of PVC, which remains one of the most important practical probems as far as applications are concerned.

The effects of UV light on the behaviour of PVC were thoroughly investigated in the last decade in order to obtain a better understanding and control of the degradation process.[1–9] Recently, Starnes[7] and Owen[9] have both presented comprehensive surveys of the work published since the early 1970s. In spite of the useful information obtained by using the more recent techniques available, it must be acknowledged that the reasons which make PVC so susceptible to UV light are still not fully elucidated. This chapter is an attempt to clarify and take stock of the current situation. Particular emphasis will be laid on the various aspects of the major photochemical processes involved, especially with regard to the nature of the initiating chromophores and the effect of such important factors as the intensity and wavelength of the incident light, the irradiation temperature and the presence of oxygen. The mechanism of the photodegradation and photooxidation of PVC will be discussed on the basis of some recent quantum yield results.

For a better understanding of the primary photochemical processes which are expected to occur by interaction of PVC with light, some basic elements of photochemistry are now briefly recalled.

3.2 KEY PROCESSES IN PHOTOCHEMISTRY

3.2.1 Absorption of light

When light interacts with a molecule, the electrons are set into motion by the oscillating electric field of the light.[10] As the incident photon exchanges its energy with the molecule, a valence electron is promoted from the highest occupied molecular orbital to one of the unoccupied molecular orbitals, with formation of an excited singlet state molecule. Each photon activates only one molecule with the requirement that the law of energy is fulfilled, i.e. $\Delta E = h\nu$, where ΔE is the difference in energy between two electronic states of a molecule, h is Planck's constant and ν is the frequency of the light. Owing to the quantised nature of electronic states, ΔE can only take discrete values. The energy of a photon is directly related to its wavelength by the equation

$$E\,(\text{kJ mol}^{-1}) = \frac{1{\cdot}195 \times 10^5}{\lambda\,(\text{nm})}$$

Thus, photons associated with 250-nm and 300-nm radiations carry energies of 478 kJ mol^{-1} and 398·3 kJ mol^{-1} respectively. If the energy of the absorbed photon, $h\nu$, is greater than the ionisation potential, i.e. the energy required to completely remove a valence electron from a molecule, then an electron is ejected and a positive molecular ion is left behind.

The main electronic absorptions consist of transitions from ground singlet states (S_0) to excited singlet states (S_i^*) since singlet–triplet transitions are forbidden by the spin-selection rule

Absorption: $S_0 + h\nu \longrightarrow S_i^* \longrightarrow S_1^* + \text{heat}$

For most photochemical reactions, only the lowest excited singlet state (S_1^*) and the lowest excited triplet state (T_1^*) have to be considered for the initiation step.[11]

3.2.2 Reactions of excited states

Excited singlet states are short-lived species, with lifetimes in the range 10^{-9}–10^{-6} s. They disappear by various competitive photophysical processes shown schematically in Fig. 3.1:

1. Internal conversion which regenerates the ground state molecule, with thermal dissipation of the absorbed energy

$$S_1^* \longrightarrow S_0 + \text{heat}$$

2. Radiative emission (fluorescence)

$$S_1^* \longrightarrow S_0 + h\nu'$$

3. Intersystem crossing to generate the first excited triplet state

$$S_1^* \longrightarrow T_1^* + \text{heat}$$

4. Energy transfer to a ground state molecule

$$S_1^* + S_0' \longrightarrow S_0 + S_1'^*$$

5. Dissociation into free radicals by intramolecular bond cleavage

$$S_1^* \longrightarrow \dot{R} + \dot{R}'$$

6. Radiation emission from the triplet state (phosphorescence)

$$T_1 \longrightarrow S_0 + h\nu''$$

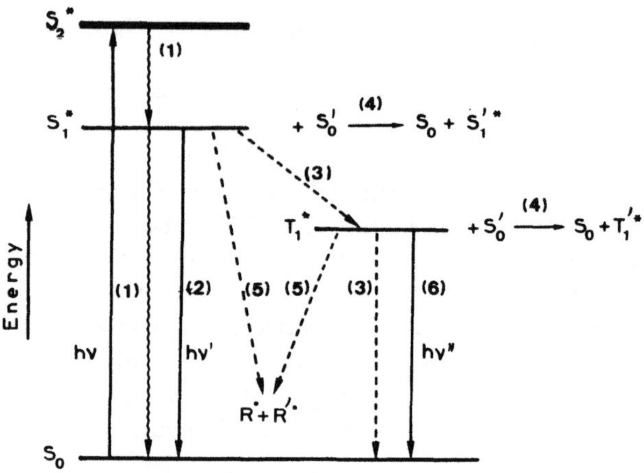

Fig. 3.1 Various deactivation pathways of electronically excited states of a molecule: (1) internal conversion, (2) fluorescence, (3) intersystem crossing, (4) energy transfer, (5) photocleavage, (6) phosphorescence.

The relative importance of these various reactions will depend on the nature of the excited chromophore, process (2) being favoured in aromatic compounds while process (3) occurs preferentially in molecules containing carbonyl groups. Reaction (5) is primarily responsible for the photodegradation of polymers since it leads to a fragmentation of the polymer chain with formation of radicals which can further propagate the degradation. This cleavage reaction most often proceeds from excited triplet states which have longer lifetimes than S_1^* states since the return to the singlet ground state is spin-forbidden.

The bond breaking reaction can only occur if the excitation energy of the molecule exceeds the bond dissociation energy or if the excited state is a repulsive one. In the case of PVC, three types of bonds are involved, with corresponding dissociation energies,[12] namely C–H (410 kJ mol^{-1}); C–C (340 kJ mol^{-1}) and C–Cl (325 kJ mol^{-1}). From energy considerations, the most likely bond to be broken in PVC is the C–Cl bond. In this process, a chloride radical is evolved together with a polymeric radical, but the polymer backbone remains intact.

3.2.3 Effect of oxygen

Since the photodegradation of polymers usually occurs in the presence of air, the effect of atmospheric oxygen must be taken into account. Molecular oxygen has a strong biradical character since its ground state electronic configuration is a triplet state where the two electrons are located predominantly on each of the two nuclei. This confers on the O_2 molecule a high reactivity toward unstable transients such as excited states and free radicals. Owing to its low-energy-lying excited singlet state ($E_s = 92\,\mathrm{kJ\,mol^{-1}}$), ground state oxygen deactivates effectively both singlet and triplet excited states by an energy transfer process, with rate constants of the order of $10^{10}\,\mathrm{M^{-1}\,s^{-1}}$ for singlet quenching and $10^{9}\,\mathrm{M^{-1}\,s^{-1}}$ for triplet quenching:[10]

$$S_1^* + O_2 \xrightarrow{k_q} S_0 + O_2^*$$

$$T_1^* + O_2 \xrightarrow{k'_q} S_0 + O_2^*$$

Although oxygen diffuses readily into amorphous polymers, its solubility is rather low ($[O_2] \sim 10^{-3}\,\mathrm{M}$ in air), so that rates of singlet quenching, $k_q[O_2]$, are of the order of $10^7\,\mathrm{s^{-1}}$. This means that S_1^* states with lifetimes shorter than $10^{-8}\,\mathrm{s}$ will not be efficiently quenched in air-saturated systems.

The oxygen molecule is also very reactive towards the alkyl radicals resulting from the photofragmentation process, with production of peroxy radicals:

$$\dot{R} + O_2 \longrightarrow R{-}O{-}\dot{O}$$

Since the rate of radical scavenging by molecular oxygen is usually within an order of magnitude of the diffusional rate, radicals with lifetimes longer than $10^{-7}\,\mathrm{s}$ react rèadily with O_2 in air-saturated systems.

3.2.4 Quantum yield

The efficiency of a given photochemical reaction is characterised by its quantum yield, Φ, defined as the number of molecules that undergo a modification for each photon absorbed by the irradiated system. Therefore Φ is always <1, except for chain reaction where it may exceed unity. The number of photons absorbed is usually expressed in Einsteins, which corresponds to 6×10^{23} photons. The quantum efficiency is then given by the ratio:

$$\Phi = \frac{\text{moles of given species formed or destroyed}}{\text{Einsteins absorbed by the system}}$$

The number of photons absorbed per unit mass or unit volume can be determined by actinometry measurements using a compound that undergoes a photochemical reaction and whose quantum yield is known.[13] For partially absorbing systems, it is necessary to evaluate both the incident light intensity, I_0, and the fraction of light absorbed by the irradiated system (f). This is the case for transparent PVC films or sheets which initially absorb only a small fraction of UV radiations of wavelength $\lambda >$ 250 nm. The value of f may increase with the irradiation time if, as happens in PVC, light-absorbing photoproducts are formed which are responsible for the discoloration of the irradiated material.

From a mechanistic point of view, it is of prime importance to evaluate the quantum yield of the different processes involved in the photodegradation of a polymer since this quantity will indicate how effectively a photon can generate products, chain scissions or crosslinks. If Φ is found to be less than unity, it can be inferred that some of the absorbed energy has been dissipated, without net chemical reaction, by a cyclic process which returns the system to its original ground state.

3.2.5 Light sources

The course of a photochemical reaction usually depends on the intensity and wavelength of the incident light. Therefore, the only strictly reliable information on the effect of sunlight on polymers will be that obtained from outdoor exposure. The ultraviolet part of the terrestrial sunlight (290–400 nm) represents only about 5% of the total radiation,[14] but it causes the most damage to polymeric materials for two main reasons: (1) UV radiation is more energetic ($300 < E < 410\,\mathrm{kJ\,mol^{-1}}$) than visible light and (2) UV radiation is more likely to be absorbed by polymers which are usually transparent to radiations >400 nm.

In order to reduce the duration of the experimentation, which may last for months or years, the light-sensitivity of polymers is often tested by using artificial light sources. The most common irradiation device is the Xenotest or Weatherometer which reproduces as faithfully as possible the conditions of natural weathering, but with an accelerating factor of 10 to 20. Xenon lamps are commonly used since their emission spectrum follows pretty well that of terrestrial sunlight. Fluorescent lamps or Pyrex-filtered medium-pressure mercury lamps which have their emission maxima at 366 nm can also be employed for irradiations in the range 300–450 nm. Since PVC has a very low absorption in this wavelength region, prolonged exposure time is still needed before some degradation can be detected. For this reason, mechanistic investigations have often been carried out by using the polychromatic light of medium-pressure mercury lamps which stretches from 248 to 578 nm. The high rate of degradation which is then observed is mainly due to the drastic increase of the intensity of the light absorbed by the irradiated polymer. Although the conclusions that can be drawn from those investigations are not questionable *per se*, they are not directly transposable to the degradation of PVC under outdoor exposure conditions.

3.3 NON-OXIDATIVE PHOTODEGRADATION OF PVC

Though atmospheric oxygen has a great influence on the mechanism by which polymeric materials deteriorate when they are exposed to natural weathering, we shall only consider in this section the primary photochemical processes which result from the interaction of PVC with UV light in the absence of oxygen. The specific effect of molecular oxygen, in relation with photo-ageing, will be discussed in Section 3.4.

When an unstabilised PVC sample is exposed to UV radiations of $\lambda >$ 250 nm, discoloration occurs rapidly due to the formation of conjugated polyene sequences. Simultaneously, large amounts of hydrogen chloride are evolved while the polymer chains undergo scissions and crosslinking, with the expected deleterious effects on the mechanical properties. Three main areas will be considered in this section, namely (1) initiation, (2) dehydrochlorination and (3) structural changes.

3.3.1 Photoinitiation

One of the outstanding questions in the photodegradation of PVC concerns the initiation step, since 'intrinsically pure' PVC does not absorb radiations of wavelength above 220 nm and certainly not terrestrial solar radiations ($\lambda > 290$ nm). The poor light stability of PVC must originate therefore from photosensitive structures which have been incorporated into the polymer during the polymerisation and/or the subsequent processing. Owing to the importance of this question, numerous studies have been devoted to the identification and removal of those functional groups in order to increase the durability of PVC. Though experimental evidence indicates that different types of structures such as carbonyl compounds, unsaturated groups or hydroperoxides can initiate the photodegradation of PVC when they are introduced into the polymer, it is still unclear which one actually initiates the photodegradation of commercial PVC. The contributions of these potential photosensitisers have been thoroughly discussed recently[7,9] and will be briefly reviewed.

(1) Unsaturation

Commercial PVC always contains alkene linkages which are essentially located at the end of the polymer chains since they arise mostly from termination or chain transfer reactions. The most probable structure for these groups is ~~CH_2—CH=CH—CH_2Cl.[15,16] The unsaturation content ranges usually from 1 to 3 double bonds per 1000 monomer units,[17] depending on the polymerisation conditions. Internal double bonds, ~~CHCl—CH=CH—CH_2~~ may represent up to 30% of the total amount of unsaturation in the undegraded material.[18]

It must be remembered, however, that isolated alkene linkages only absorb UV radiation of $\lambda < 220$ nm and therefore cannot be considered responsible for the photodegradation observed when PVC is exposed to sunlight or artificial light of $\lambda > 250$ nm. If the double bonds are conjugated, the absorption shifts steadily to longer wavelengths as the sequence length increases.[19,20] Dienes, trienes and longer polyenes which may arise during the thermal processing and storage are thus considered to be the major initially absorbing chromophores when >250 nm radiation is employed.[21] Even if their concentration in the polymer is very low, these chromophores can absorb the incident light effectively owing to their very high extinction coefficient which increases as the sequence length grows, $\varepsilon = 42\,000$ litres mol^{-1} cm^{-1} for the triene absorption at 257 nm, and 210 000 litres mol^{-1} cm^{-1} for the octaene absorption at 386 nm.[22]

Excited singlet polyenes are known to be short-lived species[10] which disappear by different routes, mainly internal conversion, *cis–trans* isomerisation, intramolecular cyclisation and fragmentation. The intersystem crossing process that generates excited triplet states was shown to be quite inefficient in those systems[10] and is not expected to intervene significantly. Among the possible reactions of excited polyenes, only the photocleavage yields free radicals able to attack the polymer and can therefore be considered as the initial step of the dehydrochlorination of PVC.

The most likely bond to be broken in PVC is the allylic C–Cl linkage, owing to its low dissociation energy, especially in the case of long polyene sequences ($n > 10$) where $E_{C\text{–}Cl}$ can apparently drop as low as 230 kJ mol^{-1}.[23] Direct photolysis of liquid allyl chlorides showed convincingly that the primary photochemical process involves essentially C–Cl homolysis.[22]

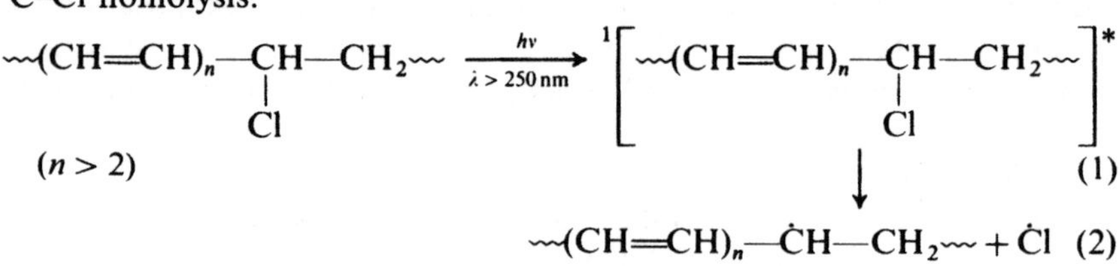

When UV radiation of $\lambda > 250\,\text{nm}$ is used, the cleavage of some C–H bonds may occur, as shown by the production of small amounts of gaseous hydrogen during the photolysis of PVC:[24]

$$^{1}\left[\sim\!\!(\text{CH}{=}\text{CH})_n{-}\underset{\text{H}}{\underset{|}{\text{CH}}}{-}\text{CHCl}\sim\right]^{*} \rightarrow \sim\!\!(\text{CH}{=}\text{CH})_n{-}\dot{\text{C}}\text{H}{-}\text{CHCl}\sim + \dot{\text{H}} \tag{3}$$

Although C–C bonds are also likely to be broken upon direct irradiation of PVC, such a process is not expected to play a significant role in the photodegradation of the solid polymer, because of the low mobility of the resulting radical fragments which have a high probability of recombination:

$$[\sim\!\!(\text{CH}{=}\text{CH})_n{-}\text{CH}_2{-}\text{CHCl}\sim]^{*} \rightleftarrows [\sim\!\!(\text{CH}{=}\text{CH})_n{-}\dot{\text{C}}\text{H}_2 + \dot{\text{C}}\text{HCl}\sim] \tag{4}$$

This back reaction is especially efficient for a polymer like PVC whose glass transition temperature lies considerably above room temperature. It will not be the case however if the polymer is irradiated in solution where macroradicals easily escape cage recombination. Indeed the quantum efficiency of the dehydrochlorination was found to be ten times larger when PVC was photolysed in solution[25] than as a film.[23]

It is now generally accepted that the photodegradation of PVC proceeds primarily by a radical-type mechanism. Both alkyl and allylic radicals have been characterised by ESR spectroscopy in PVC photolysed at low temperatures.[26–29] A contribution from some ionic species may still be possible,[30,31] particularly with regard to the observed accelerating effect of HCl[32,33] and of metal salts like $FeCl_3$ and $ZnCl_2$[32,34,35] on the rate of dehydrochlorination. The fact that typical radical scavengers, like hydroquinone or oxygen, appear to have no strong inhibiting effect on the degradation process[25,34–36] is not necessarily contradictory to a radical mechanism, provided that very short-lived species are involved.

In recent years, several experimental results have been obtained independently by different authors which all argue in favour of an alkene-photosensitised degradation of PVC. Owen and Williams[37] observed a substantial increase of the rate of dehydrochlorination upon photolysis of thermally degraded PVC films which contained large amounts of polyene structures. Similarly, Rabek *et al.*[26] found by ESR measurements that conjugated double bonds can photosensitise the formation of free-radicals in the UV degradation of PVC. Additional arguments in support of an alkene-sensitisation mechanism have been presented by Reinish *et al.*[38,39] and by Verdu *et al.*[40] Finally, the value of the quantum yield of HCl production (Φ_{HCl}) was shown to remain independent of the extent of the reaction, while the rate of dehydrochlorination increased steadily with the duration of exposure to UV light[23,41] (see Section 3.3.2). This means that the incident photons, which are increasingly absorbed by the polyenes formed, initiate the dehydrochlorination of PVC with an efficiency that remains constant throughout the photolysis. This result, together with the independence of Φ_{HCl} from the original amount of polyenes in the PVC sample[42] strongly suggests that the initially absorbing chromophores are mainly unsaturated structures already present in the commercial polymer.

(*2*) *Carbonyl groups*

Since it is very difficult to remove atmospheric oxygen completely during the industrial polymerisation of vinyl chloride, dialkyl peroxide structures (~CH_2—O—O—CHCl~) are likely to arise by copolymerisation of the monomer with O_2. The thermal decomposition of those peroxides leads, among other products,[43] to carbon monoxide, which easily polymerises with vinyl chloride. The carbonyl groups thus formed are located either within the polymer backbone:[44]

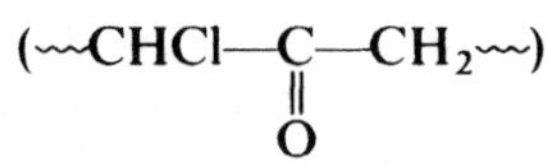

or on side groups:[45]

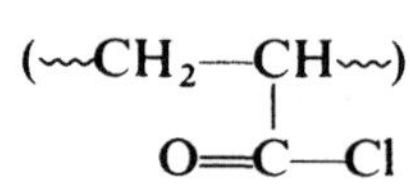

Besides ketone groups, hydroperoxides are also introduced during the processing of the polymer in the presence of air and subsequent storage.

Carbonyl groups are known to be photosensitive structures which, once they are promoted to an electronically excited state, undergo various reactions that are capable of initiating the degradation of PVC: energy transfer to the polymer, abstraction of hydrogen atoms, fragmentation into radical (Norrish type I) or non-radical (Norrish type II) products:

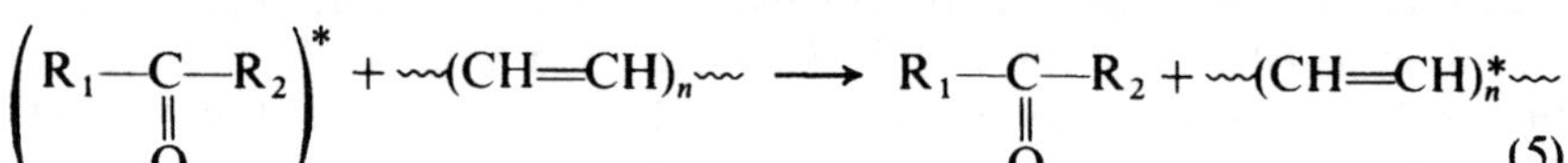

(5)

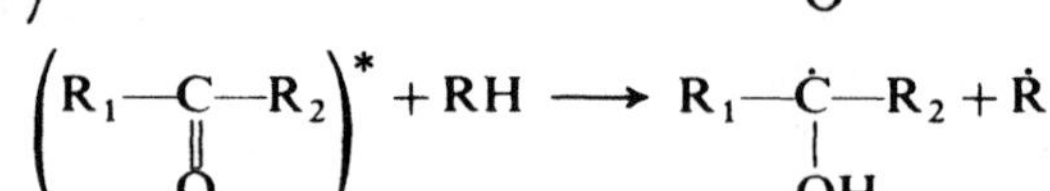

(6)

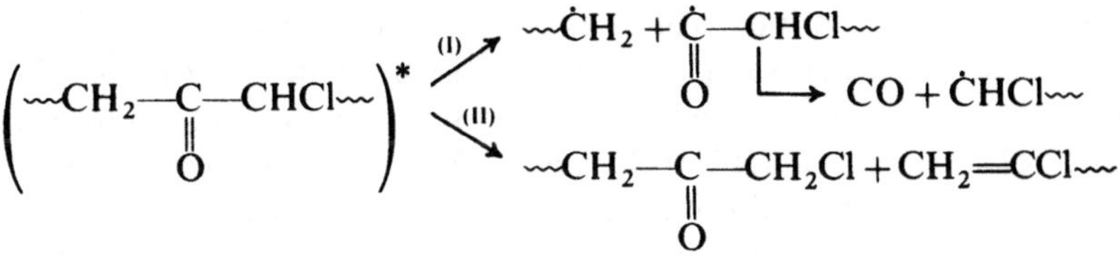

(7)

(8)

The photosensitisation of the PVC degradation by carbonyl compounds was extensively investigated in the last decade and was recently discussed in detail by Owen in a comprehensive survey.[9]

Several authors[6,44,46,47] have observed that the presence of carbonyl groups within the polymer chain, at concentrations in the range 0·05–0·2 mol %, substantially accelerates the photodegradation of PVC. The amount of HCl evolved upon the photolysis ($\lambda > 294$ nm) of a vinyl chloride–carbon monoxide copolymer can be directly related to the carbonyl content of the copolymer.[44] The light stability of PVC appears therefore to be very dependent on the presence of carbonyl chromophores which, if their concentration is sufficient to effectively absorb the incident light, will be able to initiate the dehydrochlorination reaction.

This conclusion also applies to low molecular weight ketones which may contaminate commercial PVC. Both aliphatic[48,49] and aromatic[50,51] ketones were shown to be effective sensitisers of the photodegradation of PVC, although their efficiency seems to decrease as their molecular weight increases.[52] Quantum yields as large as 0·2 were reported for the benzophenone sensitised elimination of HCl from both PVC[50] and model compounds, like *t*-butyl chloride.[53] The most usual reaction of excited benzophenone is the abstraction of a hydrogen atom from the substrate with the formation of ketyl and alkyl radicals (reaction (6)). With PVC as substrate, it will generate $\sim CH_2-\dot{C}Cl\sim$ and $\sim\dot{C}H-CHCl\sim$ radicals which, by splitting off a chlorine atom, will start the dehydrochlorination chain reaction (see Section 3.3.2).

Investigations of the ketone-sensitised photolysis of simple alkyl chlorides[52,53] also suggest that under some conditions, the excitation energy of the ketone can be transferred to the alkyl chloride which would then dissociate into radicals, and finally lead to the formation of a double bond and an HCl molecule:

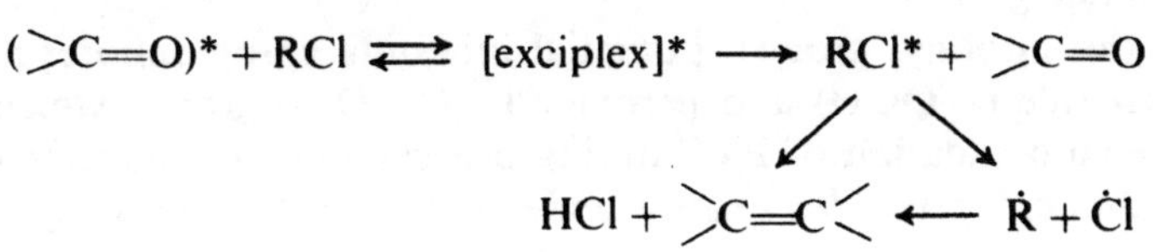

(9)

(10)

The energy transfer reaction involves essentially triplet states in the case of aromatic ketones,[53] while a singlet mechanism seems to be operative for aliphatic ketone sensitisers.[52] A similar initiation scheme was suggested for the photolysis of PVC containing C═O groups.[7] An excited complex resulting from the interaction of a carbonyl triplet and a chloromethylene unit of the polymer was postulated as the reactive intermediate. This exciplex would either dissociate into a ground state carbonyl and a vibrationally excited polymer unit that eliminates HCl, or directly induce the dehydrochlorination of PVC. The fact that molecular oxygen, an efficient quencher of triplet excited states, was found by some authors[44,54,55] to reduce the initial rate of HCl production is consistent with such a mechanism. However, once PVC starts to degrade, the presence of oxygen appears to substantially enhance the extent of the dehydrochlorination.[55,56] This suggests that carbonyl-photosensitisation is only effective at the very beginning of the irradiation. As soon as the strongly light-absorbing polyene structures appear, the alkene-sensitisation mechanism takes over and rapidly becomes the predominant initiation process. Other possible initiating chromophores are the α–β unsaturated ketones

(~CH═CH—C(═O)~)

which, owing to their high extinction coefficient in the 250–300 nm range, may compete effectively with polyenes for the absorption of the incident light.

The contribution of carbonyls to the initiation step will obviously depend on the amount of ketone groups in the starting material. This might explain why in some cases, such as with PVC films of low C═O content, no initial O_2 retarding effect could be detected during the photolysis at $\lambda >$ 254 nm.[25,41] Instead, from the outset of the irradiation, the quantum yield of HCl production was found to be 30% higher in pure oxygen than in nitrogen.[41] The lack of any O_2 inhibition effect was considered as a strong argument in favour of a degradation initiated by excited singlet alkenes which, unlike excited carbonyls, cannot be quenched by O_2 owing to their very short lifetime ($\sim 10^{-9}$ s).[10]

(3) *Peroxide groups*

Besides the carbonyl groups, potential initiating chromophores are the hydroperoxide (—OOH) and peroxide (—O—O—) groups which arise from the autooxidation of PVC during processing and storage.[57] Under UV irradiation, peroxides readily undergo a homolytic cleavage of the

O–O bond with formation of alkoxy radicals which can initiate the degradation by hydrogen abstraction from the polymer or by a β scission process (see Section 3.4.7):

$$\sim\!CH_2\!-\!\underset{\underset{O-OH}{|}}{\overset{\overset{Cl}{|}}{C}}\!-\!CH_2\!\sim \xrightarrow{h\nu} \sim\!CH_2\!-\!\underset{\underset{O^{\bullet}}{|}}{\overset{\overset{Cl}{|}}{C}}\!-\!CH_2\!\sim + {}^{\bullet}OH \qquad (11)$$

This cleavage reaction proceeds efficiently, with a quantum yield close to unity,[58] because the excited hydroperoxide molecule is in a dissociative state, i.e. as the molecule begins to vibrate after excitation there is no restoring force and the fragments fall apart.

When peroxide structures are introduced into PVC, either by polymerisation in the presence of air[40,44] or by ozonisation,[59] the polymer becomes more susceptible to subsequent photodegradation, which proves that those chromophores are indeed able to initiate this process. From their investigation on the light stability of processed PVC, Scott *et al.*[60,61] concluded that the peroxides formed during thermal processing are the major sensitisers of the photooxidation of PVC. However, olefinic unsaturation and carbonyl groups which are produced at the same time[62] may also contribute to the increased light-sensitivity of the processed polymer. Furthermore, it must be remembered that hydroperoxides have a very weak absorption in the 250–400 nm UV range. An initiation by direct photolysis of peroxide groups therefore is likely to be effective only in the very early stages of the exposure before the strongly light-absorbing polyenes appear. In the later stages, an alkene-sensitised photolysis of peroxides is considered possible,[23] by analogy with the carbonyl-photosensitised decomposition of hydroperoxides:[63,64]

$$[\sim(CH{=}CH)_n\sim]^* + POOH \longrightarrow \sim(CH{=}CH)_n\sim + P\dot{O} + \dot{O}H \qquad (12)$$

Such a process should yet be of limited importance due both to the short life time of excited polyenes and the relatively low concentration of POOH in the photooxidised PVC.[42]

(4) Structural defects

Commercial PVC is known to contain structural irregularities which may influence the photostability of the polymer. The principal defects identified so far consist of branched structures,[65,66] mostly of the type

$$\sim\!CH_2\!-\!\underset{\underset{CH_2Cl}{|}}{CH}\!-\!CH_2\!\sim$$

and

$$\sim CH_2-\underset{\displaystyle R}{\underset{|}{C}}Cl-CH_2\sim$$

where R is a side branch of variable length. Since these functional groups absorb in the far UV, like the monomer unit, they cannot be the initiating chromophores. The presence of labile tertiary hydrogen and chlorine atoms at the branch points though, is expected to increase the efficiency of both the dehydrochlorination and the oxidation. It has been shown that some of the PVC chains are terminated by allylic chloride moieties, with the following structure being most probable:[15,66,67]

$$\sim CH_2-CH{=}CH-CH_2Cl$$

These groups are very likely to be starting points for the formation of conjugated systems which will then absorb UV light and initiate the photodegradation of PVC. Recently, Gooden and Hellman[68] investigated by GPC analysis the photodegradation of PVC at solar wavelengths and found unsaturation to be mostly concentrated in low molecular weight fragments. This result strongly suggests that conjugated triene and tetraene located near the chain end are the major initiator sites.[68]

(5) *External impurities*

The presence in the polymer of traces of solvents, like tetrahydrofuran (THF), or of plasticisers, substantially increases the rate of photodegradation,[69,70,71] mostly because of the formation of some photoproducts which act as effective sensitisers of the photodehydrochlorination.[72] In addition, the increased mobility of the polymer chains will favour intermolecular propagation reactions and thus lead to shorter polyene sequences. For instance, when PVC was photolysed in solution,[25] most of the polyenes formed had sequence lengths shorter than 5, compared with values of up to 20 in the photolysis of PVC films.[41] The effect of phthalate ester plasticisers on the photodegradation of PVC was thoroughly investigated by Williams and Gerrard.[73] Both chain scission and crosslinking were shown to occur, as well as specific reactions of the plasticiser with the PVC, which leads to degradation of some of the phthalate and entrapment of the plasticiser in the crosslinked polymer. Catalyst residues and metal salts which contaminate commercial PVC were also shown to increase the light-sensitivity of the polymer by acting as sensitisers of the decomposition of peroxide groups or by generating free radicals under UV irradiation.[74,75]

The removal of all external impurities by precipitation of the polymer was found to increase the light-stability of PVC.[25,76] While the rate of degradation of the purified sample was substantially reduced, no major change was noticed in the dehydrochlorination quantum yield.[42] This means that the photons absorbed by the extraneous chromophores also participate in the initiation step. These chromophores are expected to consist mainly of low molecular weight products containing carbonyl, hydroperoxide and alkene groups.

Some of the stabilisers which are used to improve the resistance of PVC to heat and oxidation, like tin carboxylates or phenolic antioxidants, may function as efficient photosensitisers which actually accelerate the UV degradation of PVC.[77,78] Photoactive pigments, such as TiO_2 or ZnO, that are known to catalyse the photooxidation of some polyenes by generating reactive species,[79] are also likely to affect the sensitivity of PVC to UV radiation. In some recent work,[80] TiO_2 was shown to afford some protection to PVC against photodegradation by retarding the formation of carbonyl groups, hydroperoxides and polyenes. Increasing the TiO_2 pigment content in PVC increases the weathering protection by absorption of the UV light, but it also increases the probability of pigment deposition on the surface.[81] The optimum TiO_2 concentration was estimated to be 10%, based on chalking, surface erosion and weathering resistance.

(6) Conclusion on the photoinitiation of PVC

The chromophores most likely to initiate the degradation of PVC are the alkene linkages, carbonyl and peroxide groups which always contaminate commercial PVC. Their relative effectiveness in starting the dehydrochlorination depends primarily on their ability to both absorb the incident light and produce initiating species. On the basis of known values of their extinction coefficients,[10,20] at 300 nm for instance, it can be inferred that, when these chromophores are present in PVC at similar concentrations, for each photon absorbed by a peroxide group there will be about 10 excited carbonyl groups formed and of the order of 10 000 excited polyenes. These ratios are wavelength-dependent and shift even more in favour of polyene excitation as λ increases above 300 nm. The relative absorbance of these chromophores is also related to their concentration, which may vary over a large range, depending on the history of the PVC sample. In most cases, this will not be sufficient to reverse the order of importance of these photosensitising species, which is most likely to be as follows:

$$\text{alkenes} > \text{carbonyls} > \text{peroxides}$$

Whichever light-absorbing chromophores are involved in the initiation of PVC at the early stages of exposure to light, it is to be expected that the formation and photolysis of polyene products will become the predominant photochemical reaction during further irradiation of the polymer owing to both the high intrinsic absorption of these chromophores and their large rate of production.

3.3.2 Photodehydrochlorination

The main characteristic feature of the photolysis of PVC is the dehydrochlorination reaction, which develops extensively, as in the thermal degradation, and which leads to the formation of conjugated polyene sequences with concomitant evolution of HCl:

$$\sim\!CH_2\!-\!CHCl\!\sim \xrightarrow{h\nu} \sim\!(CH\!=\!CH)_n\!\sim + \, nHCl \qquad (13)$$

These polyene structures can be easily characterised by their UV

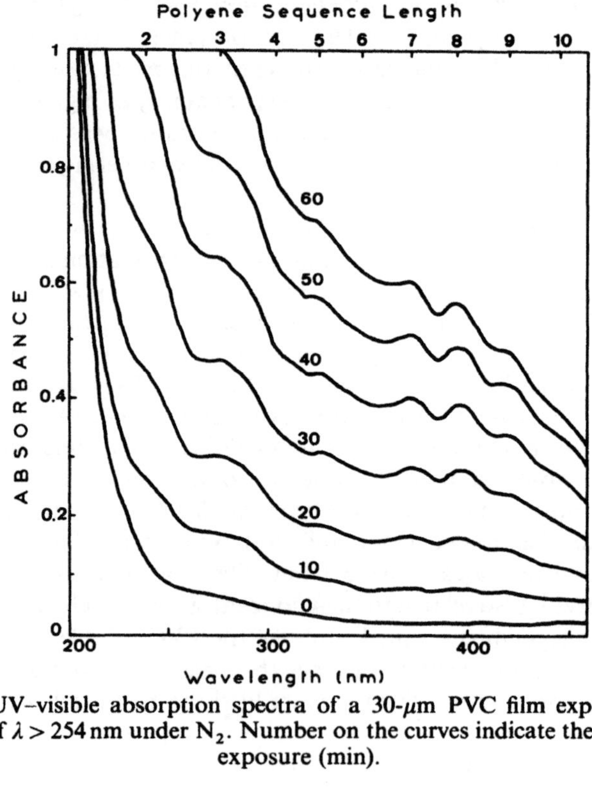

Fig. 3.2 UV–visible absorption spectra of a 30-μm PVC film exposed to UV radiation of $\lambda > 254$ nm under N_2. Number on the curves indicate the duration of exposure (min).

absorption, which ranges from 230 nm for dienes to ~400 nm for $n = 8$.[20] As the sequence length increases over 8 units, the absorption spreads into the visible region and discoloration appears. Figure 3.2 shows typical UV–visible absorption spectra of a 30-μm PVC film that was photolysed for various times up to one hour. The different maxima correspond to the strongest absorption of polyenes with sequence length ranging from 2 to 10.[19,20] For extended irradiation time, up to 20 conjugated double bonds can be formed in a single sequence, thus leading to a strongly coloured material.

(*1*) *Kinetics and quantum yield*

The kinetics of the photodehydrochlorination of PVC has been extensively investigated in recent years.[21,42,44,46,82,83,84,85] Most of the authors agree that HCl is evolved at an increasing rate in the early stages followed by a slowing down for more prolonged durations of exposure (Fig. 3.3, curve a). The reasons for this S-shaped kinetic curve have been discussed recently.[21,44] The initial auto-accelerated process is usually attributed to

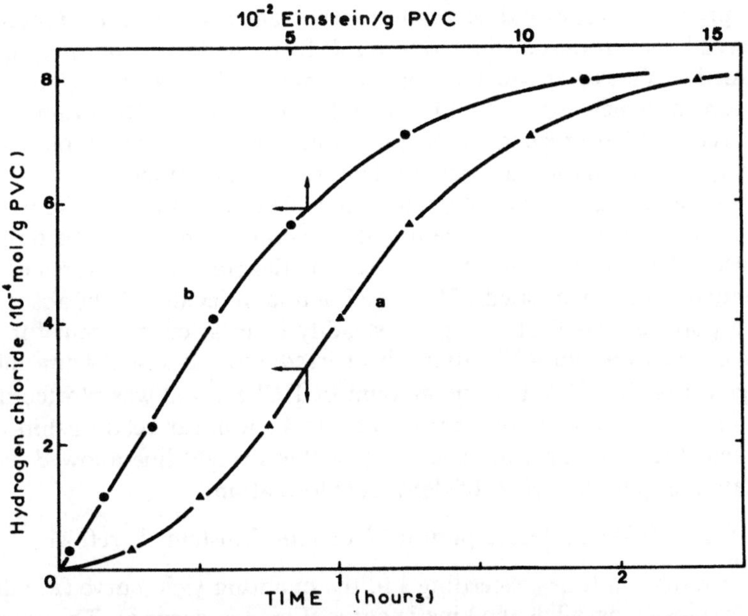

Fig. 3.3 Amount of HCl evolved in the photolysis at $\lambda > 254$ nm of a 30-μm PVC film as a function of the irradiation time (curve a) and as a function of the number of photons absorbed by the polymer (curve b).[42]

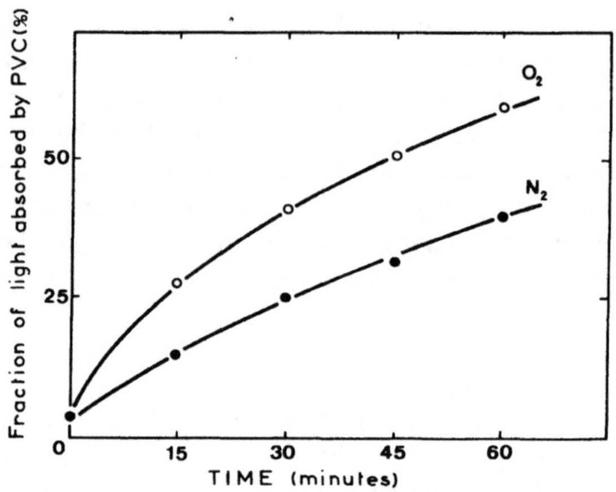

Fig. 3.4 Fraction of incident light absorbed by a 30-μm PVC film as a function of irradiation time.[23]

some catalytic effect of the HCl evolved[86] or to the photolysis of some other photoproducts that would generate reactive radicals. The auto-inhibition observed later on was considered to result from the formation of a highly degraded surface layer which absorbs the incident light and thus acts as a light-screen which protects the underlying layers.[83] Quantum yield measurements have recently shown that although the second explanation seems quite reasonable the first can be seriously questioned.[42]

By investigating the photodegradation of PVC in solution[25] and in film form,[23] we have proved that the initial accelerated kinetics is due only to increased absorption of the incident light by the polyene structures which accumulate in the irradiated PVC. The fraction of incident light absorbed by the polymer was found to grow steadily from a few per cent for the original film to about 40% after 1 h of irradiation at $\lambda > 254$ nm under nitrogen (Fig. 3.4).[23] When the amount of HCl evolved was plotted as a function of the number of photons absorbed, a linear relationship was obtained (Fig. 3.3, curve b). The slope of this straight line allowed us to evaluate the quantum yield of dehydrochlorination:

$$\Phi_{HCl} = 0{\cdot}011 \text{ molecule photon}^{-1} \text{ or mol. Einstein}^{-1} \text{ (ref. 41)}$$

For irradiation times exceeding 1 h, the quantum yield curve (Fig. 3.3, curve b) closely parallels the kinetic curve (Fig. 3.3, curve a). This means that the marked decrease in the rate of HCl production is directly related to a loss in the efficiency of the absorbed photons which are unable to induce

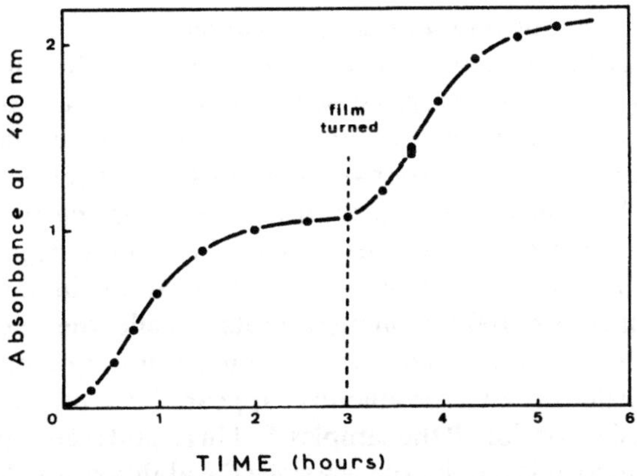

Fig. 3.5 Variation of the absorbance at 460 nm of a 30-μm PVC film as a function of the irradiation time ($\lambda > 254$ nm) in the presence of N_2 (●) and O_2 (○). The film was turned over after 3 h of exposure to light.

the dehydrochlorination of PVC. Such a drop of Φ_{HCl} was not observed in solution,[25] even for prolonged exposure, and was attributed by Gibb and MacCallum[83] to the formation, on the exposed face of the film, of an almost totally degraded layer that consists essentially of polyenes which convert the radiation energy absorbed into heat. This top layer was estimated to be only a few microns thick.[42,44]

As expected from reaction (13), the UV–visible absorbance of the PVC film and the HCl production follow similar kinetics (Fig. 3.5 and Fig. 3.3, curve a). Once the absorbance has levelled up, a second S-shaped curve can be obtained by irradiating the film on the opposite face, which appears to have suffered little degradation during the first 3 h of irradiation. Analogous curves were reported by other authors[44,83] who followed the HCl production in the photolysis of PVC films under nitrogen.

From those kinetic investigations, one can expect the photodegradation of transparent PVC films or sheets to unfold in the following way. In the early stages, the dehydrochlorination develops uniformly throughout the irradiated material but as soon as the strongly light-absorbing polyene structures appear, the process auto-accelerates and at the same time it concentrates progressively closer to the exposed face with the formation of a thin top layer of highly degraded polymer that finally prevents the bulk of the film from further degradation. The finding that the amount of HCl evolved is independent of the film thickness[38,79] is in good agreement with such a superficial process.

(2) *Influence of polymer unsaturation content*

The fact that Φ_{HCl} remains constant while the UV radiation is increasingly absorbed by the polyenes formed was considered to be a strong argument in favour of an alkene-sensitised initiation process.[23] It also implies that Φ_{HCl} does not depend on the original amount of unsaturation in the starting polymer. This conclusion was confirmed by exposing to UV light PVC samples that were either hydrogenated, purified or pre-degraded and which thus contained various amounts of unsaturation, in the range 0·05–6 double bonds per 1000 monomer units. While the rate of dehydrochlorination was directly related to the unsaturation content, the initial quantum yield of HCl production appeared to be the same, within experimental error, for all the samples.[42] This result seems quite reasonable from a photochemical point of view since the ability of an absorbed photon to initiate a given reaction is not expected to depend on the number of photons which are absorbed independently and thus on the amount of chromophores present in the system.

(3) *Influence of light intensity*

Another interesting feature of the photodehydrochlorination of PVC is that, as the discoloration develops, the fraction (f) of the incident light absorbed by the polymer increases steadily so that the absorbed light intensity, $I_a = I_0 \times f$, increases in the same way. Under given conditions,[23] a ten-fold increase in I_a occurs during the first hour of irradiation (Fig. 3.4). This apparently does not affect the quantum yield of HCl production, which remains constant during this period. Since Φ_{HCl} can be written as the ratio of the rate of dehydrochlorination, R_{HCl}, to the absorbed light intensity, $\Phi_{HCl} = R_{HCl}/I_a$, the invariance of Φ_{HCl} means that the rate of dehydrochlorination is directly proportional to the light intensity or to the rate of initiation. Such behaviour is usually observed for non-chain reactions or for first-order (linear) chain reactions, i.e. processes where the chain-propagating species is not involved in a bimolecular termination step.

Investigations[25,50,82,84] carried out at various light intensities, I_0, confirmed that the rate of photodehydrochlorination of PVC is first order (or slightly greater)[50,82] with respect to the intensity of light. This is an important result for it implies that the quantum efficiency of HCl production and polyenes formation should be the same when low light intensities are used, such as during natural weathering.

(4) *Influence of radiation wavelength*

Since the terrestrial solar spectrum does not extend below 290 nm while

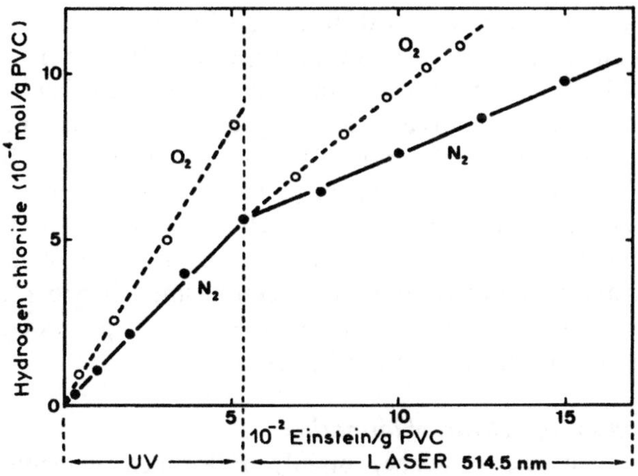

Fig. 3.6 Amount of HCl evolved in the UV and the 514·5-nm photolysis of a 30-μm PVC film as a function of the number of photons absorbed by the polymer ($I_0 = 1{\cdot}5 \times 10^{-7}$ Einstein s^{-1} cm^{-2}).[41]

most of the mechanistic studies on the photodegradation of PVC were performed with UV light of wavelength as low as 254 nm, it is of prime importance to know how the rate of degradation of PVC will depend on the wavelength of the incident light. The numerous investigations reported so far on this question[21,48,59,87,88] all point out that PVC deteriorates at a much slower rate when radiation of $\lambda > 290$ nm is employed. UV radiation ranging from 250 to 290 nm appears thus to be particularly harmful towards PVC. The observed decrease of the rate of dehydrochlorination for $\lambda > 290$ nm is expected to result either from a loss in the quantum efficiency of these less energetic photons or from a reduction of the number of photons absorbed by the PVC film, which is more transparent to radiation of $\lambda > 290$ nm—or from both factors.

In a recent study of the laser-induced degradation of PVC,[41] the ability of visible photons to induce the dehydrochlorination reaction was tested. A PVC film that had been degraded previously by UV irradiation under nitrogen was exposed to the 514·5-nm radiation of an argon-ion laser, at exactly the same light intensity. The dehydrochlorination of the polymer proceeded quite effectively (Fig. 3.6), revealing that these 'green' photons, which carry an energy of only 230 kJ mol^{-1}, are still energetic enough to induce the degradation of PVC. The value of Φ_{HCl} was about half the value measured by using >254 nm UV light. It was concluded therefore that the drastic decrease in the rate of degradation for irradiation at $\lambda > 290$ nm originates mainly from the very low absorption of PVC at this wavelength.

It should also be mentioned that in some experimental situations radiation of $\lambda > 300$ nm was unable to degrade PVC further and led instead to a redistribution of the polyene sequence lengths.[21] It has been suggested[42] that this behaviour may actually result from an efficient photobleaching of the polyenes by the HCl retained within the polymer film.[37] When HCl was thoroughly removed from the pre-degraded PVC, irradiation at $\lambda > 300$ nm led indeed to a significant dehydrochlorination of the polymer.[42] The various roles played by HCl in the photodegradation of PVC have been recently reviewed by Owen.[8]

(5) Mechanism of dehydrochlorination

Based on available evidence, mainly ESR measurements,[26-29] the elimination of HCl during the non-oxidative photodegradation of PVC is generally considered to proceed by a radical chain reaction, although some data also argue in favour of an ionic mechanism[30,89] or a concerted elimination involving some charge transfer.[89] The different types of radicals generated by the photofragmentation of the excited polyenes (see Section 3.3.1(1)) are expected to abstract hydrogen atoms from the CHCl and CH_2 groups of PVC to yield α-chloro and β-chloro-alkyl radicals, respectively (reactions (13), (14) and (15)—Scheme 1). In a rigid polymer, where the mobility of the chains is restricted, the chlorine radical evolved in reaction (2) may also abstract a hydrogen atom from the accompanying polyenyl radical (I), as well as from the adjacent CH_2 group since $\dot{C}l$ radicals were shown to be more reactive toward CH_2 than CHCl groups.[90,91] This will lead to an increase of the polyene sequence length by a non-chain process (cage reaction (16)).

The α-chloro alkyl radical (II) formed by reaction (13) is known[92] to be a less reactive species than the corresponding R—$\dot{C}$H—R radical and is expected to disappear either by abstraction of a hydrogen atom from the polymer (reaction (17)) or by a β-scission process (see Section 3.3.3(2)). The attack of PVC by the $\dot{C}l$ radicals which have escaped the cage process or by the α-chloro radicals (II) yields β-chloro alkyl radicals (III). These are short-lived species[41] which stabilise readily by splitting off a chlorine atom with the formation of a double bond (reaction (18)).[93] If the $\dot{C}l$ radical evolved reacts with the adjacent allylic CH_2 group, the 'zip' dehydrochlorination will then develop along the polymer chain and lead ultimately to polyene sequences of increasing length (reactions (19) and (20)). The $\dot{C}l$ radical can also diffuse away from its site of formation and attack a CH_2

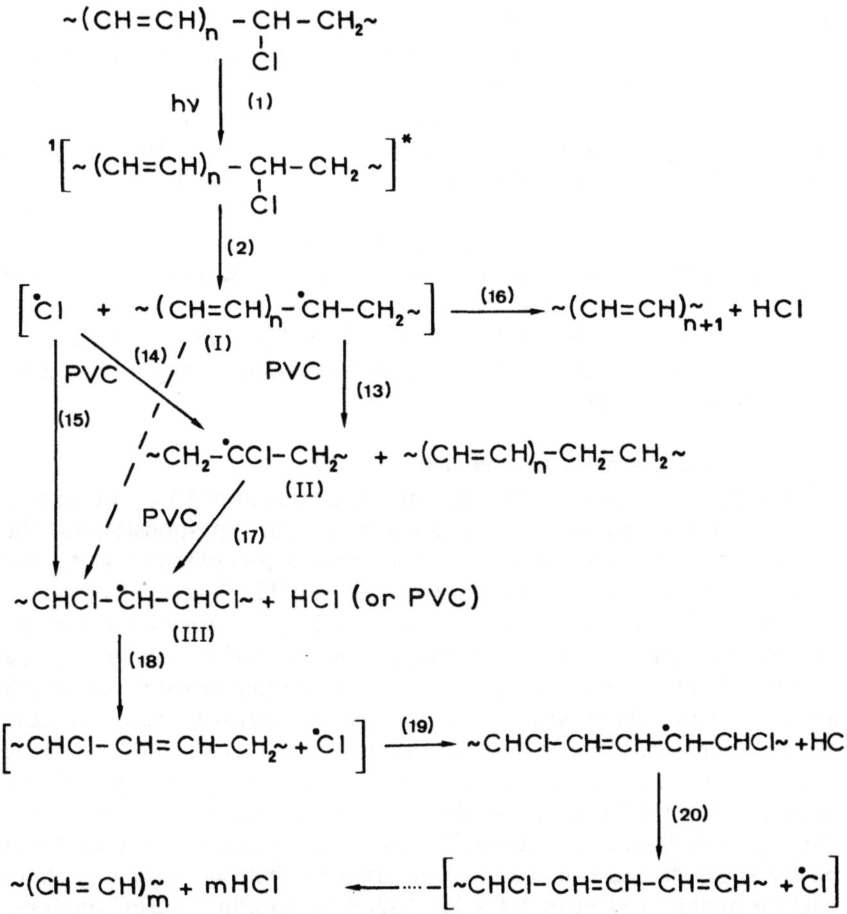

Scheme 1

group located on the same or another PVC chain. As a result, the polyene sequence will stop growing and a new conjugated double bond system will be created. The length of the polyene sequence and therefore the discoloration of photodegraded PVC is thus primarily related to the chain mobility, which depends on such factors as the temperature, the microstructure of the polymer, the presence of additives, plasticisers or solvent residues.

Interesting information regarding the mechanism of PVC photodegradation can be derived from the value of the dehydrochlorination quantum yield: $\Phi_{HCl} = 0{\cdot}011$ HCl molecule $photon^{-1}$. Since HCl is produced by a chain reaction, this result means that much less than one in every 100 photons absorbed by the polymer actually induces the dehydrochlorination of PVC. An overwhelming majority (>99%) of the excited polyenes appear thus to disappear by routes other than the photofragmentation reaction which is initially responsible for the deterioration of the physical and mechanical properties of the polymer. The most probable reactions of the excited singlet polyenes are the internal conversion, the *cis–trans* isomerisation of the conjugated double bond system and intramolecular reactions leading to cyclic structures.[94] Aromatic structures were indeed detected in thermally[95] as well as in chemically and photochemically degraded PVC where strongly fluorescent species were characterised.[74,96]

3.3.3 Crosslinking and chain scission

UV radiation is known to affect the molecular weight of PVC by inducing both chain scissions and crosslinks which are partly responsible for the change in the mechanical properties and surface aspect of the light-exposed polymer. While these processes were thoroughly investigated in the photooxidation of PVC (see Section 3.4.3), little information is reported on the corresponding effects on the photodegradation of this polymer in the absence of oxygen. Most authors[48,97–100] agree that crosslinking, which generates new intermolecular bonds between polymer chains, usually occurs under these conditions. A sharp increase of the PVC intrinsic viscosity was observed from the very beginning of the irradiation with the subsequent formation of an insoluble gel.[42] By contrast, when PVC was photolysed in deaerated solution,[25] chain scission predominated, as shown by the rapid decrease of the intrinsic viscosity, with an estimated chain scission quantum yield of $1{\cdot}1 \times 10^{-3}$ scission $photon^{-1}$. GPC analysis, however, revealed a broadening of the molecular weight distribution toward both ends of the chromatogram which indicated that crosslinking was also taking place.

(1) Quantum yields

The theoretical aspect of polymer crosslinking was developed by Charlesby[101] for the degradation of polymers by high-energy radiation and can be extended to photoinduced reactions. By measuring the fraction of the polymer which remains soluble after various times of exposure to UV light, it is possible to evaluate the quantum yields of crosslinking, Φ_{CL}, and chain scission Φ_S, by using the following equation:[102]

$$S + S^{1/2} = \frac{\Phi_S}{2\Phi_{CL}} + \frac{1}{\Phi_{CL}\bar{M}_{\omega_0}I_a t} \quad (21)$$

where S is the fraction of the PVC film that remains soluble after irradiation time t, $\bar{M}_{\omega_0}$ is the weight-average molecular weight of the unirradiated polymer and I_a is the absorbed light intensity expressed in Einsteins per second per gram of polymer. As expected from this equation, a linear relation was obtained when $S + S^{1/2}$ was plotted against the reciprocal number of photons absorbed ($r = I_a t$) (Fig. 3.7).[42] The slope of this straight line is inversely proportional to Φ_{CL} which can thus be estimated as $\Phi_{CL} = 1{\cdot}4 \times 10^{-3}$ crosslinks per photon.

The fact that the straight line obtained does not pass through the origin indicates that a competitive chain scission process also develops during the photolysis of PVC under nitrogen. UV radiation of $\lambda > 254$ nm is thus capable of cleaving the PVC chains, independently of any oxidation process. From the value of the intercept, 1·1, and by using eqn (21), Φ_S was

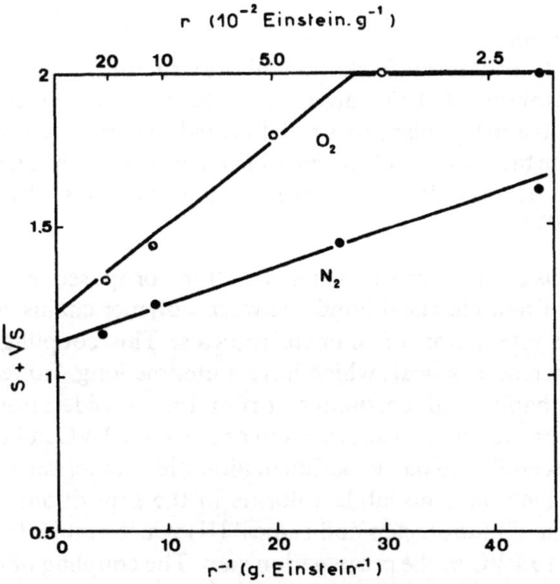

Fig. 3.7 Plots of $S + S^{1/2}$ against the reciprocal number of photons absorbed by the PVC film exposed to UV radiation ($\lambda > 254$ nm) in the presence of N_2 (●) or pure O_2 (○).[42]

calculated to be $3{\cdot}1 \times 10^{-3}$ scissions per photon. Since there are on average less than four main chain scissions for each intermolecular bond formed ($\Phi_S < 4\Phi_{CL}$), crosslinking predominates and leads to a partially insoluble polymer. As the photodegradation proceeds, the gel fraction increases steadily to finally reach, after prolonged exposure, a limit value of 55%.[42] A substantial fraction of the heavily N_2-irradiated PVC thus remains soluble because polymer chains are cleaved and crosslinked at the same time.

The relative importance of the major processes which occur during the photolysis of PVC in the absence of oxygen can be estimated from these quantum yield values. Under given experimental conditions,[42] for each crosslink formed there were two cleavages of the polymer backbone and eight HCl molecules evolved together with eight double bonds appearing on the PVC chain as conjugated polyene sequences. These figures clearly indicate that, besides the widely described dehydrochlorination reaction, UV-degraded PVC also suffers important modifications of its physical structure as a result of main chain scission and crosslink formation which both occur with remarkably high efficiencies, even in the absence of O_2.

(2) Mechanism

The presently available information on the structural changes taking place in the photolysis of PVC appears to be too incomplete to allow a comprehensive mechanism to be elaborated. Therefore, this section will only describe briefly some of the possible reactions which can be considered in order to account for the crosslinks and scissions observed in UV-irradiated PVC.

Crosslinking. The most usual reaction proposed to explain the formation of new chemical bonds between polymer chains consists of the bimolecular interaction of in-chain radicals. This coupling process will involve only macroradicals which have a lifetime long enough to ensure a definite probability of encounter. From the considerations developed previously for the dehydrochlorination process, the PVC radicals which are the most susceptible to participation in biomolecular termination reactions are those which have no labile chlorine in the β position, i.e. radical (I) formed in the initiation step and radical (II) which results from hydrogen abstraction to PVC in the propagation step. The coupling of these radicals leads to the formation of intermolecular bridges:

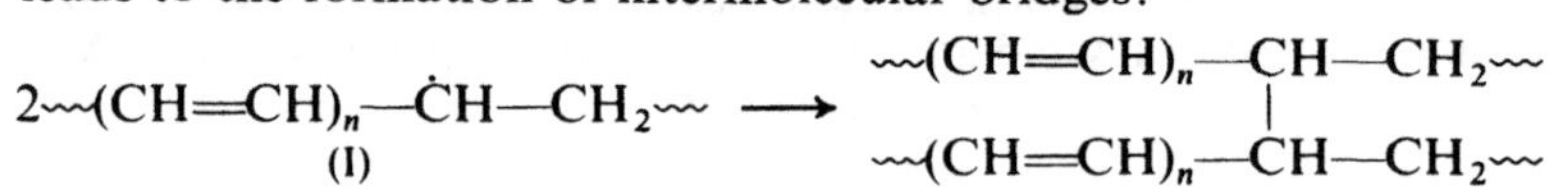

(I) (22)

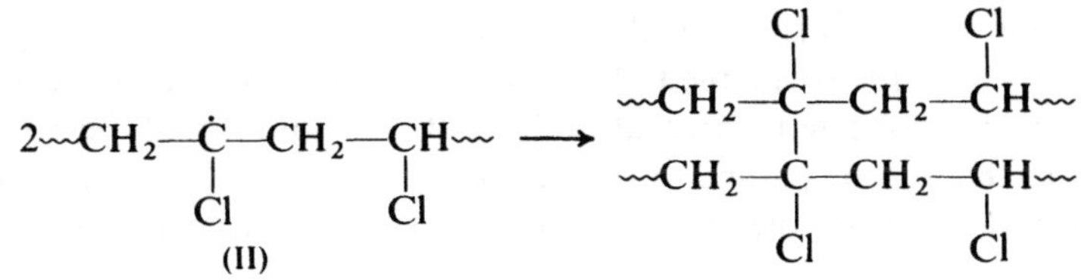

(23)

Cross terminations between radicals (I) and (II) are also likely to occur. Such a reaction may become particularly efficient if it involves radicals generated in close proximity, as for instance those resulting from the fragmentation of an excited singlet polyene (reactions (2) and (14)).

An additional crosslinking process was suggested by Rabek *et al.*[26,103] in order to account for the gel formation in PVC photolysed at low temperature (77 K) where the molecular motion of the macroradicals is drastically reduced. It consists of the addition of an alkyl or polyenyl radical on to a double bond of a polyene sequence:

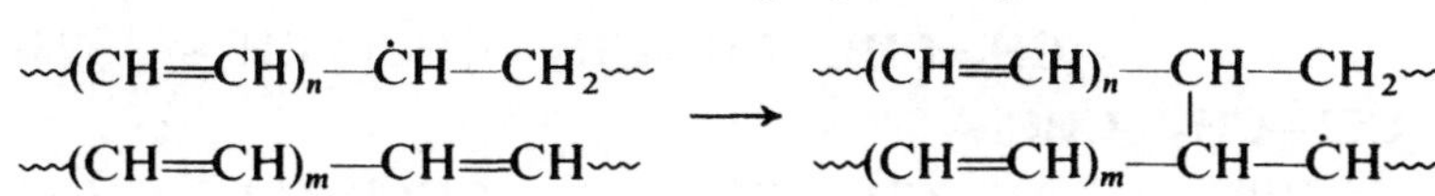

(24)

A similar reaction involving carbonium ions, resulting from HCl protonation of polyenes, was considered to be responsible for the crosslinking observed in the photolysis of PVC.[104] Since the efficiency of such an electron or ion transfer reaction depends directly on the concentration of the acceptor molecule, this crosslinking mechanism is expected to take place mainly in the later stages of the degradation, once sufficient amounts of polyenes have accumulated in the PVC film. Indeed, the strongly light-absorbing surface layer which appears in heavily degraded PVC was found to consist mostly of insoluble crosslinked material.[105]

Chain scission. The cleavage of the PVC backbone seems unlikely to result from the breaking of C–C bonds by the direct effect of UV light since, in a rigid polymer, the recombination of the two resulting radical fragments will rapidly take place. Furthermore, Φ_S was found to be substantially larger for PVC irradiated as a film[42] than in solution.[25] Since cage recombination is much less efficient in the liquid state, the opposite trend would be expected if direct C–C bond cleavage were to play a significant part in the fragmentation of the PVC chain.

In UV-irradiated polymeric materials, main-chain scissions often result from the photocleavage of backbone carbonyl groups according to Norrish type I or type II mechanisms. In the photolysis of PVC under nitrogen where no C=O groups are formed, this reaction could only involve the few carbonyl functions originally present in the PVC film and would therefore occur only at the very beginning of the irradiation. As PVC chains appear to be cleaved throughout the exposure to light, another explanation must be found.

One of the possibilities is to consider that the stabilisation of alkyl or polyenyl radicals can occur by a β scission involving not only C–Cl bonds (reaction (18)) but also C–C bonds, as suggested recently for the photodegradation of chlorinated PVC.[106] Since the β-chloro radical (III) is known to stabilise preferentially by splitting off rapidly the β-chlorine atom, the most likely candidates for such a reaction are radicals (I) and (II) which carry no chlorine in the β position:[93]

$$\underset{\text{(I)}}{\sim(CH{=}CH)_n{-}\dot{C}H{-}CH_2{-}CHCl\sim} \longrightarrow \sim(CH{=}CH)_n{-}CH{=}CH_2 + \dot{C}HCl{-}CH_2\sim \qquad (25)$$

$$\underset{\text{(II)}}{\sim CH_2{-}\dot{C}Cl{-}CH_2{-}CHCl\sim} \longrightarrow \sim CH_2{-}CCl{=}CH_2 + \dot{C}HCl{-}CH_2\sim \qquad (26)$$

According to this mechanism, the same kind of radicals would be responsible for both chain scission and crosslink formation. The fact that the GPC chromatogram of PVC showed no significant shift of the maximum after photolysis under nitrogen[25] indicates that the two competitive processes originate from the same species,[107] which is in good agreement with the suggested mechanism.

The major primary processes that occur when PVC is exposed to UV radiation in the absence of oxygen can finally be represented by Scheme 2:

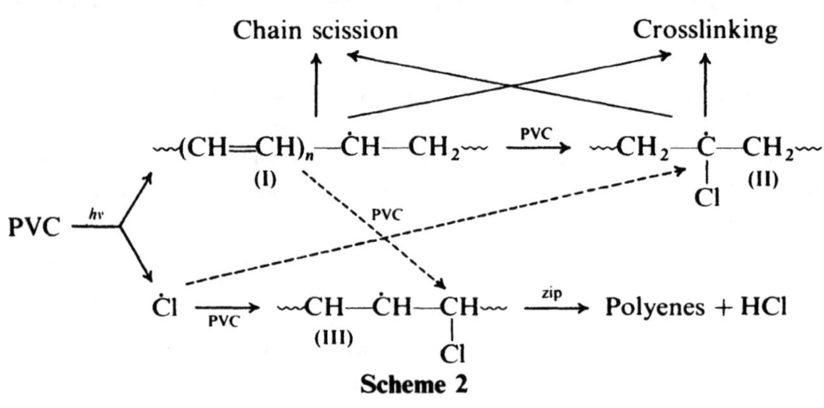

Scheme 2

Structural changes in the photodegraded PVC molecule are expected to result mostly from reactions of radicals (I) and (II), while the $\dot{C}l$ radical and the short-lived β-chloro radical (III) are mainly responsible for the discoloration of the polymer by starting the zip-dehydrochlorination.

3.4 PHOTOOXIDATION OF PVC

Since polymeric materials used for outdoor applications are always surrounded by air when they are exposed to the sunlight, it is of prime importance to know how the presence of oxygen will interfere with the various photochemical processes involved and thus affect the durability of the irradiated polymers.

As mentioned in Section 3.2.3, molecular oxygen may readily react with the unstable transients arising by UV irradiation, either by deactivating the excited states (quenching reaction) or by combining with free radicals to form peroxy ($P\dot{O}_2$) radicals (scavenging reaction):

Quenching: $$A^* + O_2 \rightarrow A + O_2^*$$

Scavenging: $$\dot{P} + O_2 \rightarrow P\dot{O}_2$$

The relative importance of these two reactions during the photodegradation of PVC in the presence of oxygen will depend primarily on the lifetime of the transients formed. In air-saturated systems, these reactions are only likely to compete with the other routes of disappearance for transient species with lifetime exceeding 10^{-7} s.[10]

By deactivating the excited polyenes formed by UV irradiation of PVC, the O_2 quenching reaction should reduce the extent of the zip-dehydrochlorination and therefore improve the light stability of PVC. Conversely, the scavenging reaction is expected to start an autooxidation process and thus to be responsible for the production of such photosensitive structures as hydroperoxide or carbonyl groups. Besides ruptures of the polymer chain, peroxide crosslinks occur in the photooxidation of PVC as a result of the various reactions of the $P\dot{O}_2$ radicals. All these chemical changes lead ultimately to a deterioration in the mechanical properties and embrittlement of the UV-oxidised material.

In this section, the effects of oxygen on each of these processes will be considered in turn. On the basis of recent quantitative data, an overall reaction scheme of the photooxidation of PVC will finally be presented which is assumed to be valid at least during the early stages of the irradiation.

3.4.1 Zip-dehydrochlorination

Most of the authors who have investigated the photooxidation of PVC[26,43,54,55,56] agree that dehydrochlorination still occurs, but the extent of this reaction appears to depend strongly on the intensity and wavelength of the incident light as well as on the temperature of the sample. It still remains unclear whether atmospheric oxygen actually accelerates or reduces the efficiency of this reaction. Since the energy absorbed by the PVC film may vary greatly from one sample to another, depending on its history, and since it increases considerably upon irradiation as a result of polyene formation, a direct comparison of the rate of HCl production is meaningless. The only way to unambiguously answer this question is by measuring the quantum yield of the dehydrochlorination (Φ_{HCl}).

(1) Kinetics and quantum yield

When a PVC film is exposed to UV radiation of $\lambda > 254$ nm in the presence of oxygen at a light intensity of $1{\cdot}5 \times 10^{-7}$ Einstein s^{-1} cm^{-2}, discoloration develops rapidly and HCl is evolved at a rate which increases with irradiation time (Fig. 3.8, curve a). This accelerated dehydrochlorination was shown[23] to result from an enhanced absorption of the incident light by the irradiated polymer, as in the photolysis of PVC under nitrogen. By plotting the total amount of HCl evolved (n_{HCl}) as a function of the number of photons absorbed by the PVC film (r), the accelerated kinetics turn into a linear relationship, where $n_{HCl} = \Phi_{HCl} r$ (Fig. 3.8,

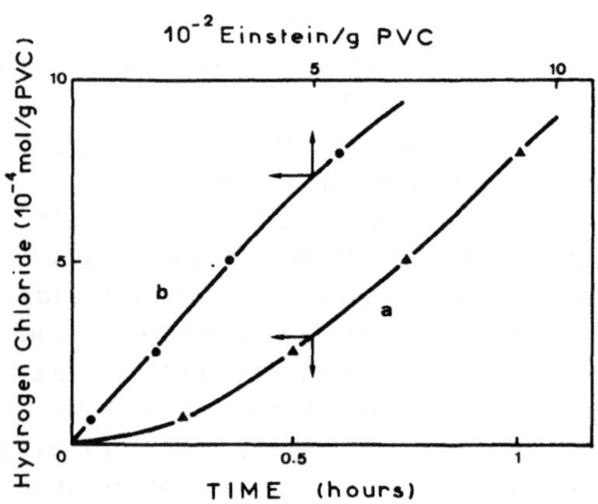

Fig. 3.8 Amount of HCl evolved in the photooxidation of a 30-μm PVC film as a function of the irradiation time (curve a) and as a function of the number of photons absorbed by the polymer (curve b).

curve b). From the slope of the straight line obtained, the quantum yield of HCl production was estimated to be 0·015 mol. Einstein^{-1}, a value which is 36% higher than that of Φ_{HCl} in the photolysis of PVC under nitrogen.[42] This result clearly demonstrates that the presence of oxygen actually increases the quantum efficiency of the zip-dehydrochlorination, under the given conditions.

From the lack of any O_2 inhibition effect it can be inferred that the excited singlet polyenes formed by UV irradiation at $\lambda > 254$ nm do not live long enough to be effectively quenched by atmospheric oxygen, i.e. their lifetime should be less than 10^{-8} s, in full agreement with previous observations.[10,108] The formation of singlet oxygen by energy transfer from the excited polyene structures to ground state oxygen, which was suggested to occur in the photooxidation of PVC,[55,109] appears therefore to be highly questionable. Furthermore, this result also implies that the β-chloropolyenyl radical (III) which, by splitting off its β chlorine, initiates the chain dehydrochlorination of PVC, is not scavengeable by the oxygen dissolved in the polymer, i.e. the stabilisation of this radical by $\dot{C}l$ removal and formation of a double bond occurs within less than 10^{-7} s. It will therefore be very difficult to scavenge β-chlororadicals and therefore to stop the zip-dehydrochlorination once it has started.

Since the presently available evidence suggests that molecular oxygen does not interfere significantly in the primary photochemical processes leading to dehydrochlorination when PVC is exposed to intense UV radiation of $\lambda > 254$ nm, one can expect reactions (1) and (2) as well as the sequence of reactions (15), (18), (19) and (20) proposed previously (Scheme 1, Section 3.3.2(5)) to remain valid for irradiation carried out in the presence of air.

(2) Photobleaching of polyenes by oxygen

It is well known that, under some conditions, the presence of oxygen actually reduces the extent of the discoloration in UV-irradiated PVC.[87,110] This bleaching process becomes particularly efficient when radiation of low intensity and of wavelength above 300 nm is used.[76,111] Irradiation of predegraded PVC at $\lambda > 315$ nm[54] or 514·5 nm[108] was shown to induce a very effective photobleaching by O_2 of preformed polyenes. Because of this polyene destruction reaction, discoloration usually develops at a slow pace when PVC films or sheets are exposed to sunlight during natural weathering.[112]

The observed effect of the radiation wavelength on the bleaching efficiency can be accounted for by assuming that the lifetime of the excited singlet polyenes increases as the conjugated double bond sequence grows, thus allowing O_2 to quench some of these excited states. In addition, the destruction of the polyenes can also result from a reaction of O_2 with the more stable polyenyl radicals (I) to yield cyclic peroxides, as suggested by Nagy *et al.*:[113]

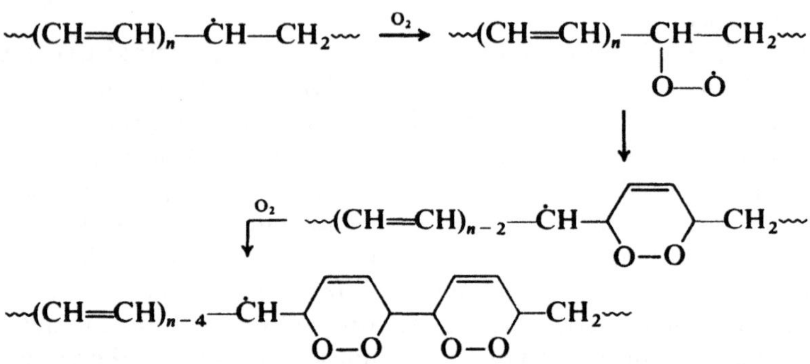

(27)

(28)

(29)

If the resulting polyenyl radical reacts again with oxygen, it will lead to a step-by-step shortening of the polyene sequence and the formation of adjacent cyclic peroxide structures.

It should be mentioned that polyene bleaching by atmospheric oxygen also takes place in the absence of any radiation.[108] This dark-bleaching was considered to result from a reaction of ground state oxygen with conjugated polyene systems[114] or with free electrons delocalised in the longer polyenes.[55] Both photo- and dark-bleachings by O_2 play an important role during the natural weathering of PVC by substantially reducing the rate of discoloration. The apparently poor effectiveness of visible radiation toward PVC is thus a result primarily of the fact that these radiations are very little absorbed by the bleached polymer.

In order to know whether photobleaching by O_2 occurs under the experimental conditions used, it is necessary to follow the variation of the PVC film absorbance upon photolysis as a function of the amount of HCl evolved, both in the presence and in the absence of oxygen. Figure 3.9 shows the plot obtained by using data from the work described in Section 3.4.1(1).[23] When UV radiation of $\lambda > 254$ nm is employed, the film absorbance, for a given amount of HCl evolved, is only 10% less in the presence of O_2 than in the presence of N_2, thus indicating that O_2 photobleaching is not proceeding effectively under these conditions. In contrast, when laser radiation of 514·5 nm is used, a marked difference appears between the two curves (Fig. 3.9). The drop of the absorbance in

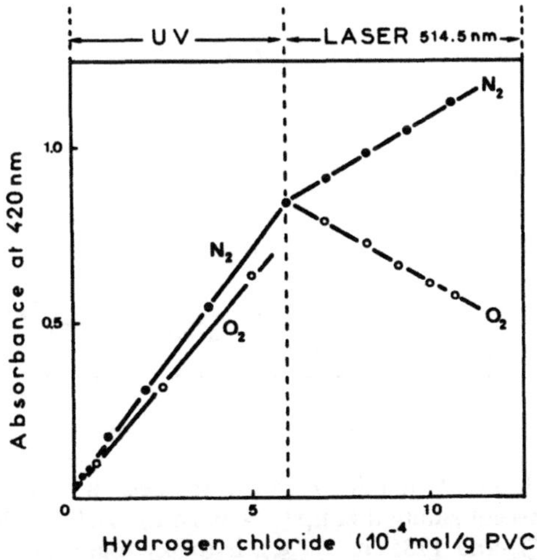

Fig. 3.9 Variation of the absorbance at 420 nm of the irradiated PVC film as a function of the amount of HCl evolved in the UV and 514·5-nm photolysis of PVC in the presence of N_2 (●) or pure O_2 (○).

the presence of O_2 clearly reveals that efficient photobleaching is taking place. The quantum yield of this reaction was found to be significantly higher than Φ_{HCl}, therefore causing a fast bleaching of the pre-degraded PVC film.[108] Interestingly, in the latter case, the actual rate of HCl production appeared to be lower for PVC irradiated under O_2 than under N_2, although the quantum yield Φ_{HCl} was still 40% higher in the presence of O_2.[41] The reduced rate of degradation in air therefore does not result from a drop in the photon efficiency but only from a decrease in the amount of light absorbed by the bleached polymer.

3.4.2 Oxidation products

Different types of oxygenated structures appear in the PVC chain on photolysis in the presence of oxygen. They consist mainly of OH and C=O groups which have been characterised by their infra-red absorption in the range 3200–3600 cm^{-1} and 1700–1800 cm^{-1}, respectively. Numerous investigations have been reported on this subject.[24,25,26,56,59,60,70,76,87,103,115-118]

The very broad hydroxyl band can be attributed to hydroperoxide structures

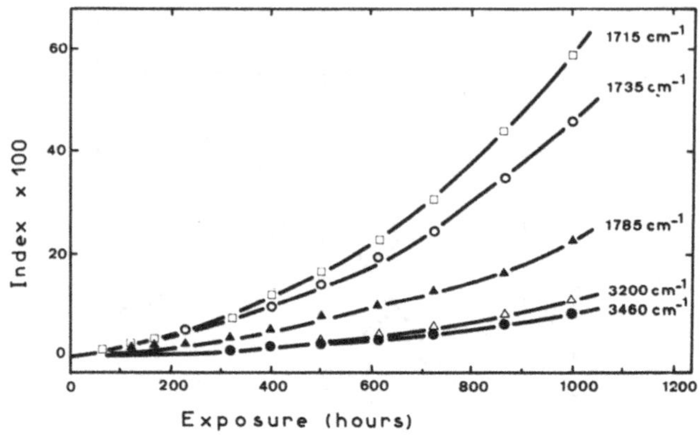

Fig. 3.10 Dependence of functional group indices on exposure for PVC films exposed to a fluorescent sunlight lamp ($\lambda > 300$ nm): carbonyl band at 1715 cm^{-1} (□), 1735 cm^{-1} (○) and 1785 cm^{-1} (▲), hydroxyl band at 3200 cm^{-1} (△) and 3460 cm^{-1} (●).[115]

which were quantitatively determined by chemical titration,[23] and to some alcohol functions, although the presence of these functions in photo-oxidised PVC has not yet been clearly demonstrated. The carbonyl absorption consists of a composite band centred around 1720 cm^{-1} with two prominent shoulders at 1735 and 1780 cm^{-1} which were assigned to different kinds of functional groups containing the C═O structure, namely α- and β-chloro carboxylic acids (1715 cm^{-1}),[115] α-[116] and β-[24,56] chloroketones (1725–1740 cm^{-1}), acid chlorides,[115] peracids[24] and lactones[56] (1775–1785 cm^{-1}). The presence of carboxylic acids was inferred from the substantial decrease of the absorption of the carbonyl band at 1715 cm^{-1} after treatment of the aged film with triethylamine.[115]

The kinetics of production of all of these oxidation products usually exhibits an auto-accelerating character,[56,87,115] like for the production of hydrogen chloride and conjugated polyenes. Figure 3.10 shows the dependence of the various infrared bands on UV exposure at $\lambda > 300$ nm for cast PVC films, as reported by Scott and Tahan.[115] The auto-accelerating effect is apparently less pronounced for those PVC samples where discoloration develops less intensively due to an efficient bleaching of the polyenes by oxygen.[54,76,116] This observation suggests that, like the kinetics of dehydrochlorination, the observed increase of the rates $R_{C=O}$ and R_{POOH} at which these oxygenated structures appear mainly results from an enhanced absorption of the incident light by the increasingly coloured

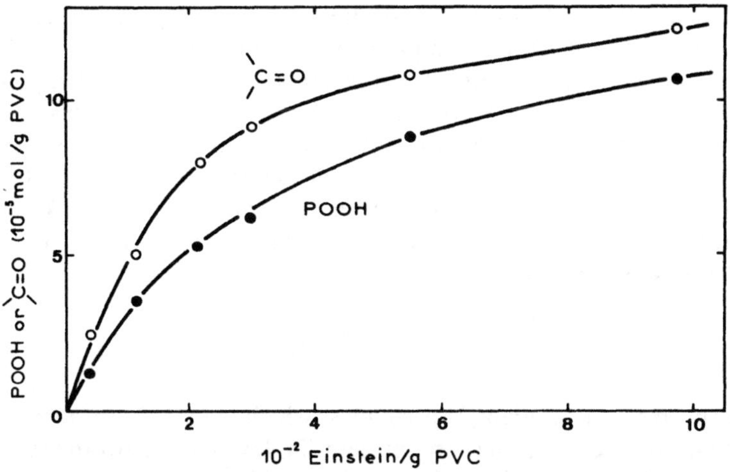

Fig. 3.11 Concentrations of carbonyl and hydroperoxide groups as a function of the number of photons absorbed by the polymer in the photooxidation of PVC films at $\lambda > 254\,nm$.[23]

PVC film. Quantum yield measurements have fully confirmed this explanation.[23]

For extended irradiation time, both rates $R_{C=O}$ and R_{POOH} were found to decrease steadily, down to almost zero[35,65] (Fig. 3.11). Such an important rate reduction results essentially from the build-up of strongly UV-absorbing species in the surface layers of the polymer which is thus protected from further photooxidation. This was demonstrated by surface analysis of a highly photooxidised PVC film by using X-ray photo-electronic spectroscopy (ESCA)[76] and multiple internal reflection (MIR) infra-red spectroscopy.[42] The carbonyl concentration was shown to be much higher in the superficial layer directly exposed to UV radiation than in the interior of the film. A careful kinetic examination[42] revealed that the slowing down of $R_{C=O}$ and R_{POOH} does not parallel the slowing down of R_{HCl}, as would be expected if this screen effect were the sole reason, but that it begins at an earlier time. This indicates that an additional mechanism is responsible for the rate reduction, most probably a polyene-sensitised photodecomposition of carbonyl and hydroperoxide groups,[23] since these photosensitive chromophores are unlikely to be directly excited in the presence of large concentrations of strongly UV-absorbing polyenes.

Small amounts of volatile oxidation products, mainly CO_2, CO and H_2O, were also shown to be formed during the photooxidation of PVC.[24]

Only in the latter stages of the irradiation do these products contribute significantly to the O_2 consumption by the polymer. They must thus be considered as resulting mainly from secondary reactions involving the two major oxidation products, carbonyl and hydroperoxide groups.

The fact that photooxidised PVC becomes partly insoluble in THF reveals that intermolecular crosslinking is taking place (see next section). In the presence of a sufficient supply of oxygen, these crosslinks consist essentially of peroxide bridges between polymer chains with the following most probable structure:[42]

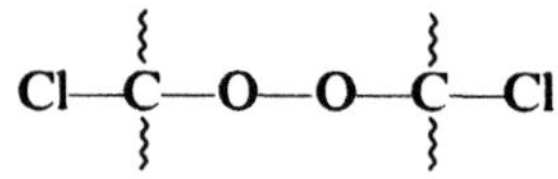

As expected, these dialkyl peroxides which result from termination reactions of $P\dot{O}_2$ radicals are present in lower concentrations in the oxidised polymer than the hydroperoxides which are produced by a chain reaction:

$$\text{Termination} \qquad P\dot{O}_2 + P\dot{O}_2 \longrightarrow POOP + O_2$$

$$\text{Propagation} \qquad P\dot{O}_2 + PH \longrightarrow POOH + \dot{P}$$

$$\dot{P} \xrightarrow{O_2} P\dot{O}_2$$

Though hydroperoxide and carbonyl groups are usually considered to be the main oxygen-containing structures, direct evidence that could be obtained by O_2 uptake measurements has not been provided so far. In the case of the γ-induced oxidation of PVC, these oxygenated structures were found to account for over 70% of the oxygen consumption.[119]

3.4.3 Change in the molecular weight of PVC

PVC chains undergo an efficient scission process when they are exposed to UV radiation of $\lambda > 254$ nm in the presence of O_2, as shown by the decrease of the molecular weight and subsequent embrittlement of the photo-oxidised film.[23,48,69,76,97,115] This process was considered to be one of the main causes of the observed loss of the mechanical properties of PVC sheets exposed to natural weathering.

The average number of main chain scissions, $\bar{X}$, can be evaluated from the number-average molecular weight of the original and of the degraded polymer, $\bar{M}_{n_0}$ and $\bar{M}_n$, respectively, determined by GPC or intrinsic viscosity measurements, by using the following equation:

$$\bar{X} = (\bar{M}_{n_0}/\bar{M}_n) - 1$$

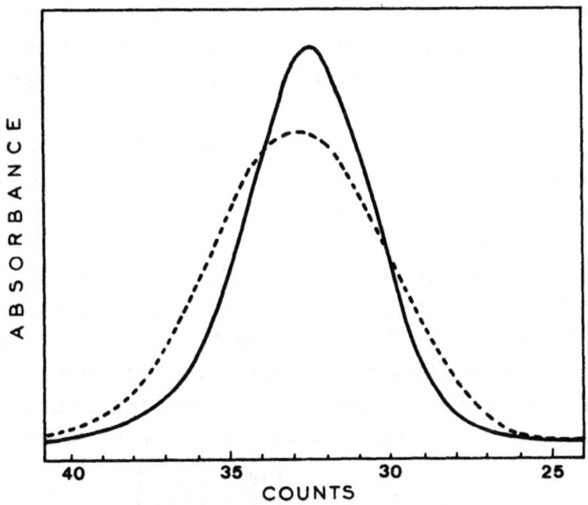

Fig. 3.12 GPC molecular weight distribution of PVC samples. ——, Before degradation; ---, after 1 h of UV irradiation in the presence of O_2.

From the linear relationship between $\bar{X}$ and the number r of photons absorbed per gram of PVC ($\bar{X} = \Phi_s r \bar{M}_{n_0}$ or $1/\bar{M}_n = 1/\bar{M}_{n_0} + \Phi_s r$), the quantum yield of chain scission, Φ_s, was estimated to be $1{\cdot}7 \times 10^{-4}$ scissions per photon absorbed.[23] GPC chromatograms of the oxidised PVC clearly show a broadening of the molecular weight distribution toward both the lower and the higher molecular weights (Fig. 3.12). The ratio $\bar{M}_w/\bar{M}_n$, initially close to 2, was found to increase linearly with the UV exposure time.[115] This effect was attributed to a cross-linking process which develops simultaneously with the chain scission reaction,[26,27,69,97,98,120] so that the calculated value of Φ_s must be an underestimate.

During the photooxidation of PVC, Mori *et al.*[76] observed that the number of main chain scissions increased linearly with the carbonyl concentration. From this finding it was inferred that the cut ends of the main chain may probably be oxidised. Though this seems likely to be true, it should be pointed out that the linear correlation between the number of chain scissions and the carbonyl content is a condition, indeed necessary, but not sufficient to decide in favour of such a process. It only means that both the backbone cleavages and the C=O groups arise with similar kinetic laws, but through reactions which can, *a priori*, be totally independent. For example, an analogous linear relationship could be obtained by plotting the number of chain scissions against the amount of hydrogen chloride evolved,[23] although there is certainly no direct connection between these two processes. The same argument also applies for the linear correlation observed between the carboxyl index and the colour formation,[118] but neither should this be regarded as convincing evidence that conjugated carbonyl groups are responsible for the discoloration of photooxidised PVC films.[110]

As the duration of the exposure to UV light is increased, a fraction of the PVC film becomes insoluble in hot tetrahydrofuran, thus indicating that intermolecular crosslinking is taking place. An overall theory has been developed by Charlesby[101] and Saito[121] to deal with simultaneous crosslinking and degradation of polymers with random initial molecular weight distribution. The change of the weight-average molecular weight, $\bar{M}_w$, with the absorbed dose, r, follows a simple expression:

$$\frac{1}{\bar{M}_w} = \frac{1}{\bar{M}_{w_0}} + \left[\frac{\Phi_s}{2} - 2\Phi_{CL}\right] r$$

where Φ_{CL} is the quantum yield of crosslinking.

Three limiting cases can be considered (Fig. 3.13):

1. If $\Phi_s > 4\Phi_{CL}$, the polymer will be predominantly degraded and $\bar{M}_w$ decreases accordingly.
2. If $\Phi_s = 4\Phi_{CL}$, chain scission and crosslinking exactly compensate each other and $\bar{M}_w$ does not change with the absorbed dose.
3. If $\Phi_s < 4\Phi_{CL}$, crosslinking is predominant and a three-dimensional network appears at the gel dose corresponding to $1/\bar{M}_w = 0$.

In the photooxidation of PVC, there is a gradual shift from a situation where chain scission is initially controlling the degradation ($\Phi_s > 4\Phi_{CL}$) to a situation where crosslinking takes over, with formation of an insoluble gel ($\Phi_s < 4\Phi_{CL}$).[69] Measurements of the weight fraction (s) of the soluble part of the film in this second period allow quantum yields of chain scission and of crosslinking to be evaluated from a plot of $s + s^{1/2}$ against the reciprocal dose, by using the Charlesby–Pinner equation[102] (Fig. 3.7). Values of Φ_s ($1{\cdot}4 \times 10^{-3}$) and Φ_{CL} (6×10^{-4}) were both found to be about half those obtained in the photolysis of PVC under nitrogen.[42] In highly photooxidised PVC, nearly 50 % of the polymer remains soluble, as shown by the value (1·2) of the y-axis intercept of the straight line $s + s^{1/2} = f(r^{-1})$. This indicates that the main-chain scission process is still efficient in the latter stages of the irradiation since, for each crosslink formed, two PVC chains are broken.

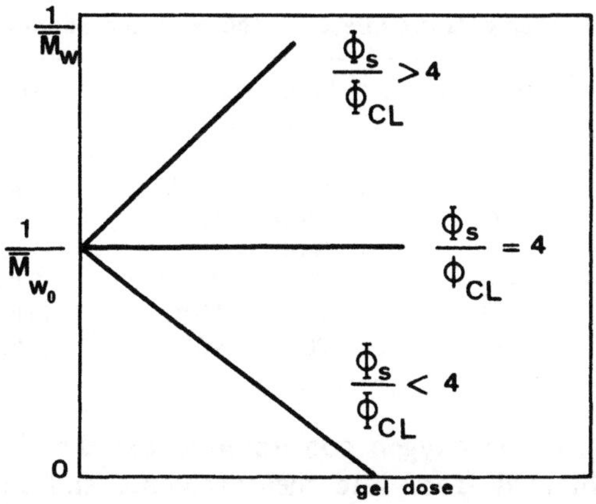

Fig. 3.13 Reciprocal weight-average molecular weight as a function of the absorbed dose for different values of the ratio Φ_S/Φ_{CL}.[101]

3.4.4 Effect of temperature

The temperature of a PVC sheet exposed to sunlight may rise up to 80 °C, especially once the red-brown colour appears. At this temperature, polyene formation develops much more intensively, both in the presence[24,54,56,122] and in the absence[83,84,123] of oxygen. Braun and Bezdadea[56] observed, for instance, a three-fold increase in the rate of dehydrochlorination of PVC films when the irradiation temperature was raised from 20 to 90 °C.

Since the absorption of light is not a temperature-dependent process, variations of the quantum yields with the temperature are often related to a change of the physical state and thus the molecular mobility, which causes a higher reaction probability. This explains the rapid increase of the rates of dehydrochlorination and oxidation of PVC that are observed[122] in the vicinity of the glass transition temperature (~78 °C). Segmental mobility will also influence the separation of the polymer fragments resulting from the photocleavage of excited states. The efficiency of the cage recombination of these fragments is greatly affected by an increase of the temperature which favours radical escape and thus enhances the extent of the degradation of PVC.

As shown in Table 3.1 the values of the activation energy (which should more correctly be called temperature coefficient) for the HCl elimination range from 8 to 18 kJ mol^{-1} and are said to be independent of the oxygen concentration.[54] In the photooxidation of PVC slightly higher activation energies were measured by following either the carbonyl group content

Table 3.1 Temperature effects on the photodegradation of PVC

Process	*Atmosphere*	*Temperature range*	*Activation energy*	*Ref.*
Photodehydro-chlorination	N_2	−40 to +70 °C	18 kJ mol^{-1}	39
	N_2	0 to 60 °C	14 kJ mol^{-1}	84
	N_2	20 to 90 °C	8·3 kJ mol^{-1}	82
	O_2	20 to 90 °C	8 kJ mol^{-1}	54
O_2 uptake	O_2	14 to 50 °C	33 kJ mol^{-1}	24
Carbonyl group	O_2	20 to 90 °C	21 kJ mol^{-1}	54

(21 kJ mol^{-1})[54] or the oxygen consumption (33 kJ mol^{-1}).[24] All these values were obtained from rate measurements and may actually be overestimates since the high rates that are measured at elevated temperatures correspond to more advanced stages of the degradation where the coloured polymer is strongly absorbing the incident light. The true temperature effect on the quantum yields is therefore likely to be less pronounced than is suggested by these figures.

An interesting feature observed by Mitani and Ogata[123] in the photolysis of PVC *in vacuo* is the finding that the enhanced polyene formation above the glass transition temperature results essentially from an increase of the average polyene length, $\bar{n}$ (Fig. 3.14). At room temperature, the polyene distribution remains constant during the early stages of the irradiation, with an average polyene length between four and five, depending on the syndiotacticity of the original sample. As the irradiation temperature is increased above 80 °C, $\bar{n}$ grows rapidly up to values of 20. The long conjugated polyene sequences absorb in the visible range and are thus responsible for the deep dark-red coloration of PVC films photolysed at 90 °C.[82] The considerable increase of $\bar{n}$ above T_g was ascribed to a free rotation of the chain backbone that favours the propagation reaction of the long polyenes.[123] In the same work,[123] the photodehydrochlorination was shown to generate mostly *trans* polyenes which arise predominantly from the racemic dyads of *trans–trans* conformation in the PVC chain.

When PVC films are irradiated at high temperatures in the presence of oxygen, photobleaching of the polyenes formed proceeds efficiently and leads to a less coloured material.[54] As a result, large amounts of carbonyl groups were detected in PVC photooxidised at 80 °C.[122] In addition, the average molecular weight of the polymer was sharply reduced while at the same time the $\bar{M}_w$ distribution became broader.[122] This indicates that an

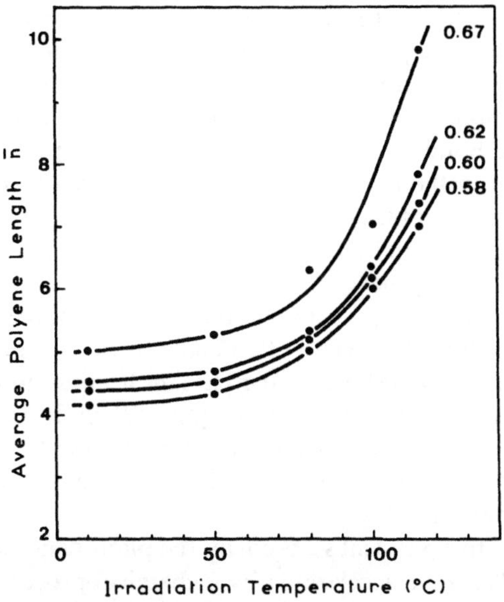

Fig. 3.14 Variation of the average polyene length as a function of the irradiation temperature in the photolysis of PVC *in vacuo*. Numbers on the curves refer to the fraction of syndiotactic dyads.[123]

increase in temperature enhances both the chain scission and the crosslinking processes and will thus have dramatic effects on the mechanical properties and the durability of the aged polymer.

In this connection, it should be realised that the thermal processing of PVC at temperatures of 170–210 °C, *prior* to the UV exposure, also reduces the light stability of PVC by generating oxygenated structures, mostly hydroperoxides, and conjugated polyenes which effectively initiate the photodegradation.[60,124,125] This difficulty can be partly overcome by incorporating into the polymer a thermal stabiliser, such as tin maleate, which by reacting with both HCl and allylic chlorines eliminates the production of polyenes and peroxides during the thermal treatment[126] and thus improves the light-resistance of processed PVC.

3.4.5 Effects of solvents and additives

The susceptibility of PVC to UV light is substantially enhanced by the presence of traces of solvents which are always difficult to remove completely from cast films. In the presence of oxygen, both the dehydrochlorination rate and the oxidation rate are increased by residual solvent,[69,70,127] while in the absence of oxygen the relative concentration of the shorter polyene products increases.[128] This effect was attributed to the ability of the remaining solvent to act as a plasticiser and, by reducing the glass transition temperature of the polymer, to allow an easy rotation around the C–C bonds of the longer polyenes leading to a loss of planarity and a reduction in the length of conjugation.[128] Such an explanation is, however, in direct contrast to the one proposed to account for the temperature effect where free rotation of the chain was assumed to favour the growth of the polyenes.[123]

The increase of the segmental mobility of the polymer chain is also expected to facilitate the escape of radicals from cage reactions (16) and (19). The net result will be both an increase in the initiation efficiency and a reduction in the polyene sequence length. When PVC is irradiated in solution, where polymer chains can move freely, a ten-fold increase of Φ_{HCl} is indeed observed while at the same time the average polyene length is considerably shortened.[25,42]

In addition to this physical state effect, the photodegradation of solvent-containing PVC can be accelerated by a chemical process, since the light-induced oxidation of the solvent or of the added plasticiser may generate initiating chromophoric groups.[73] This is particularly true for THF which, unlike DCE, participates actively in the overall oxidation process and yields, in particular, crosslinked structures by reacting with the PVC polyenes.[129] The main products of THF photooxidation are α-hydroxy tetrahydrofuran, γ-butyrolactone and α-hydroperoxy tetrahydrofuran (THF—OOH).[130] The latter is thermally and photochemically unstable and markedly accelerates the dehydrochlorination of PVC by producing alkoxy radicals

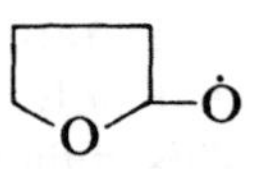

which are very effective for initiating the degradation process.[127] Both chain scission and crosslinking develop more intensively in THF—OOH-containing PVC films which thus suffer an enhanced deterioration upon exposure to UV light.[127]

The presence of ferric chloride[131,132] and of Fe containing organic compounds[74] that often contaminate PVC products also contribute to lower the light stability of these materials. Concentrations of Fe^{2+} and Fe^{3+} as low as 0·001 % were shown to be sufficient to significantly increase the rate of oxidation and of chain scissions and crosslinks formation during outdoor or artificial ageing of PVC.[74] The removal of extraneous impurities, mostly catalyst residues and metal salts, by purification of the polymer substantially reduces the rate of PVC photodegradation.[42,76]

Though water has apparently no direct effect on the photooxidation of PVC,[26] acidic rain, moisture and air pollutants may accelerate the deterioration of PVC sheets during outdoor exposure by forming photochemically active species.[133] Furthermore, the highly degraded surface layer which acts as a UV screen becomes brittle upon prolonged exposure and will be washed away by natural weathering thus causing the bulk of the PVC sheet to be gradually degraded. Surface coating of PVC materials with UV-curable epoxy-acrylates was recently shown to provide an efficient protection against photodegradation.[134]

Since plasticised PVC is usually more susceptible to photooxidation than rigid PVC, an efficient stabilisation is required if such material is to be used for outdoor applications. It should be remembered that some phenolic compounds which are often employed as antioxidants can strongly photosensitise the dehydrochlorination of PVC[78] most probably through an excited charge-transfer complex.[135] Similarly, some heat stabilisers were found to actually accelerate the photooxidation of PVC under certain conditions[77,136] (see Chapter 5).

3.4.6 Changes in physical and mechanical properties

As a result of the various chemical reactions occurring upon exposure to UV light, PVC undergoes important modifications of its physical and mechanical characteristics. Considerable work has been devoted to this topic, mostly in order to reduce the detrimental effects of natural weathering on the durability of PVC materials used outdoors.[137-142]

For some polymers, the ultimate tensile strength was shown to vary linearly with the reciprocal number-average molecular weight of the photooxidised sample,[143] in agreement with the empirical equation derived by Flory.[144] This seems not to be the case in the photodegradation of PVC,[140] probably because chain scissions and crosslinks are formed simultaneously. Both the ultimate tensile strength and the elongation decrease with exposure time leading finally to the embrittlement of the aged material. In the outdoor exposure tests, PVC samples showed first a ductile behaviour at fracture for irradiation times up to two years and then a brittle fracture for exposures over three years.[141] Typically the elongation at break was reduced to one-quarter of its original value after about one year of natural weathering.

From the shape of the curve obtained by plotting the ultimate tensile strength against elongation, it was inferred that main-chain scission is the predominant process responsible for PVC photodegradation[140] and that the scissions are localised in microdomains that constitute the starting point of the rupture. By exposing PVC sheets outdoors at a temperature of 45 °C, Scott and Tahan[145] observed a linear decrease of the impact strength with irradiation time that closely parallels the reduction in molecular weight. The shift of T_g to lower temperatures and the reduced intensity of the β-relaxation were attributed to a plasticisation of the polymer by the low-molecular weight photoproducts resulting from an efficient bond cleavage process. Changes in some of the mechanical properties of aged PVC were also shown to depend on the gel content of the polymer[146] and are thus strongly influenced by the extent of the crosslinking reaction.

The optical and surface properties of PVC sheets exposed to UV radiation are severely modified since both the dehydrochlorination and the oxidation process develop primarily in the top layer of the irradiated material. Loss of brilliance and transparency as well as colour changes deeply affect the appearance of the degraded polymer.[112,147] A correlation was found between mechanical properties and surface characteristics by performing microstructure analyses of the superficial layer of PVC samples subjected to natural or accelerated ageing.[148] Another sensitive technique which can be used to follow the superficial photodegradation of PVC films involves determining the interfacial free energy by contact angle measurements. The increase in wettability observed with H-bonding liquids appears to be strongly related to the formation of polar groups on the film surface.[149] A new analytical method was developed recently by Jacob[150] in order to evaluate the thickness of the superficially degraded layer in aged PVC materials. It is based on a microscopic examination of the crazes, cracks and dislocations which appear after stretching the sample up to rupture. The thickness of the degraded layer was found to be typically in the range 10–20 μm for outdoor exposure durations exceeding one year. The chemical modifications occurring in this top layer during natural or artificial weathering of plasticised PVC were followed quantitatively by MIR infrared spectroscopy.[151]

Recently, Roux and Eurin[152] proposed an indentation test which allowed them to predict fairly accurately the time of embrittlement of rigid PVC extruded profiles exposed to weathering. The durability of the thin PVC sheet was evaluated by monitoring the variations of the brittle temperature, T_B, which was determined by using an indentation test with constant ram speed. The changes in T_B upon artificial and natural weathering followed similar laws and, by extrapolation of T_B values from short weathering times, the mechanical durability of PVC could be defined to close approximation.

3.4.7 Mechanism of the photooxidation

Although many mechanistic studies have been reported on the photooxidation of PVC,[23,24,26,56,87,109,115,117] the precise reactions by which PVC deteriorates upon exposure to UV light in the presence of O_2 are far from being completely elucidated. All the reaction schemes suggested have as a common factor the need to involve the alkyl peroxy radicals ($P\dot{O}_2$) which are considered to be responsible for the formation of the various functional groups detected in the oxidised polymer. The purpose of this section is not to review the numerous mechanisms postulated so far, but rather to present a feasible reaction scheme which may satisfactorily account for the quantitative information recently obtained regarding the major primary products (HCl, polyenes, hydroperoxide and carbonyl groups) and the structural changes taking place (main-chain scissions and crosslinks) when PVC is exposed to UV radiations in air.[153]

In the presence of a sufficient supply of oxygen and with intense radiation of $\lambda > 254$ nm, quantum yield measurements indicate that, for each scission of the polymer backbone, 11 molecules of HCl are evolved while, at the same time, three carbonyl groups, two hydroperoxide groups and 0·4 intermolecular crosslinks appear on the PVC chain.[23] All these products can be considered as resulting from primary reactions, owing to the observed constancy of the quantum yields with the duration of exposure, at least in the early stages of the photooxidation.

(1) Initiating radicals

The primary photophysical processes occurring by interaction of UV light with PVC were described previously (Section 3.3.2(5)). Due to their short lifetime, excited singlet polyenes are not likely to be quenched by O_2, so that the main effect of atmospheric oxygen will be its rapid reaction with alkyl radicals to create peroxy radicals ($P\dot{O}_2$) on the polymer chain. The characteristic signal of $P\dot{O}_2$ radicals was observed by esr spectroscopy in PVC photolysed at low temperature in the presence of O_2.[15,80] Upon heating the irradiated sample, the concentration of $P\dot{O}_2$ radicals was found to first increase in the temperature range -150 to $-90\,°C$, as the reaction of polymeric radicals with O_2 becomes easier, and then to decrease due to the subsequent reactions of the $P\dot{O}_2$ radicals.

From the lack of any inhibition effect of O_2 on the quantum efficiency of the PVC dehydrochlorination it can be inferred that the chain carrier radical (III) (~~(CH=CH)$_n$—ĊH—CHCl~~) is not scavengeable by O_2. Therefore, the alkyl radicals most likely to react with O_2 are those which bear no easily removable chlorine atom in the β-position, i.e. radical (I) (~~(CH=CH)$_n$—ĊH—CH_2~~) and radical (II) (~~CH_2—ĊCl—CH_2~~) which generate secondary and tertiary peroxy radicals on the PVC chain:

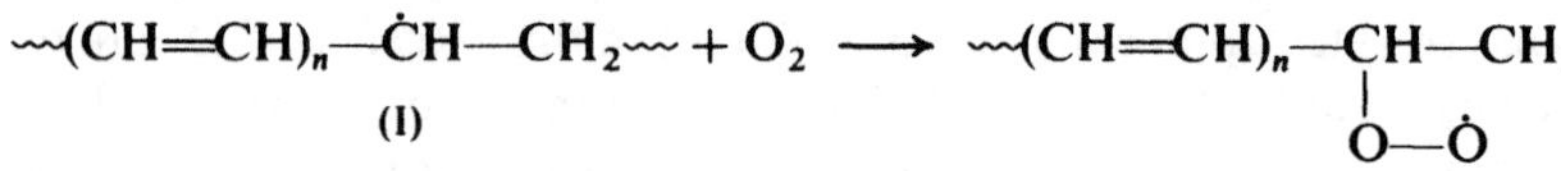
(30)

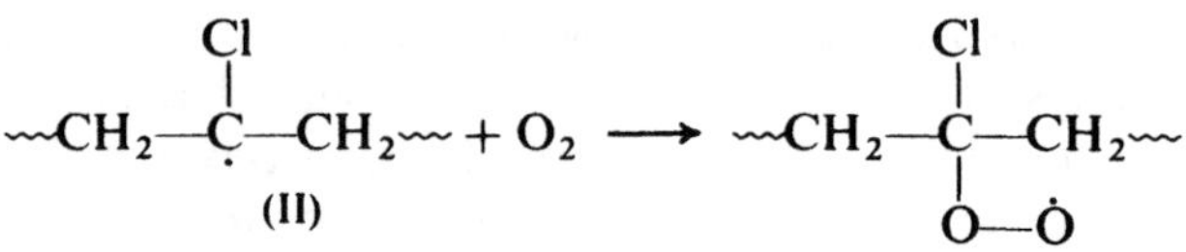
(31)

These radicals are considered to be the main initiating species of the PVC oxidation and the direct precursors of the oxygenated structures detected in the aged polymer.

(2) *Reactions of peroxy radicals*
P$\dot{O}_2$ radicals are known to abstract hydrogen atoms easily from the CHCl or CH_2 groups of the polymer to give hydroperoxide groups and regenerate a propagating radical (II) or (III) (with $n = 0$), respectively:

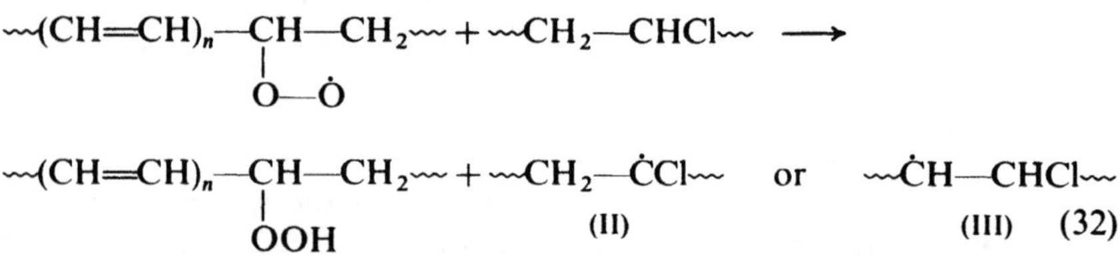
(32)

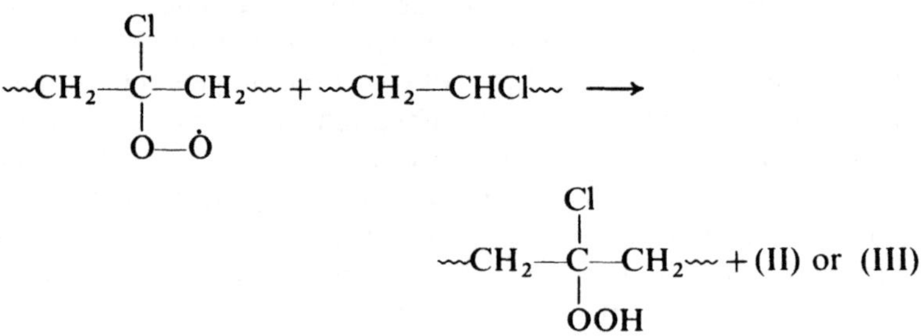
(33)

Since it is a chain reaction, hydroperoxidation will develop most extensively when low intensity radiation is employed, as for instance in natural weathering.

When two alkyl peroxy radicals are formed in close vicinity to each other, they are likely to interact to yield, by a termination process, either an alcohol and a ketone group in the case of primary and secondary $P\dot{O}_2$ radicals:[154]

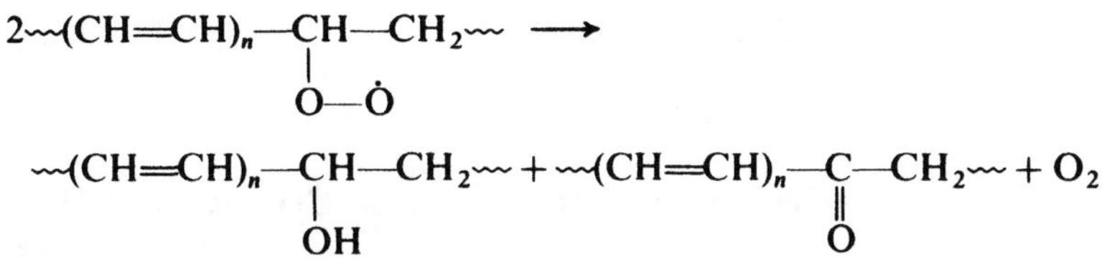

(34)

or a dialkyl peroxide (usually via an unstable tetroxide) in the case of tertiary $P\dot{O}_2$ radicals:

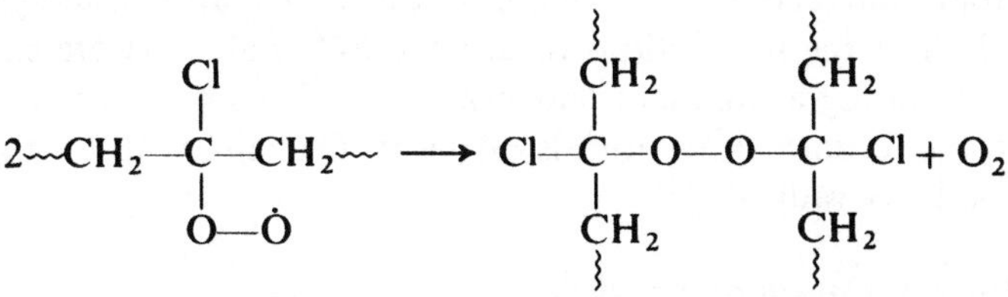

(35)

The formation of such intermolecular peroxide bridges may account for the crosslinking detected by GPC in the early stages of the photooxidation of PVC. In the latter stages, once polyenes have accumulated in large amounts, the addition of $P\dot{O}_2$ radicals onto double bonds is also likely to contribute to the partial insolubilisation of the photooxidised polymer:

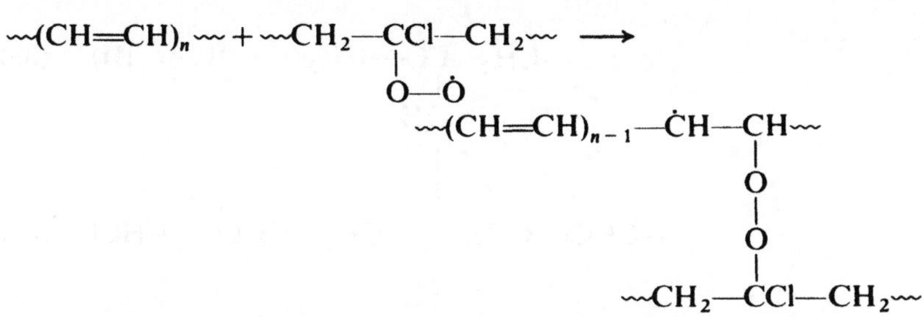

(36)

The resulting crosslinked radical will be further oxidised and react essentially like radical (I).

Under some conditions, the interaction of tertiary $P\dot{O}_2$ radicals can be a non-terminating process which generates free alkoxy radicals:[155]

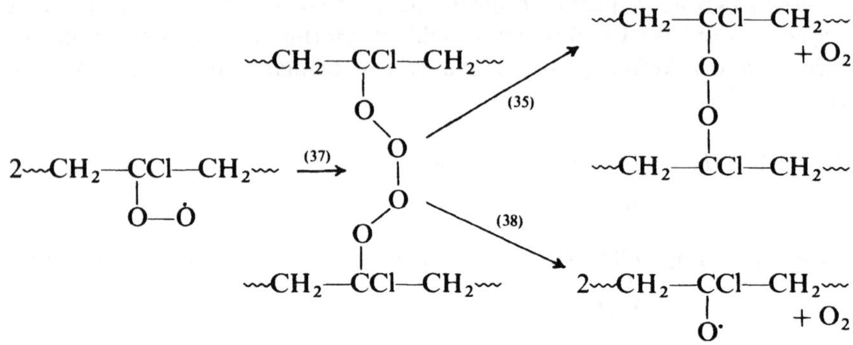

This means that a significant fraction of the alkoxy radicals resulting from the decomposition of the very unstable tetroxide is able to escape the cage recombination reaction (35). In the γ-induced oxidation of polypropylene, about half of the interactions of tertiary $P\dot{O}_2$ radicals were shown to be non-terminating at ambient temperature.[156] This was assumed to apply also to the tertiary $P\dot{O}_2$ radicals of the PVC chain, with formation of β-chloro-alkoxy radicals.[119]

(3) Reactions of alkoxy radicals

The most probable reaction of these alkoxy radicals is the hydrogen abstraction from the PVC molecule to yield a chain carrier radical (II or III) and an unstable α-chloro-alcohol which stabilises readily into a ketone by elimination of an HCl molecule:

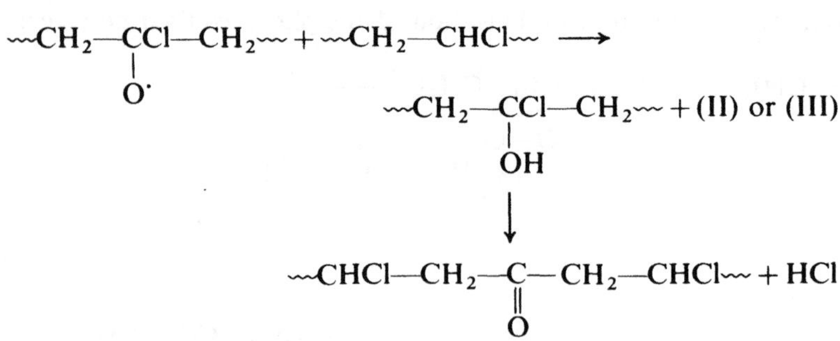

(39)

(40)

In addition, tertiary alkoxy radicals are known to undergo an easy β-scission process involving one of the two equivalent C–C bonds:[157]

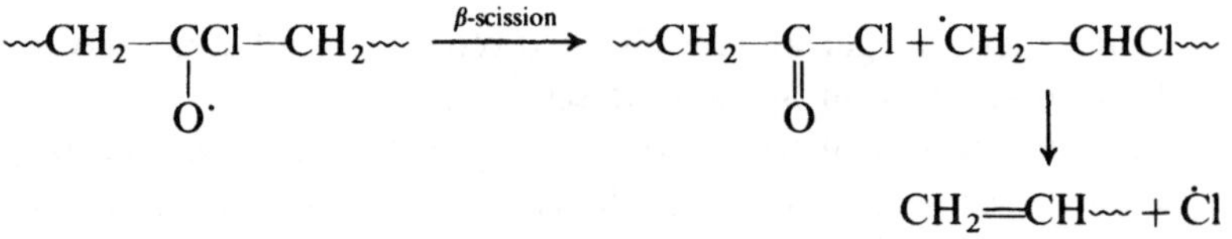

(41)

(42)

Reaction (41) may explain both the backbone scissions observed in photooxidised PVC and the presence of acid chlorides and/or carboxylic acids detected by infrared spectroscopy.[115] For alkoxy radicals which bear a halogen atom in the β-position to the unpaired electron, stabilisation can also occur by splitting off this atom,[93] much like reaction (18) of radical (III):

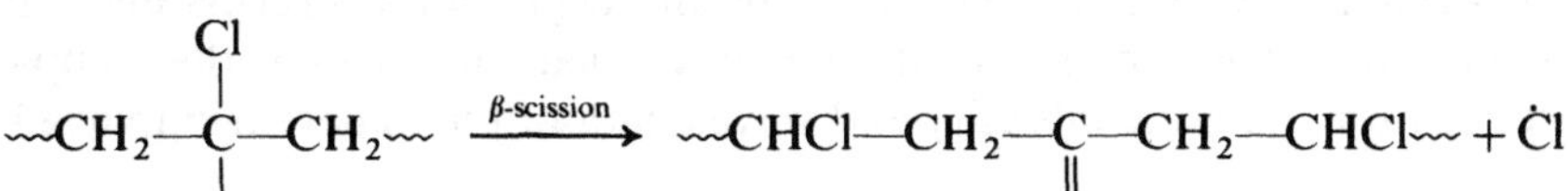

(43)

It is noticeable that reactions (39) + (40) and (43) lead to the formation of β,β'-dichloroketones, thus accounting for the relatively large concentration of these carbonyl groups in aged PVC. The additional production of reactive chlorine radicals through reactions (41) + (42) and (43) was considered to be mainly responsible for the higher Φ_{HCl} observed when PVC was photolysed in the presence of O_2 instead of N_2.[23]

This reaction scheme, which bears a formal resemblance to the mechanism of the γ-oxidation of PVC,[119] is not supposed to settle definitely the controversy about the mechanism of photooxidation of PVC. It only constitutes a feasible route to explain the formation of the main primary oxidation products detected, i.e. products resulting from reactions which involve solely reactive intermediates and not the photoproducts formed, like carbonyl or hydroperoxide groups.

(*4*) *Secondary reactions*

In addition to these primary processes, subsequent reactions are also expected to take place in the later stages of the PVC degradation, once photoproducts have accumulated to a sufficient concentration to enable them to compete for radicals produced in the primary reactions. Two important types of secondary reaction are considered to play a predominant role in the photooxidation of PVC:

1. The photo- and dark-bleaching of the conjugated polyenes by both

atmospheric oxygen and the HCl evolved which lead ultimately to a lower discoloration of the oxidised polymer.

2. The polyene-sensitised photodecomposition of the carbonyl groups according to Norrish type I and II processes, and photodecomposition of the hydroperoxide groups, which yield alkoxy radicals. This will result in the formation of final oxidation products, like carboxylic acids, and gaseous products (CO_2, CO, H_2O), as well as some additional main-chain scissions and crosslinks.

The relative importance of these primary and secondary reactions will depend on such factors as the microstructure of the PVC sample, its permeability to oxygen, the extent of the degradation process and the conditions at which the irradiation is carried out, mainly the sample temperature and the intensity and wavelength of the radiation employed.

(5) *Reaction scheme*

The overall mechanism of the photooxidation of PVC can finally be represented by the following reaction scheme (Fig. 3.15):

The upper cycle corresponds to the zip-dehydrochlorination which is apparently not very much affected by the presence of O_2 and involves as chain carrier species the chlorine radical and the short lived β-chloroalkyl (or polyenyl) radical (III), designated here by $\dot{P}_\beta$. The net result of this chain reaction is the production of HCl and conjugated polyene sequences which are responsible for the discoloration of the irradiated PVC.

The oxidation process is represented by the lower cycle which includes several chain reactions. The major species are the O_2-scavengeable radicals (I)

$$(\sim(\mathrm{CH{=}CH})_n\!-\!\dot{\mathrm{C}}\mathrm{H}\!-\!\mathrm{CH_2}\sim)$$

and (II)

$$(\sim\mathrm{CH_2}\!-\!\dot{\mathrm{C}}\mathrm{Cl}\!-\!\mathrm{CH_2}\sim),$$

designated here by $\dot{P}$; through the intermediate peroxy ($P\dot{O}_2$) and alkoxy ($P\dot{O}$) radicals, they generate oxygenated structures on the polymer backbone (mostly carbonyl and hydroperoxide groups), and lead also to the fragmentation of the PVC chain and to the formation of peroxide crosslinks. The subsequent photoreactions of these oxidation products are represented by dashed arrows.

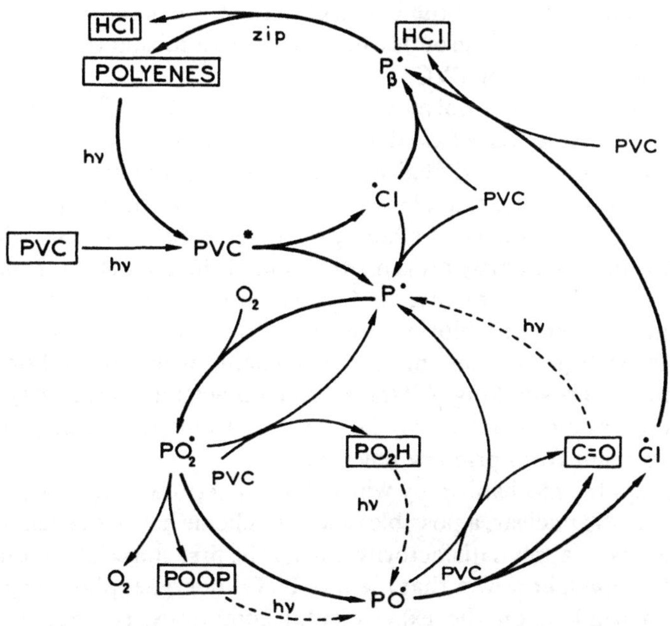

Fig. 3.15 Schematic representation of the various processes occurring during the UV photolysis of PVC in the presence of O_2. Final products are in boxes.

$\dot{P}$: ~(CH=CH)$_n$—ĊH—CH$_2$~ and ~CH$_2$—ĊCl—CH$_2$~

$\dot{P}_\beta$: ~(CH=CH)$_n$—ĊH—CHCl~

3.5 CONCLUSIONS

In spite of the useful information provided by recent studies on the photodegradation of PVC, the problem of the initiation step has not yet been definitively settled. The initial active sensitisers are most probably alkene linkages, hydroperoxides and carbonyl groups which are always present as impurities in commercial and processed PVC. As soon as the strongly light-absorbing polyenes appear, these chromophores assume the predominant role of sensitisation. The excited polyenes formed can apparently stabilise by easily splitting off an allylic chlorine atom, and thus initiate both the zip-dehydrochlorination through the Ċl radical and the oxidation process through the polymer backbone radical. The main cause of the great sensitivity of PVC to UV light seems therefore to originate from the presence of these allylic chlorines, both in the starting polymeric material and as a result of the UV degradation. Replacement of the allylic chlorines by more stable groups usually leads to a substantial improvement of the photostability of PVC.

The long conjugated polyene sequences, ~(CH=CH)$_n$~, which arise from the intramolecular chain dehydrochlorination are responsible for the discoloration of the irradiated polymer. As degradation proceeds, these chromophores concentrate closer to the surface of the exposed material, leading to the appearance of a highly degraded superficial layer which acts as a UV screen and partly prevents the bulk of the PVC sheet from further deterioration. In the presence of O_2, oxidation develops simultaneously with the zip-dehydrochlorination with formation of carbonyl and hydroperoxide groups on the polymer chains which in addition suffer cleavages and crosslinking. All these products accumulate linearly with the number of photons absorbed, i.e. with constant quantum yields, and therefore result from primary reactions.

Although the mechanism by which PVC deteriorates upon exposure to UV light is still unclear, a possible reaction scheme has been suggested here which seems to apply satisfactorily during the first stages of irradiation. It must be realised, however, that the way PVC undergoes photodegradation is very dependent on the experimental conditions, so that no unique mechanism is expected to operate in all cases. Clearly, additional work is required in the future in order to elucidate the many different processes occurring during natural weathering of PVC and to increase the durability and thus extend the field of application of this already widely used polymer.

REFERENCES

1. Owen, E. D. *A.C.S. Symp. Ser.*, **25** (1976), 208.
2. Close, L. G., Gilbert, R. D. and Fornes, R. E. *Polym. Plast. Technol. Eng.*, **8** (1977), 177.
3. Wirth, H. O. and Andreas, H. *Pure Appl. Chem.*, **49** (1977), 627.
4. Starnes, W. H., Jr. *Adv. Chem. Ser.*, **169** (1978), 309.
5. McKellar, J. F. and Allen, N. S. *Photochemistry of Man-Made Polymers*, Applied Science Publishers Ltd, London, 1979, 95.
6. Braun, D. *Pure Appl. Chem.*, **53** (1981), 549.
7. Starnes, W. H., Jr. *A.C.S. Symp. Ser.*, **151** (1981), 197.
8. Owen, E. D. *A.C.S. Symp. Ser.*, **151** (1981), 217.
9. Owen, E. D. In: *Developments in Polymer Photochemistry*, N. S. Allen (Ed.), Applied Science Publishers Ltd, London, 1982, 165.
10. Turro, N. *Modern Molecular Photochemistry*, Benjamin Cummings, Menlo Park, California, 1978.
11. Kasha, M. *Disc. Farad. Soc.*, **9** (1950), 14.

12. Calvert, J. G. and Pitts, J. N. *Photochemistry*, John Wiley, New York, 1966.
13. Murov, S. L. *Handbook of Photochemistry*, Marcel Dekker, New York, 1973.
14. Searle, N. Z. and Hirt, R. C. *J. Opt. Soc. Amer.*, **55** (1965), 1413.
15. Starnes, W. H., Jr., Schilling, F. C., Abbas, K. B., Cais, R. E. and Bovey, F. A. *Macromol.*, **12** (1979), 556.
16. Caraculacu, A., Buruiana, E. C. and Robila, G. *J. Polym. Sci., Polym. Chem. Ed.*, **16** (1978), 2741.
17. Boissel, J. *J. Appl. Polym. Sci.*, **21** (1977), 855.
18. Braun, D., Bezdadea, E., Holzer, G. and Sonderhof, D. *Angew. Makromol. Chem.*, **104** (1982), 145.
19. Sondheimer, F., Ben Efraim, D. A. and Wolovsky, R. *J. Amer. Chem. Soc.*, **83** (1961), 1675.
20. Daniels, V. D. and Rees, H. H. *J. Polym. Sci., Polym. Chem. Ed.*, **12** (1974), 2115.
21. Gibb, W. H. and MacCallum, J. R. *Europ. Polym. J.*, **10** (1974), 529.
22. Phillips, R. W. and Volman, D. H. *J. Amer. Chem. Soc.*, **91** (1969), 3418.
23. Decker, C. and Balandier, M. *Polym. Photochem.*, **1** (1981), 221.
24. Kwei, K. P. S. *J. Polym. Sci., Polym. Chem. Ed.*, **7** (1969), 1075.
25. Balandier, M. and Decker, C. *Eur. Polym. J.*, **14** (1978), 995.
26. Rabek, J. F., Canbäck, G., Lucky, J. and Rånby, B. *J. Polym. Sci., Polym. Chem. Ed.*, **14** (1976), 1447.
27. Rånby, B., Rabek, J. F. and Canbäck, G. J. *J. Macromol. Sci. Chem.*, **12** (1978), 587.
28. Yang, N. L., Liutkus, J. and Haubenstock, H. *A.C.S. Symp. Ser.*, **142** (1980), 35.
29. Torikai, A., Tsuruta, H. and Fueki, K. *Polym. Photochem.*, **2** (1982), 227.
30. Mayer, Z. *J. Macromol. Sci., Rev. Macromol. Chem.*, **10** (1974), 263.
31. Starnes, W. H., Jr. and Edelson, D. *Macromol.*, **12** (1979), 797.
32. Riechi, A., Grimm, A. and Mucke, H. *Kunststoffe*, **52** (1962), 265.
33. Geddes, W. C. *Eur. Polym. J.*, **3** (1967), 267.
34. Hartman, A. *Kolloid Z.*, **149** (1956), 67.
35. Baum, B. *Soc. Plast. Eng. J.*, **17** (1961), 71.
36. Winkler, D. E. *J. Polym. Sci.*, **35** (1959), 3.
37. Owen, E. D. and Williams, J. I. *J. Polym. Sci., Polym. Chem. Ed.*, **12** (1974), 1933.
38. Reinish, R. F., Gloria, H. R. and Wilson, D. E. *A.C.S. Div. Polym. Chem. Preprints*, **7** (1966), 372.
39. Reinish, R. F., Gloria, H. R. and Androes, G. M. In: *Photochemistry of Macromolecules*, R. F. Reinish (Ed.), Plenum, New York, 1970, 185.
40. Verdu, J., Michel, A. and Sonderhof, D. *Eur. Polym. J.*, **16** (1980), 689.
41. Decker, C. and Balandier, M. *J. Photochem.*, **15** (1981), 213.
42. Decker, C. and Balandier, M. *Eur. Polym. J.*, **18** (1982), 1085.
43. Bauer, J. and Sabel, A. *Angew. Makromol. Chem.*, **47** (1975), 15.
44. Braun, D. and Wolf, M. *Angew. Makromol. Chem.*, **70** (1978), 71.
45. Ratti, L., Visani, F. and Ragazzani, M. *Eur. Polym. J.*, **9** (1973), 429.
46. Kawai, W. and Ichihashi, T. *J. Polym. Sci., Polym. Chem. Ed.*, **12** (1974), 201, 1041.
47. Osawa, Z. and Aiba, M. *Polym. Photochem.*, **2** (1982), 397.

48. Kenyon, A. S. *Nat. Bur. Stand.* (*US*) *Circ.*, **525** (1953), 81.
49. Kagiya, U. T., Takemoto, K. and Hagiwara, M. *J. Appl. Polym. Sci. Appl. Polym. Symp.*, **35** (1979), 95.
50. Owen, E. D. and Bailey, R. J. *J. Polym. Sci., Polym. Chem. Ed.*, **10** (1972), 113.
51. Owen, E. D. and Williams, J. I. *J. Polym. Sci., Polym. Chem. Ed.*, **11** (1973), 905.
52. Golub, M. A. *J. Phys. Chem.*, **75** (1971), 1168.
53. Harriman, A. and Rockett, B. W. *J. Chem. Soc. Perkin 2*, **5** (1974), 485.
54. Braun, D. and Kull, S. *Angew. Makromol. Chem.*, **86** (1980), 171.
55. Gibb, W. H. and MacCullum, J. R. *Eur. Polym. J.*, **10** (1974), 533.
56. Braun, D. and Bezdadea, E. *Angew. Makromol. Chem.*, **99** (1981), 55.
57. Carlsson, D. J. and Wiles, D. M. *J. Radiation Curing*, **2** (1975), 2.
58. Carlsson, D. J. and Wiles, D. M. *J. Macromol. Sci., Rev. Macromol. Chem.*, **C14** (1976), 65.
59. Marechal, J. C. *J. Macromol. Sci. Chem.*, **12** (1978), 609.
60. Scott, G., Taman, M. and Vyvoda, J. *Eur. Polym. J.*, **14** (1978), 1021.
61. Cooray, B. B. and Scott, G. *Polym. Deg. Stab.*, **3** (1981), 127.
62. Scott, G. *Adv. Chem. Ser.*, **169** (1978), 30.
63. Ng, H. G. and Guillet, J. E. *Macromol.*, **11** (1978), 937.
64. Geuskens, G. and David, C. *Pure Appl. Chem.*, **51** (1979), 233.
65. Starnes, W. H., Jr., Schilling, F. C., Abbås, K. B., Plitz, I. M., Hartless, R. L. and Bovey, F. A. *Macromol.*, **12** (1979), 13.
66. Starnes, W. H., Jr., Schilling, F. C., Abbås, K. B., Cais, R. E. and Bovey, F. A. *Macromol.*, **12** (1979), 556.
67. Starnes, W. H., Jr., Schilling, F. C., Plitz, I. M., Cais, R. E., Freed, D. J. and Bovey, F. A. *Preprints Third. Int. Symp. on Poly*(*vinyl chloride*), Cleveland, 1980, 58.
68. Gooden, R. and Hellman, Y. *Proc. 28th Macro. IUPAC Symposium*, Amherst, 1982, 303.
69. Kamal, M. R., El-Kaissy, M. M. and Avedesian, M. M. *J. Appl. Polym. Sci.*, **16** (1972), 83.
70. Rabek, J. F., Shur, Y. J. and Rånby, B. *J. Polym. Sci., Polym. Chem. Ed.*, **13** (1975), 1285.
71. Kaminska, A. and Kaczmarek, H. *Polymery* (*Warsaw*), **27** (1982), 209.
72. Rabek, J. F., Skowronski, T. A. and Rånby, B. *Polymer*, **21** (1980), 226.
73. Williams, G. E. and Gerrard, D. L. *J. Polym. Sci., Polym. Chem. Ed.*, **21** (1983), 1491.
74. Joffe, Z. and Rånby, B. *J. Appl. Sci., Appl. Polym. Symp.*, **35** (1979), 307.
75. Owen, E. D. *Organ. Coat. Appl. Polym. Polym. Sci. Proc.*, **46** (1982), 641.
76. Mori, F., Koyama, M. and Oki, Y. *Angew. Makromol. Chem.*, **64** (1977), 89.
77. Rabek, J. F., Canbäck, G. and Rånby, B. *J. Appl. Polym. Sci.*, **21** (1977), 2211.
78. Foster, R. J., Whitling, P. H., Mellor, J. M. and Phillips, D. *J. Appl. Polym. Sci.*, **22** (1978), 1129.
79. Allen, N. S., McKellar, J. F., Phillips, G. O. and Wood, D. G. M. *J. Polym. Sci., Polym. Chem. Ed.*, **12** (1974), 241.
80. Davidson, R. S. and Meek, R. R. *Polym. Photochem.*, **2** (1982), 1.
81. Ferguson, W. J. *Kunststoffe*, **73** (1983), 36.

82. Braun, D. and Kull, S. *Angew. Makromol. Chem.*, **85** (1980), 79.
83. Gibb, W. H. and MacCallum, J. R. *Eur. Polym. J.*, **7** (1971), 1231.
84. Gibb, W. H. and MacCallum, J. R. *Eur. Polym. J.*, **8** (1972), 1223.
85. Braun, D., Bezdadea, E. and Holzer, G. *Angew. Makro. Chem.*, **106** (1982), 47.
86. Corsato-Arnaldi, A., Palma, G. and Talamini, G. *Mat. Plast.*, **32** (1966), 50.
87. Verdu, J. *J. Macromol. Sci. Chem.*, **12** (1978), 551.
88. Braun, D. and Wolf, M. *Kunst. Forschrittsber.*, **2** (1976), 13.
89. Starnes, W. H., Jr. and Edelson, D. *Macromol.*, **12** (1979), 797.
90. Miller, A. A. *J. Phys. Chem.*, **63** (1959), 1755.
91. Decker, C., Balandier, M. and Faure, J. *Preprints IUPAC 27th Int. Symp. Macromol., Strasbourg*, Vol. I (1981), 445.
92. Kharash, M. S. *J. Amer. Chem. Soc.*, **73** (1951), 632.
93. Friedlina, R. Kh. Chapter 6 in: *Advances in Free-Radical Chemistry*, Vol. 1, G. H. Williams (Ed.), Logos, London, 1965.
94. Owen, E. D. and Read, R. L. *J. Polym. Sci., Polym. Chem. Ed.*, **17** (1979), 2719.
95. Brauman, S. and Chen, I. J. *J. Polym. Sci., Polym. Chem. Ed.*, **19** (1981), 495.
96. Owen, E. D. and Read, R. L. *Eur. Polym. J.*, **15** (1979), 41.
97. Kwei, K. P. S. *J. Appl. Polym. Sci.*, **12** (1968), 1543.
98. Sobue, H., Tabata, Y. and Tajima, Y. *J. Polym. Sci.*, **27** (1958), 596.
99. Sobue, H. and Tajima, Y. *J. Chem. Ind. Tokyo*, **62** (1959), 1149.
100. Ulinska, A. and Kaminska, A. *Polimery*, **11** (1966), 522.
101. Charlesby, A. *Atomic Radiation and Polymers*, Pergamon, Oxford, 1960.
102. Charlesby, A. and Pinner, S. H. *Proc. Roy. Soc. London*, **A249** (1959), 367.
103. Rånby, B., Rabek, J. F. and Canbäck, G. *J. Macromol. Sci. Chem.*, **12** (1978), 587.
104. Owen, E. D., Pasha, I. and Moayyedi, F. *J. Appl. Polym. Sci.*, **25** (1980), 2331.
105. Bailey, R. J. Ph.D. Thesis, University of Wales, 1972.
106. Decker, C. and Balandier, M. *Makromol. Chem.*, **183** (1982), 1263.
107. David, C. and Baeyens-Volant, D. *Eur. Polym. J.*, **14** (1978), 29.
108. Decker, C. and Balandier, M. *J. Photochem.*, **15** (1981), 221.
109. Rabek, J. F., Rånby, B., Östensson, B. and Fodin, P. *J. Appl. Polym. Sci.*, **24** (1979), 2407.
110. Martin, K. G. and Tilley, R. I. *Br. Polym. J.*, **3** (1971), 36.
111. Kohn, P., Marechal, C. and Verdu, J. *J. Anal. Chem.*, **51** (1979), 1000.
112. Marks, G. C. and Butters, G. *J. Macromol. Sci. Chem.*, **A12** (1978), 569.
113. Nagy, T. G., Kelen, T., Turcsanyi, B. and Tudos, F. *J. Polym. Sci., Polym. Chem. Ed.*, **15** (1977), 853.
114. Pochan, J. M., Gibson, H. W. and Bailey, F. C. *J. Polym. Sci., Polym. Lett. Ed.*, **18** (1980), 447.
115. Scott, G. and Tahan, M. *Eur. Polym. J.*, **11** (1975), 535.
116. Martin, K. G. and Tilley, R. I. *Br. Polym. J.*, **1** (1969), 213.
117. Rabek, J. F., Canbäck, G. and Rånby, B. *J. Appl. Polym. Sci., Appl. Polym. Symp.*, **35** (1979), 299.
118. Scott, G. and Tahan, M. *Eur. Polym. J.*, **13** (1977), 989.
119. Decker, C. *J. Appl. Polym. Sci.*, **20** (1976), 3321.
120. Kryszewski, M. and Mucha, M. *J. Polym. Sci., Polym. Phys. Ed.*, **5** (1967), 1095.

121. Saito, O. *J. Phys. Soc. Jpn.*, **13** (1968), 198, 1451, 1465.
122. Mori, F., Koyama, M. and Oki, Y. *Angew. Makromol. Chem.*, **75** (1979), 113.
123. Mitani, K. and Ogata, T. *J. Appl. Polym. Sci.*, **18** (1974), 3205.
124. Scott, G. *Adv. Chem. Ser.*, **169** (1978), 30.
125. Cooray, B. B. and Scott, G. *Polym. Deg. Stab.*, **3** (1981), 145.
126. Scott, G., Tahan, M. and Vyvoda, J. *Eur. Polym. J.*, **15** (1979), 51.
127. Rabek, J. F., Skowronski, T. A. and Rañby, B. *Polymer*, **21** (1980), 226.
128. Gibb, W. H. and MacCallum, J. R. *Eur. Polym. J.*, **9** (1973), 771.
129. Osawa, Z. and Aiba, M. *Polym. Photochem.*, **2** (1982), 339.
130. Stenberg, V. I., Wang, C. T. and Kulevsky, N. *J. Org. Chem.*, **35** (1970), 1774.
131. Rånby, B. and Rabek, J. F. *J. Appl. Polym. Sci., Appl. Polym. Symp.*, **35** (1979), 243.
132. Oki, Y., Mori, F. and Tanaka, A. *Kinzohn Hyomen Gijutsu*, **24** (1973), 329.
133. Verdu, J. *Trib. CEBEDEAU*, **27** (1974), 360.
134. Decker, C. J. *Appl. Polym. Sci.*, **28** (1983), 97.
135. Hirayama, S., Foster, R. J., Mellor, J. M., Whitling, P. H, Grant, K. R. and Phillips, D. *Eur. Polym. J.*, **14** (1978), 679.
136. Bellenger, V., Verdu, J. and Carette, L. B. *Preprints, Third Int. Symp. on Poly(vinyl chloride)*, Cleveland, 1980, 297.
137. Manfredi, M. and Pacelli, E. *Mat. Plast. Elast.*, **36** (1970), 647.
138. Aoki, H., Yoshida, A. and Suzuki, T. *Zairyo*, **21** (1972), 309.
139. Roux, G. and Huet, J. M. *Materiaux et Techniques* (April 1979), 139.
140. Pabiot, J. and Verdu, J. *Polym. Eng. Sci.*, **21** (1981), 32.
141. Kubota, H., Suzuki, S., Nishimura, O., Hattori, S., Yoshikawa, K. and Shirota, T. *Kenkyu Hokoku—Kogyo Gijustsuin*, **2** (1981), 23.
142. Levai, G., Ocskay, G. and Meszlenyi, A. *Kunst. Fortschr. Ber.*, **7** (1980), 23.
143. Geuskens, G. and David, C. *Pure Appl. Chem.*, **51** (1979), 2385.
144. Flory, P. J. *J. Amer. Chem. Soc.*, **67** (1945), 2049.
145. Scott, G. and Tahan, M. *Eur. Polym. J.*, **13** (1977), 997.
146. Summers, J. W., Ismer, J. D. and Rabinovitch, E. R. *Polym. Eng. Sci.*, **20** (1980), 155.
147. Roux, G., Revirand, G., Rochas, G. and Giraud, D. *Cahier Centre Sci. Techn. Bâtiment*, **214** (1980), cah. 1677.
148. Nishimura, O., Kubota, H. and Suzuki, S. *Hokkaido Kogyo Kaihatsu Shikensho Hokoku*, **24** (1981), 115, 145.
149. Price, T. R. and Fox, R. B. *A.C.S. Organ. Coat. Plast. Chem.*, **26** (1966), 242.
150. Jacob, R. *G.F.P., Meeting on Aging and Stabilization of PVC*, Lyons, 1981.
151. Hill, C. A. S. *J. Appl. Polym. Sci.*, **27** (1982), 3313.
152. Roux, G. and Eurin, P. A. *J. Macromol. Sci. Phys.*, **B20** (1981), 505.
153. Decker, C. *Eur. Polym. J.*, **20** (1984), 149.
154. Russell, G. A. *J. Amer. Chem. Soc.*, **79** (1957), 3871.
155. Allara, D. L., Mill, T., Hendry, D. G. and Mayo, F. R. *Adv. Chem. Ser.*, **76** (1968), 40.
156. Decker, C. and Mayo, F. R. *J. Polym. Sci., Polym. Chem. Ed.*, **11** (1973), 2847.
157. Bell, E. R., Raley, J. H., Rust, F. F., Seybold, F. H. and Vaughan, N. E. *Disc. Faraday Soc.*, **10** (1951), 242.

Chapter 4

SPECTROSCOPIC AND CHEMICAL INVESTIGATIONS OF PVC STRUCTURE

W. F. MADDAMS

The British Petroleum Company plc, Sunbury-on-Thames, Middlesex, UK

4.1 INTRODUCTION

The substantial progress achieved during the past two decades in elucidating the molecular structure of PVC owes much to four spectroscopic techniques. At the outset, infrared spectroscopy proved valuable for the characterisation of the configurational and conformational chain isomers present and, more recently, ^{1}H- and ^{13}C-NMR spectroscopy have provided the means for quantitative measurements of tacticity. As there are increasing indications that a knowledge of the configurational and conformational states of PVC chains are important for an understanding of the propagation stage of the degradation process,[1-4] the availability of such information is very useful.

Similarly, infrared spectroscopy and ^{1}H- and ^{13}C-NMR spectroscopy have been harnessed very effectively for the identification of structural defects, most notably chain branching and unsaturation. Chemical methods have also played an important part, by providing information on the concentration of labile chlorine atoms, total and internal unsaturation, and peroxide/hydroperoxide groups. This opens the way to a fuller understanding of the initiation stage of the degradation process.

The earlier work on the characterisation of the conjugated polyene chains, formed by the sequential elimination of hydrogen chloride during degradation, was based largely on ultraviolet spectroscopy. More recently, resonance Raman spectroscopy has proved to be a versatile and valuable method for those polyenes containing more than about 10 conjugated double bonds. Raman spectroscopy is also making an increasing impact as a convenient method for obtaining the vibrational spectrum of PVC, because of its ability to accommodate samples in a range of shapes and sizes, thereby avoiding the preparation of the very thin films required for infrared spectroscopy.

The application of these techniques, and the results obtained, will be considered under three headings. They are the characterisation of configurational and conformational isomers, the study of structural defects such as chain branching, unsaturation, labile chlorine and carbonyl groups, and the examination of degraded samples.

4.2 CONFIGURATIONAL AND CONFORMATIONAL ISOMERISM

In the very early work on the structure of the PVC chain a planar zigzag structure was tacitly assumed, probably partly by analogy with polyethylene but also because of the X-ray diffraction study of Fuller.[5] The realisation that this represented a very oversimplified view emerged for two very different reasons: the considerable changes in the infrared spectrum as a function of polymerisation temperature and the development of the concept of stereoregular polymers, and the discovery of catalysts that lead to the formation of such polymers. The subsequent period of 25 years has been very fruitful and has led to a rather detailed qualitative understanding of configurational and conformational isomerism in the PVC chain. The picture is less complete quantitatively but is, nevertheless, sufficiently clear for some appreciation of the role of these two types of isomerism in the propagation stage of the degradation process.

The early infrared work demonstrated conclusively that the changes in the carbon–chlorine stretching region as a function of polymerisation temperature relate to changes in concentration of both the configurational and conformational isomers. It has subsequently proved difficult to separate the overlapping bands characteristic of these various isomers with sufficient certainty to justify a quantitative interpretation of the spectrum. Hence, the approach that will be adopted is to consider the information to be obtained by infrared spectroscopy rather than to attempt disjunct accounts of the characterisation of the two types of isomers. In a continuation of this approach, the information to be obtained by Raman and NMR spectroscopy will then be detailed.

4.2.1 Infrared spectroscopic studies

The first tentative suggestion that the various bands appearing between 600 and 700 cm^{-1}, arising from the C–Cl stretching modes, might be indicative of rotational isomers of the polymer chain came from Krimm and Liang[6] in 1956. This suggestion received some support from the work of Mizushima *et al.*[7] who, from studies on simple chlorinated paraffins, concluded that the bands at 610, 634 and 692 cm^{-1} are consistent with the Cl and CH group in the polymer chain in the *trans* and *gauche* conformations.

However, the first clear indications of the value of infrared spectroscopy for characterising the configurational and conformational states of the PVC chain, and the differences in polymers made in various conditions, came from the work of Grisenthwaite and Hunter[8,9] and Krimm *et al.*[10] Measurements on polymers prepared at temperatures between +170 °C and −100 °C showed a marked change in the spectra; the lower the polymerisation temperature the less intense is the band at 692 cm^{-1} by comparison with those at 615 and 635 cm^{-1}, as shown in Fig. 4.1. This led to the suggestion that the 692 cm^{-1} band is to be associated with disordered units in the polymer chain.

It also became evident from measurements on the polymers prepared at the lowest temperatures which, on the basis of energetic considerations and the intensity of the 692 cm^{-1} band, should contain the highest concentration of ordered chains, that there are two closely adjacent modes at 605 and 615 cm^{-1}. Krimm *et al.*[10] assigned the first of these to the C–Cl stretching mode of the ordered chain, and were the first to suggest that this

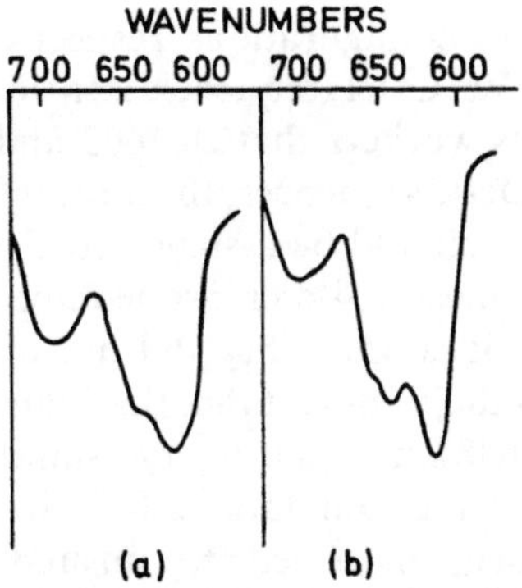

Fig. 4.1 The infrared spectrum of polyvinyl chloride in the region characteristic for the carbon–chlorine stretching vibration: (a) polymer prepared at +50 °C, (b) polymer prepared at −78 °C.

would have the syndiotactic configuration. They were less specific about the 615 cm^{-1} band, associating it with disordered portions of the chain.

The position at this juncture, as summarised by Shimanouchi *et al.*[11] was, therefore, that four distinct C–Cl stretching bands in the region 600–700 cm^{-1} were recognised. The pair at 603 and 638 cm^{-1} (various workers reported slightly different values for what are clearly the same bands) were associated with planar syndiotactic chains, the one at 615 cm^{-1} less certainly with syndiotactic chains of unknown conformation, and the one at 690 cm^{-1} with an isotactic structure. It was realised that this would have to be a bent or folded structure, possibly a three-fold helix. The assignment of the 603 and 638 cm^{-1} bands to the B_1 (antisymmetric) and A_1 (symmetric) stretching modes was indicated by the normal coordinate calculations of Shimanouchi and Tasumi[12] on a planar syndiotactic PVC chain.

There was a rapid breakthrough during the period 1960–65, which led to an almost complete qualitative understanding of the C–Cl stretching region of the spectrum in terms of frequencies characteristic for C–Cl units in specific configurational states. This was partly the result of additional work by the Japanese spectroscopists at Tokyo University but, more particularly, from the collaborative work of Professor Krimm at the University of Michigan and staff at the B. F. Goodrich Co. Research Center at Brecksville, Ohio. Although the problem was approached from several directions, studies on simple model chloroparaffins provided the key to the situation.

Shimanouchi and Tasumi[13] measured the spectra of meso- and dl-2,4-dichloropentanes, which can be regarded as vinyl chloride 'dimer', with isotactic and syndiotactic configurations, respectively, and compared their spectra with the four PVC C–Cl frequencies. In addition to confirming the conclusions of previous workers that the 603 and 638 cm^{-1} bands characterise planar syndiotactic sequences, the most interesting conclusion was that the 690 cm^{-1} band should be assigned to C–Cl units in the isotactic configuration, because meso-2,4-dichloropentane has a band hereabouts whereas the dl-isomer does not. They did not discuss the nature of the 616 cm^{-1} band beyond the comment that this is an amorphous band which splits into the 603 and 638 cm^{-1} pair in crystalline regions. This imprecise statement probably laid the foundations for careless comments by later workers which, in turn, have led to misunderstandings about the determination of crystallinity from the infrared spectrum.

Shipman *et al.*[14] examined the spectra of a range of secondary alkyl chlorides, covering the energetically probable conformers that are likely to occur, both in these low molecular weight model compounds and in the syndiotactic and isotactic configurations of PVC. This work is better appreciated if the notation used to define the various conformers is discussed. It is possible to use the familiar *trans* and *gauche* notation but the alternative system, developed by Shipman *et al.* and an extension of the notation of Mizushima *et al.*,[7] is more convenient and flexible.

The letters P, S and T are used to denote primary, secondary and tertiary chlorides, respectively; in the present context, only secondary chlorides are of interest. The conformations are specified by reference to the chlorine atom. The subscript H specifies that it is *trans* to a hydrogen atom and C that it is *trans* to a carbon atom. The simplest secondary chloride, 2-chloropropane, can exist in only one conformation, S_{HH}. However, 2-chlorobutane and higher 2-chlorohomologues can exist in three rotationally (conformationally) isomeric forms, S_{HH}, S'_{HH} and S_{CH}. In one of the two S_{HH} conformations the first four carbon atoms are planar while in the other they are non-planar. This bent conformer is designated S'_{HH}. Other more energetically improbable rotational isomers can be postulated but are unlikely to occur in practice.

From a study of the spectra of 2-chloropropane, 2-chlorobutane, 2,2-dimethyl-3-chlorobutane and 3-chloro-2,2,4,4-tetramethyl pentane, Shipman *et al.*[14] were able to give frequency ranges for the three types of conformations, as follows: S_{HH} 608–615 cm^{-1}, S'_{HH} 627–637 cm^{-1} and S_{CH} 655–674 cm^{-1}. Shipman *et al.*[14] noted, wisely, that whereas the relative concentrations of the rotational isomers in the secondary alkyl chlorides studied are determined by the temperature and the energy differences between them, in the case of PVC, additional constraints due to solid state forces may result in higher concentrations of some conformers than would be anticipated from the solution studies on model compounds.

By a systematic application of these characteristic conformational frequencies, and by a comparison of the spectra of PVC, PVC-αd_1, PVC-βd_1, PVC-$\alpha\beta d_2$ and PVC-d_3, Krimm *et al.*,[15] in a detailed and important paper, were able to explain the salient spectral features in the 600–700 cm^{-1} region. The peak at about 690 cm^{-1} is clearly an S_{CH} mode and this is possible from both bent syndiotactic and isotactic configurations. On the basis of the work of Shimanouchi *et al.*,[16] who observed that this band moves from 693 to 685 cm^{-1} when the temperature is raised, Krimm *et al.*[15] concluded that two overlapping peaks are present. They assigned the one at 693 cm^{-1} to the S_{CH} vibration of syndiotactic units and the one at 685 cm^{-1} to the corresponding vibration of isotactic units.

Comparison of the spectra of PVC samples (and those prepared from the

Table 4.1 Assignments for the various bands in the C–Cl stretching region of the infrared spectrum of PVC, from the results of Krimm and co-workers[17]

Frequency (cm^{-1})	*Conformation*	*Configuration*
601	S_{HH} *TTTT*	Long, planar syndiotactic sequences
612	S_{HH} *TTTT*	Short, planar syndiotactic sequences
622	S'_{HH} (*TG*)(*TT*)(*G*T*)	Energetically somewhat unfavourable bent isotactic structure
635	S'_{HH} (*TG*)(*TG**)(*TG*)	Energetically unfavourable bent isotactic structure
639	S_{HH} *TTTT*	Long, planar syndiotactic sequences
685	S_{CH} (*TG*)(*TG*)(*TG*)	Most probable bent isotactic structure
693	S_{CH} *TTG*G** or *TTTG**	Bent syndiotactic structure

In the *trans*, *gauche* nomenclature, *G** represents the alternative *gauche* conformation, staggered through 120° from *trans* in the opposite direction from the *G* conformation.

four deuterated monomers) made via the urea canal complexes at −78 °C, and prepared conventionally at +50 °C, shows that there is additional absorption at 615 cm^{-1}. This is assigned to the S_{HH} mode of short syndiotactic sequences, and is in line with the value observed for low molecular weight model compounds. The splitting of this mode into the two observed components at 604 and 640 cm^{-1} occurs only when there is a long sequence of monomer units in the planar syndiotactic configuration in crystalline regions, where there are two polymer chains passing through the orthorhombic unit cell. This splitting is predicted by group theory which, however, gives no indication of the required number of monomer units. Finally, there is evidence for additional absorption at 638 cm^{-1}, obtained from a careful comparison of the band shape in this vicinity among samples polymerised at different temperatures. This band becomes more intense as the order in the sample decreases, and this is consistent with its assignment to an S'_{HH} mode of an isotactic structure that is energetically less favourable than the one responsible for the absorption at 685 cm^{-1}.

In a subsequent note, Krimm *et al.*,[17] using a spectrometer with a higher resolution, reported an additional band at 622 cm^{-1}. They assigned this to the S_{HH} conformation of a bent isotactic structure. Hence, their work as a whole indicated the presence of seven bands, assignable to various conformations of the syndiotactic and isotactic configurations, as indicated in Table 4.1.

As noted above, Krimm *et al.* obtained their evidence for the existence of some of these seven bands from a careful visual comparison of the shapes of the overlapped bands. Subsequently, mathematical methods have been developed, which will be discussed in more detail below when quantitative studies are considered, to aid the resolution of overlapping peaks. These methods are of two types: curve fitting and derivative spectroscopy. With the former, the overall system of overlapping bands is matched, to a 'goodness of fit' criterion, by summing a series of individual peaks. Derivative spectroscopy, using even-order derivatives, gives peaks at the same positions as those in the conventional spectrum, but of appreciably smaller width, approximately one-third for the second derivative and one-fifth for the fourth. Hence, the effective resolution is improved and overlapped peaks are more readily discernible.

Pohl and Hummel[18,19] used a particularly simple form of curve fitting, involving a graphical procedure which gave interesting results. In particular, they detected an additional band at 647 cm^{-1} which, on very circumstantial evidence, they associated with a high-energy bent syndiotactic structure. However, this assignment seems improbable as its intensity, in the case of the urea canal complex polymer, was found to be greater than that of the 693 cm^{-1} band, which characterises what Krimm *et al.* regard as the most energetically favourable of the bent syndiotactic structures. Pohl and Hummel used their graphical resolution method to interpret measurements made over a range of temperatures. They separated the broad band at about 690 cm^{-1} into two components, at 676 and 696 cm^{-1}, and found that the intensity of the former varies significantly with temperature whereas that of the latter does not. Using the argument that the S_{CH} form of the isotactic structure will be stable with increasing temperature but the concentration of the S_{CH} form of the syndiotactic structure will decrease, they assigned the 696 cm^{-1} band to the former, thus reversing the assignments of Krimm *et al.*[17]

More recently, computer curve fitting has been used by Baruya *et al.*[20] and by Robinson *et al.*,[21] respectively, to interpret the infrared and Raman spectra of various PVC samples in the C–Cl stretching region, and the results confirm the findings of Pohl and Hummel. There is very strong evidence for the presence of a band at about 647 cm^{-1}, from an S'_{HH} structure of the *TTTG* conformation, for the reversed assignments of the two S_{CH} bands, and for the presence of an additional band at 704 cm^{-1} in the Raman spectrum.[21] This has been assigned to an S_{CH} conformation of the syndiotactic structure and as it is appreciably stronger than one appearing at 680 cm^{-1} (the 676 cm^{-1} band of Pohl and Hummel) in all the

Table 4.2 Revised and more detailed assignments for the various bands in the C–Cl stretching region of the vibrational spectrum of PVC, based on results from curve fitting and derivative spectroscopy

Frequency (cm^{-1})	*Conformation*	*Configuration*
601 (608 Raman)	S_{HH} *TTTT* long	Long, planar syndiotactic sequences
614	S_{HH} *TTTT* short	Short, planar, syndiotactic sequences
623	S'_{HH} (*TG*) (*TT*) (*G*T*)	Energetically somewhat unfavourable bent isotactic structure
634	S'_{HH} (*TG*)(*TG**)(*TG*)	Energetically unfavourable bent isotactic structure
638	S_{HH} *TTTT* long	Long, planar syndiotactic sequences
647	S^{1}_{HH} *TTTG*	Energetically favourable bent syndiotactic structure
680	S_{CH} *TTG*G**	Bent syndiotactic structure
692	S_{CH} (*TG*)(*TG*)(*TG*)	Most probable bent isotactic structure
704	S_{CH} *TTTG**	Bent syndiotactic structure of intermediate energy

samples examined, it is probably of lower energy. On this basis it has been assigned to the *TTTG** conformation and the *TTG*G** structure is associated with the 680 cm^{-1} band. This gives a total of nine peaks, as indicated in Table 4.2. Useful confirmatory evidence for the presence of most of these bands has come from the work of Maddams and Tooke[22] and Maddams and Southon,[23] who obtained second and fourth derivative infrared spectra of a commercial polymer and of the urea clathrate material.

The assignment of the various C–Cl stretching bands to specific configurational and conformational structures is an example of the application of the concept of group frequencies, so widely used in the interpretation of the infrared spectra of organic compounds. The correctness of this approach is not in doubt in the case of model compounds such as 2,4-dichloropentane and 2,4,6-trichloroheptane.[24,25,26] However, measurements on low molecular weight model compounds in solution may not be wholly applicable to polymer chains examined in the solid state. An interesting and important new approach to the problem has been made by

Opaskar and Krimm.[27] They consider first the vibrational modes of an infinitely long planar syndiotactic PVC chain. Their analysis shows that of the infinite number of modes, the only ones optically allowed will be those in which the phase of the motion is the same in each unit of the repeating structure; these are the factor-group modes. For phase differences other than zero the oscillating dipole moment averages out to zero when summed over the entire chain. Such vibrational modes are infrared inactive. When this concept is applied to the C–Cl stretching mode the predicted bands are at 607 and 634 cm^{-1}, in tolerably good agreement with the experimental values of 601 and 639 cm^{-1}. There is a small, but finite, probability that will be absorption at intermediate frequencies, arising from modes where the phase differences are not exactly zero.

Opaskar and Krimm then calculated the effect of introducing an occasional defect into the infinitely long chain; the defect may be a different type of monomer unit or the same type of monomer unit in a different configurational or conformational state. They reached a number of interesting conclusions. If the natural vibrational frequency of the defect lies outside the lattice band of the regular polymer it will be active and will give a band which can be regarded as a localised vibration with a group frequency. If this defect is loosely coupled to the remainder of the chain it will have very little influence on it. However, if it is strongly coupled it may activate some of the weakly allowed lattice modes of the main chain. If the defect frequency falls inside the lattice band its infrared activity will depend on the strength of the coupling. If this is weak the defect will absorb at its natural frequency. When the coupling is more than about one-tenth of the normal interaction between monomer units the intensity of this characteristic absorption dwindles to zero. With still stronger coupling it will activate all the possible lattice band vibrations equally and the absorption will therefore be determined by the probability of these occurring. With strong coupling one or two defect frequencies will appear just outside the limits of the lattice band.

Opaskar and Krimm applied these ideas qualitatively to the C–Cl stretching region of PVC samples with a significant concentration of defects, that is all polymers except those prepared under special conditions designed to give a substantial degree of order. They concluded that the S_{CH} mode at 693 cm^{-1} is a localised defect and the band is therefore a good group frequency. The situation with respect to the various configurational and conformational modes lying between the A_1 and B_1 modes of the long syndiotactic chain is less certain. Very probably, all the modes in the lattice band are excited, although they will not absorb equally strongly. Hence, there will be a good deal of additional absorption between 601 and 639 cm^{-1} and this may not be assignable to discrete modes. If this happens, the concept of group frequencies will be of limited value and as this situation is likely to occur most markedly with the most disordered polymers—the commercial type of materials—the realistic interpretation of the spectra, particularly in quantitative terms, may be open to question.

The other important theoretical study is that of Rubčič and Zerbi[28,29] who, although considering the behaviour of defect structures in an otherwise ordered chain, were concerned with the need to correlate their findings with the experimental data. They calculated density of states plots for various types and lengths of defect structures inserted into an ordered chain and a very reasonable correlation with measured spectra emerged for some of these structures. This work, more so than that of Opaskar and Krimm, gives pragmatic credence to the use of group frequencies. Furthermore, anticipating the results from quantitative curve fitting presented in Section 4.2.2, the fact that the spectra of polymers over a range of temperatures may be fitted with the same set of component peaks in different proportions suggests that the group frequency concept is applicable.

The fact that the various bands in the C–Cl stretching region are characteristic for both configurational and conformational isomers poses a fundamental problem for tacticity measurements. It is clearly necessary to sum the concentrations for the various conformational forms of the syndiotactic and isotactic configurations in order to obtain the required information. The problems involved will be considered in Section 4.2.2, because the majority of the serious work on this topic relates to Raman spectroscopic measurements. An alternative approach to the estimation of tacticity is to use the CH_2 deformation mode at about 1430 cm^{-1}. In their early work, Krimm *et al.*[10] noted that the observed band consists of two overlapping peaks. As the polymerisation temperature is lowered the component at 1430 cm^{-1} becomes weaker relative to the one at 1423 cm^{-1}. This led them to conclude that the latter is associated with syndiotactic units and the former with disordered portions of the chain.

This observation was followed up in some detail by Germar *et al.*,[30] who showed that there are two CH_2 deformation modes, at 1428 and 1434 cm^{-1}, when the spectrum is measured in solution in tetrachloroethane. They considered two possible conformers of the isotactic structure, one with chlorine *trans* to hydrogen and *trans* to chlorine, the other with chlorine *trans* to carbon and chlorine *trans* to hydrogen. They reasoned that these two conformers have mirror symmetry and are therefore of equal energy. In the case of the syndiotactic structure they concluded that the two probable conformers, one with both the chlorine atoms *trans* to hydrogen atoms (denoted by S_1), and the other with both the chlorine atoms *trans* to carbon atoms (denoted by S_2), cannot be of equal energy. From the variations of relative intensity with polymerisation temperature it follows that the band at 1428 cm^{-1} must be associated with S_1 and the other at 1434 cm^{-1} with S_2. They also concluded that the two isotactic conformers absorb at 1434 cm^{-1}.

If the temperature of the solution is changed, the proportions of the two syndiotactic conformers will also change, as will the ratio of the absorbance values at 1428 and 1434 cm^{-1}. If this ratio is measured at two temperatures, T_1 and T_2, and the values are denoted by λ_1 and λ_2, then

$$\frac{\lambda_1 + 1}{\lambda_2 + 1} = \frac{1 + \exp(-\Delta E/RT_1)}{1 + \exp(-\Delta E/RT_2)}$$

Thus, the energy difference, ΔE, may be found, and if this is substituted into the equation

$$\alpha = \frac{1 + \exp(-\Delta E/RT)}{1 + \lambda}$$

the degree of syndiotacticity, α, may be found. Germar *et al.*[30] used the method to examine six polymers, prepared at various temperatures. A material polymerised at +30 °C, and hence not greatly different from commercial polymers, was found to be 54 % syndiotactic; one prepared at −70 °C gave a value of 80 %. Nakajima *et al.*[31] also used the method and examined samples polymerised in the temperature range +90 to −75 °C and obtained syndiotactic contents between 51 and 77 %.

The values obtained for the less syndiotactic materials are in substantial agreement with those from NMR spectroscopic studies (Section 4.2.3), whereas those for low temperature polymers are erroneously high. This suggests the presence of a systematic error, the probable cause of which has recently been demonstrated by Maddams and Tooke,[22] using fourth-derivative spectroscopy. The band sharpening achieved by this approach led to the splitting of the CH_2 band into three components, one of which is broader than the other two and probably consists of two markedly overlapping components. The use of this complex band for the estimation of syndiotacticity is, therefore, far from straightforward and it must be concluded that the vibrational spectrum as a whole is not suitable for this purpose.

4.2.2 Raman spectroscopic studies

The various configurational and conformational isomers of PVC are of rather low symmetry and, therefore, as expected, the various vibrational modes manifest in the infrared spectrum also appear in the Raman spectrum. Their relative intensities differ, considerably in some cases, because the intensity of a given vibrational mode in infrared absorption is determined by the change in dipole moment whereas it is the change in polarisability that is relevant for Raman scattering.

The importance of the Raman spectrum of PVC, which is becoming increasingly recognised, stems from the fact that it provides more realistic and flexible conditions for the examination of samples. The infrared spectrum is usually examined with a thin film, whose thickness is approximately 0·025 mm. Such films are prepared by hot pressing or casting from solution. Both methods are open to considerable objection on the grounds that the conformational equilibrium in a particular polymer specimen, which may reflect some mechanical or thermal treatment to which it has been subjected, has been disturbed. The alternative method for sample preparation involving alkali halide cold-sintered discs, involves some degree of grinding and it has long been recognised that this leads to conformational changes.[32] Smaller differences between the infrared spectra of the urea clathrate polymer, run as a mull and as a film pressed at 200 °C have also been noted.[33] By contrast, Raman spectra may be obtained from polymer specimens in powder or bulk form and, with the latter, considerable latitude is permissible with the sample dimensions. The only potential practical problem is sample fluorescence which can vary considerably and which, at worst, is a serious hindrance.

The early work of Koenig and Druesedow,[34] on a commercial polymer and on the urea clathrate material, did not reveal the wealth of detail that had emerged from the more extensive infrared studies of Krimm and co-workers, but it did suggest that the ν (C–Cl) region was capable of an equally fruitful interpretation. They found a value of 607 cm^{-1} for the antisymmetric (B_1) stretching mode of long planar syndiotactic sequences, significantly different from the corresponding value in the infrared spectrum. This point was not taken up, but the value was confirmed later by Robinson *et al.*,[35] who pointed out that it was a consequence of the correlation splitting arising from the interaction of the two chains in the orthorhombic unit cell. The best known example of this is the doublet at 720 and 730 cm^{-1} in the infrared spectrum of crystalline polyethylene. In this instance the splitting leads to two bands which are both infrared-active.

However, Moore and Krimm[36] predicted that the interactions between the two syndiotactic PVC chains in the orthorhombic unit cell would lead to a splitting of the symmetric (A_1) and antisymmetric (B_1) C–Cl stretching modes; the predicted separations were 5 cm^{-1} and 2 cm^{-1}, respectively. Furthermore, one component of each pair should be infrared-active and the other one Raman-active. Hence, there should be differences of 5 cm^{-1} and 2 cm^{-1} between the measured infrared and Raman frequencies of the two bands in the spectrum of the urea clathrate polymer. The experimental values[35] of 0 and 7 cm^{-1}, while confirming the existence of correlation splitting, indicate that the force field used for the calculations by Moore and Krimm is inexact.

As noted above, Robinson *et al.*[21] used the mathematical technique of curve fitting to interpret the Raman spectrum of commercial polymers in the ν (C–Cl) region in terms of nine configurational and conformational peaks. Curve fitting is the process of representing a complex of overlapping bands as the sum of a series of component peaks, using a least-squares non-linear fitting procedure in which the 'goodness of fit' between the measured profile and the sum of the component peaks is maximised. Three parameters are required to define each of the peaks: its position, shape and half-width. Furthermore, as it is possible to improve the goodness of fit almost indefinitely by the addition of more small peaks, some prior knowledge of the total number of peaks present is highly desirable. These are the fundamental limitations of curve fitting, stressed by Morrow and Cody,[37] Vandeginste and De Galan,[38] and Maddams.[39]

The work of Baker *et al.*[40–42] showed that many infrared and Raman bands are a good approximation to the Lorentzian profile and this band shape was used by Robinson *et al.*[21] in their curve fitting work on Raman spectra. It is possible to obtain a reasonable estimate of the number of component peaks by the use of second- and fourth-derivative spectra,[22,23] but Robinson *et al.* followed an alternative approach. They used curve fitting to separate the component peaks in the simple, well-defined spectrum of the urea clathrate polymer. This fit was then used to interpret the spectrum of a low temperature polymer, by adding the minimum number of peaks required to obtain an acceptable 'goodness of fit'. By an extension of this approach fits were obtained for the spectra of commercial polymers, using nine peaks. Although this approach does not necessarily eliminate systematic errors the results can be viewed with more confidence than those from an *a priori* nine-peak fit.

Although quantitative results from curve fitting must be viewed with caution, the results of Robinson *et al.*,[21] typical examples of which are given in Table 4.3, follow a coherent pattern with respect to the

Table 4.3 Results from the curve fitting of the Raman spectra, in the C–Cl stretching region, of various PVC samples

Peak position (cm^{-1}):	608·3	614·0	622·9	634·4	637·6	646·5	679·5	692·4	704·3
Half-width (cm^{-1}):	12·2	13·6	17·6	14·6	12·2	19·6	34·1	23·8	25·4
Polymer									
Commercial, suspension	4·4	7·9	9·5	11·9	7·3	14·6	7·6	18·5	16·1
Commercial, mass	4·3	7·4	10·0	11·0	7·1	15·0	8·2	17·6	16·8
Prepared at +5°C	4·6	7·7	10·3	8·3	7·7	19·3	7·1	16·6	16·0
Prepared at −30°C	5·2	8·3	10·2	6·0	8·6	22·9	4·9	16·7	13·6
TR371CH, low molecular weight	12·0	5·0	4·6	16·4	20·0	8·8	10·2	9·6	12·2
Urea clathrate	23·9	0·7	0·2	0·1	39·9	24·7	0	7·0	2·5

polymerisation conditions. For example, the area of the 608·3 and 637·6 cm^{-1} bands characteristic for the planar syndiotactic structure, increases with decreasing polymerisation temperature, and with the two samples, TR 371 CH and urea clathrate, prepared using special conditions. This trend is as anticipated. Similarly, with these last two materials, the intensities of most of the bands arising from isotactic structures is low and it can be inferred that they are of comparatively high syndiotacticity. It is also clear that the proportion of planar syndiotactic material present in commercial polymers is low. It may be noted that the half-widths of the various peaks in the intervals 600–650 cm^{-1} and 670–710 cm^{-1} are not equal, contrary to the assumptions of Pohl and Hummel,[18,19] whose simple graphical curve fitting of infrared spectra will, therefore, only give semi-quantitative results.

Although the values given in Table 4.3 are very useful for following the concentration trend of a given configurational/conformational isomer among a series of samples, they are probably not a good measure of the relative concentrations of the various isomers in a particular sample. They represent fractional peak areas and in order to convert them into isomer concentrations it is necessary to have the appropriate extinction coefficients in the case of infrared spectra and the corresponding calibration factors for the Raman spectra. These are not available, and it has been common practice[18,19,21] to assume that they are equal. Work with the infrared spectra of model compounds such as 2-chlorobutane and 2,4-dichloropentane[43] shows that this is not the case. Furthermore, calibration factors derived from solution measurements on low molecular weight model compounds may not be wholly applicable to PVC spectra run in the solid state. Nevertheless, as the vibrational spectrum provides the only method capable of yielding information on the conformational isomers present in PVC this approach will undoubtedly continue to receive attention and should, ultimately, provide reasonable quantitative information. This will undoubtedly be of value for obtaining a fuller understanding of the degradation mechanism of PVC.

4.2.3 NMR spectroscopic studies

The rapid development of NMR spectroscopy during the past two decades, and the present wide availability of high performance instruments, has transformed tacticity measurements on polymers. The pioneering work of Bovey and Tiers[44] on poly(methyl methacrylate) proved, in retrospect, to be atypically simple. In general, the ^{1}H (proton) spectra of vinyl polymers are difficult to interpret, and PVC is no exception, but they have provided

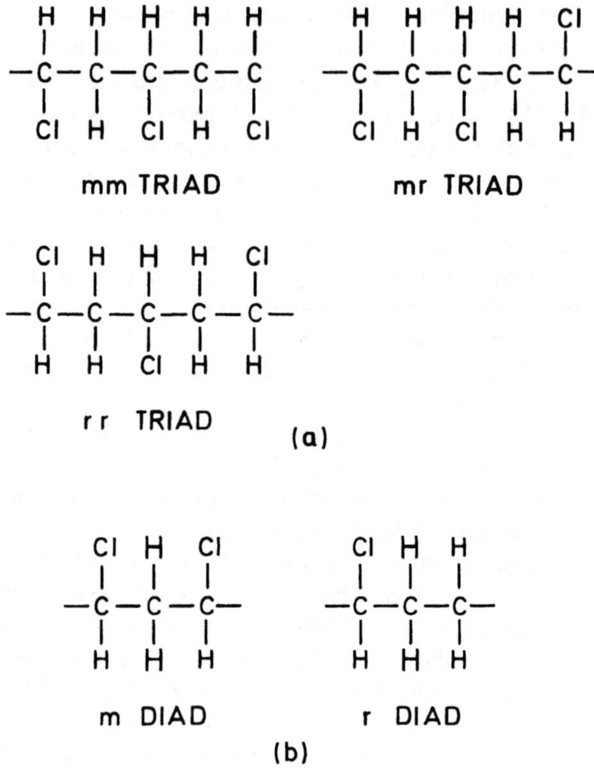

Fig. 4.2 Stereoregularity in polyvinyl chloride: (a) the mm, mr and rr triad structures, (b) the m and r diad structures.

much information. This class of polymer has two types of hydrogen atoms, methine and methylene, attached to the backbone carbon atoms, and the signals appear at different frequencies because the two types are in different structural environments. Furthermore, there is a secondary effect, the result of spin–spin coupling with adjacent hydrogen atoms, that leads to fine structure in the two signals. In the case of the methine hydrogen atom, coupling occurs with each of the four hydrogen atoms on the adjacent methylene groups to give a quintet, whereas with the two methylene protons coupling can only occur with two adjacent hydrogen atoms, one on each of the flanking methine groups, to give a triplet.

When tactic placements are taken into consideration (Fig. 4.2(a)), it is clear that a particular methine proton can occur in three different structural environments, mm, mr and rr triads, depending on the stereoplacement of the chlorine atoms but the methylene protons are present in two structural forms, m and r diads (Fig. 4.2(b)). Hence, the methine proton signal should consist of three quintets and that of the methylene proton two triplets. The results of the early measurements, made at comparatively low resolution, seemed at variance with these expectations and resulted in considerable controversy. This issue has been well documented by Sewell[45] and by Shimanouchi[46] and will, therefore, not be discussed in depth.

Johnsen[47] observed that the methylene proton signal is sensitive to differences in stereoregularity and treated it as two overlapping triplets, the one at higher field being assigned to syndiotactic units. However, Tincher[48] pointed out that the two methylene protons in isotactic units should be non-equivalent and so the overall signal should be of greater complexity than two triplets. He simulated the measured spectra by computer calculations and was able to obtain tacticity values. The problem was resolved by the examination of deuterated polymers, the use of spin decoupling, and by studying the spectra of simple model compounds. The measurements of Yoshino and Komiyama[49] on poly(α-d vinyl chloride) gave a single signal for the meso(isotactic) protons rather than the quartet expected if they are non-equivalent. The use of spin decoupling, an instrumental technique for eliminating the coupling between various types of hydrogen atoms, gave two single peaks; one for each of the two configurational environments of the methylene protons.[50] Finally, although the two methylene protons of meso-2,4-dichloropentane give different resonance frequencies and are therefore non-equivalent,[51] there is near-equivalence in the case of syndiotactic 2,4,6-trichloroheptane.[26] Hence, the observed equivalence in the case of PVC, which must be accidental, is not unreasonable. In the case of the methine proton signal the problem was to account for the fact that only seven peaks were discernible in experimental spectra rather than the 15 that might be anticipated. This is the result, not surprisingly, of a marked overlap of the three groups of signals and when the decoupled spectrum is measured a triplet is found.[50]

This clarification of the peak assignments, at a time when instrumental performances were improving steadily, opened the way to quantitative tacticity measurements. One consequence of the superior instrumentation was enhanced resolution, sufficient to discern fine structure in both the methylene and methine proton signals, the result of next-nearest neighbour interactions. For example, the resonance frequencies of the methylene protons are determined primarily by their immediate chemical environment, particularly the chlorine atoms on the two adjacent carbon atoms. However, the chlorine atoms next further removed on either side exert a small secondary effect. It is therefore necessary to consider tetrad units to account for this fine structure; there are six such units, rrr, mmm, rmr, mrm, rrm and mmr, in the usual notation.

Heatley and Bovey[52] measured the undecoupled methylene proton spectrum at high resolution, and although it is of appreciable complexity they were able to computer-fit it satisfactorily in terms of the six tetrads. In the case of the methine proton the next nearest neighbour interaction involves pentad units and there are 10 of these, rrrr, rrrm, mrrm, rrmr, rrmm, mrmr, mrmm, rmmr, rmmm and mmmm. Abdel-Alim[53] published a decoupled high resolution methine proton spectrum showing eight peaks and has concluded that overlap occurs between the signals from the rrmm and mrmr pentads, and also the rrmr and mrmm pentads.

Although the assignments of these various signals represent the cumulative efforts of a number of workers, the problems associated with quantitative tacticity measurements have attracted comparatively little attention. The most careful study is probably that of Robinson *et al.*[54] They made a systematic examination of the factors which determine the precision and accuracy, using comparatively low resolution spectra (60 and 90 Hz), and both decoupled and undecoupled methine and methylene proton spectra. Although the spectrometer resolution was insufficient to resolve fully the fine structure from next-nearest neighbour interactions, information on this point was obtained by computer curve fitting. The reproducibility of the tacticity values from different solutions of a given polymer proved significantly worse than that for a series of runs on a particular solution, showing that sampling errors are larger than those from instrumental sources. Incomplete dissolution of the polymer, particularly in the case of materials prepared at low temperatures, may be one pertinent factor. In order to improve the precision, Robinson *et al.* recommended that tacticity values should be derived from both methine and methylene proton spectra and that replicate runs should be made on more than one solution of a given material. In these circumstances, a precision of $\pm 2\%$ on a value in the range 55–65 % should be achievable. With the improvements in instrumentation since this work was undertaken, there should be no difficulty in surpassing this figure.

The larger chemical shifts in and simplicity of ^{1}H-decoupled ^{13}C-spectra more than offset the lack of sensitivity, which necessitates spectral accumulation over a period of hours; hence, such spectra are now being widely used for tacticity determinations on a range of polymers. The published work on PVC[55–58] shows that there is no difficulty in obtaining good quality high resolution spectra in which resonances characteristic for tetrad and pentad structures of methylene and methine carbons, respectively, are readily resolved. Such a spectrum, run at 50·1 MHz, is shown in Fig. 4.3. Although there was initially some disagreement over the assignments, those of Carman *et al.*[55] are now generally accepted and provide a sound basis for tacticity determinations. Furthermore, Schaefer[59] has shown that good quality spectra may be obtained from gels, the state in which the more syndiotactic materials probably occur if they do not enter completely into solution.

It has been shown recently that the amount of fine structure obtained in the methine and methylene carbon signals is, for a given instrumental resolution, solvent-dependent. Liu *et al.*[60] found that *o*-dichlorobenzene containing about 7% of dimethyl sulphoxide gave a very clear resolution of the mrm and mmr tetrads, whereas they are badly overlapped in pure *o*-dichlorobenzene. Elgert and Kosfeld[61] made measurements in a range of solvents and found that chloroalkanes and ethers gave the highest resolution, particularly in the methine carbon signal. In particular, with dioxane as solvent, the rr and mr signals are split into five components and the one from mm triads into six. These workers also recommended a ceiling temperature of 375 K to minimise decomposition. Ethers, collectively, are not particularly good solvents for PVC and as the degree of splitting achieved by Elgert and Kosfeld is not necessary for tacticity determinations, it is likely that *o*-dichlorobenzene will continue to find favour as a solvent. Typically, spectra are run at 400 K using 10% solutions and although blackening of the solutions occurs during the period of 20 h that the spectral accumulation measurements are in progress, the amount of thermal dehydrochlorination is probably no greater than a fraction of one per cent.

The only systematic study of the precision and accuracy of ^{13}C tacticity measurements is that of King *et al.*,[62] who examined a range of polymers, including one polymerised at -30 °C. They found good agreement between the values derived from the methine and methylene carbon signals and the precision achieved, using a frequency of 50·1 MHz, was approximately ± 1% of the value being measured, e.g. a typical commercial polymer would have a syndiotacticity of 55 $\pm$ 0·5%. Small, but probably significant, differences were noted between a mass polymer and three prepared by the suspension route; the latter had syndiotacticities 1–2% greater. The increase in syndiotacticity in passing to a polymer prepared at $+5$ °C was several times the uncertainty in the measurements.

These results are in good agreement with those of Pham *et al.*[63] given in Table 4.4. From the variation of syndiotacticity with polymerisation

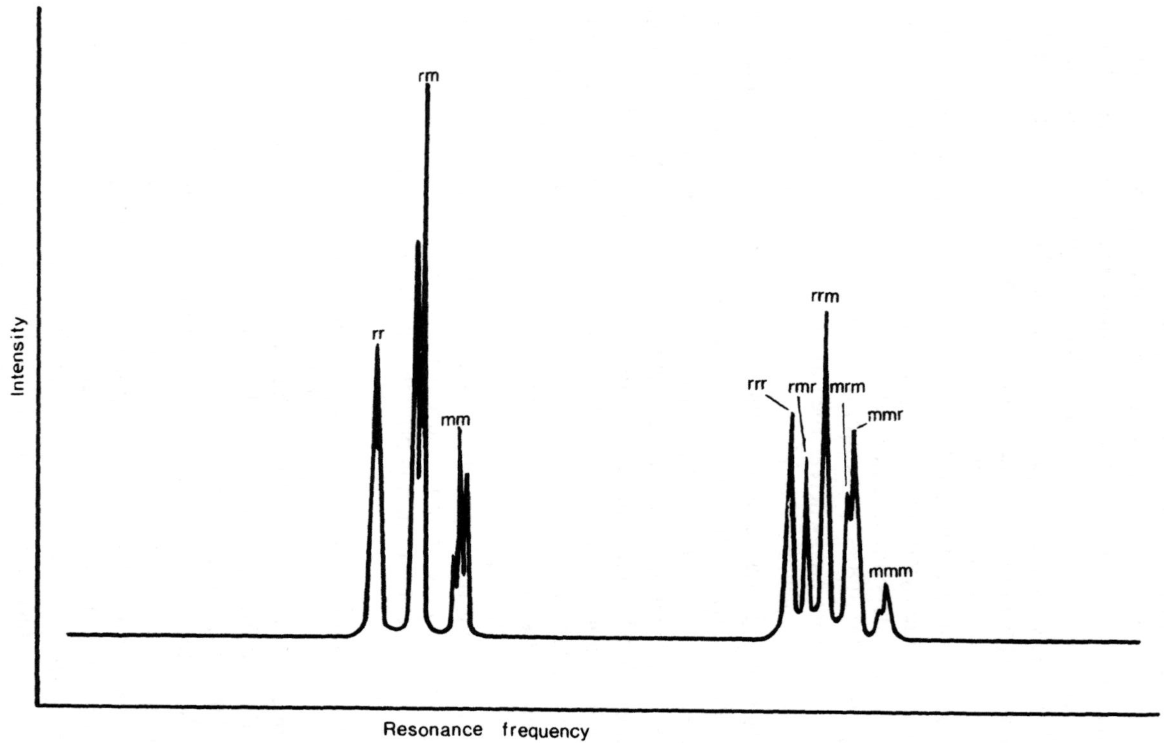

Fig. 4.3 The ^{13}C-NMR spectrum of polyvinyl chloride. The group of peaks on the left characterises the triad structures involving the α (methine) carbon atom and the group on the right characterises the tetrad structures from next-nearest neighbour interactions of the β (methylene) carbon atom.

temperature it is possible to calculate a value for the enthalpy difference, ΔH, in favour of syndiotactic placement. This is about 2500 J mol^{-1} and it may be used to calculate the syndiotacticity for any polymerisation temperature from the equation $S_t = S_0 \exp(-(\Delta H/kt))$. This approach is not valid in the case of polymerisations involving stereospecific catalysts, such as Grignard reagents, which give rather greater syndiotacticities and for which the polymerisation temperature is probably of secondary importance. Cais and Brown[64] have recently reported syndiotacticity values for polymers prepared at several temperatures and have claimed a precision of about $\pm 0{\cdot}5\%$ in the syndiotacticity values. However, they report a value of 0·604 for a polymer prepared at −78 °C and one of 0·584 for a −45 °C material. These values are appreciably lower than would be anticipated from the results of King *et al.*[62] and Pham *et al.*;[63] the presence of a systematic error must be suspected. Predictably, Cais and Brown report a markedly lower value, 1100 ± 20 J mol^{-1}, for ΔH.

Although the fine structure present in the ^{1}H and ^{13}C spectra gives information on the configuration of sequences of five monomer units, this number is significantly less than the sequence length commonly associated with the establishment of three dimensional order and, hence, crystallinity in polymers. Furthermore, it is unlikely that improvements in NMR techniques will provide configurational sequence information beyond six or seven units. However, it is possible to infer the behaviour of longer sequences in a statistical sense by means of information derived from the shorter sequences, together with a knowledge of the type of chain propagation involved.

The simplest situation is that of Bernoullian propagation, where the way in which a monomer unit adds on is not influenced by the stereochemistry of the end of the growing chain, but is a random process. If P_m/r denotes the probability that a monomer unit adds in r fashion to an m chain end and P_r/m is the probability that a monomer unit adds in m fashion to an r chain end then, for Bernoullian propagation, it follows that $P_{m/r} + P_{r/m} = 1$. It can also be shown that $P_{m/r}/P_{r/r} = 1$, as is $F_s/P_{r/m}$, the persistence ratio of Coleman and Fox,[65,66] where F_s is the fractional syndiotacticity. These relations may therefore be used to test if a particular polymer has Bernoullian configurational statistics.

On the other hand, the propagation step may be influenced by the stereochemistry of the growing chain end. There are then two conditional probabilities, and first-order Markov statistics are applicable. It is also possible to envisage a chain propagation mechanism in which both the ultimate and penultimate units influence the way in which the monomer

Table 4.4 The syndiotacticity of polyvinyl chloride as a function of polymerisation temperature

Polymerisation temperature (°*C*)	*Fractional syndiotacticity*
+55	0·55
+25	0·57
0	0·60
−30	0·64
−50	0·66
−76	0·68

adds; this gives four conditional probabilities, a situation covered by second-order Markov statistics. The mathematical relationships governing the various probabilities have been discussed in depth by Bovey.[67] For present purposes it suffices to note the type of sequence information, obtained from NMR tacticity measurements, required in order to test a particular model. It is necessary to measure the concentrations of mm, mr and rr units in order to test for random propagation and, hence, the applicability of Bernoullian statistics. This, therefore, involves the characterisation of triad units. If this type of propagation does not occur and it is required to test for the applicability of first-order Markov statistics, it is readily demonstrable that tetrad sequence concentrations are required.

A number of workers have used one or more of the three relations noted above as a test for the applicability of Bernoullian propagation statistics for various PVC samples without, perhaps, due regard for the effect of experimental errors. With a measured value that differs somewhat from unity it is necessary to assess the measuring errors in order to decide if the difference is significant. Martinez *et al.*,[68] using ^{13}C spectroscopy, regarded values outside the limits 0·95–1·05 as significant and, in the particular circumstances, this was probably a realistic criterion. With the recent careful work of King *et al.*,[62] the limits are closer, probably 0·975–1·025. On this basis, they found that a polymer prepared at +5 °C lay just outside these limits and that one prepared at −30 °C was definitely non-Bernoullian. The results of Liu *et al.*[60] indicate that at temperatures above about −10 °C the polymerisation mechanism is Bernoullian. However, the work of Cais and Brown[64] suggests that the propagation is Bernoullian at polymerisation temperatures as low as −78 °C. In view of this disagreement, which would not be expected in the light of the stated precision of their measurements, there is the need for additional careful experimental work.

When the type of propagation statistics has been established it is possible to calculate the probability of a particular type of sequence for a given degree of syndiotacticity or isotacticity. For example, Fordham[69] has shown that with Bernoullian propagation statistics the fraction of a polymer having a syndiotactic sequence length n, denoted by F_s^n, is given by the equation $F_s^n = n\alpha^2\beta^n$, where α and β are the probabilities of isotactic and syndiotactic placement, respectively. This equation has been utilised by several workers who have been interested in the probability of occurrence of longer syndiotactic sequences in relation to the problem of accounting for the crystallinity of PVC. In the present context the ability to calculate different types of sequence lengths may prove of increasing interest in the future, in view of the indications that are now emerging for a relationship between the configurational and conformational states of the PVC chain and the propagation stage of the degradation process.[1-4]

It was noted in Sections 4.2.2 and 4.2.3 that bands in the C–Cl stretching region of the infrared and Raman spectra of PVC are characteristic for individual conformational isomers but that it is difficult to obtain quantitative information because the appropriate calibration factors are not available. Although this quantitative problem does not arise in the case of NMR spectroscopy, the technique is of limited value for the examination of the conformational states of the PVC chain, for a different reason.

Any molecule having conformational isomers is in a state of dynamic equilibrium, the proportions of the individual isomers being determined by the temperature and the enthalpy differences between these isomers. There is rapid interconversion between them, involving rotation about one or more bonds, such as C–C, on a time scale of approximately 10^{-8}–10^{-9} s. It follows that if individual isomers are to be characterised the time scale of the measuring process must be rapid by comparison. This criterion is met in the case of vibrational spectra but not with ^{1}H and ^{13}C-NMR spectra where the time scale is several orders of magnitude slower. The spectra therefore characterise time-averaged conformations weighted according to the proportions of the isomers involved. The spectral differences between isomers are determined primarily by differences in the spatial arrangements of the hydrogen atoms, giving differences in the spin–spin coupling. The observed coupling constant for a particular pair of hydrogen atoms is $J_{obs} = \sum C_i J_i$, where C_i and J_i are the mole fraction and coupling constant of a particular conformer, respectively.

Despite this time-averaging process, the measurement of coupling constants has provided useful conformational information for the model compounds 2,4-dichloropentane and 2,4,6-trichloroheptane. In the former case, Schneider *et al.*[70] have shown from ^{1}H spectra that with *meso*-2,4-dichloropentane the *TG* conformer predominates but with the dl-isomer a significant amount of the *GG* conformer is present in addition to the *TT* form. The change in the coupling constant with temperature suggests an enthalpy difference of about 2·5 kcal mol^{-1} between the two forms. Doskocilova *et al.*[25] have obtained reasonably detailed information for the three configurational isomers of 2,4,6-trichloroheptane. At about 300 K the syndiotactic isomer consists of about 85% of the *TTTT* conformer and 15% of the *TTGG* form. The isotactic isomer has 80% of the *TGTG* form and 20% of the *GTGT* conformer. With the heterotactic isomer there are approximately equal concentrations of the *TTGG* and *GGTG* conformers. In view of the obviously greater conformational complexity of PVC, as instanced by the vibrational spectrum in the C–Cl stretching region, the amount of information obtainable by NMR spectroscopy will not be great. Nevertheless, the results of Heatley and Bovey[52] for a commercial type of polymer suggest that the conformational behaviour is similar to that already noted for 2,4,6-trichloroheptane.

4.3 STRUCTURAL DEFECTS

The growth in knowledge of the structural defects in PVC during the past decade has been even more striking than the comparable advances in the understanding of the configurational and conformational isomerism of the chain during the past 20 years. Although the first work on short chain branching dates back to 1953 the detailed studies are very recent, and this is also true of the characterisation of unsaturation and labile chlorine atoms. The published work up to 1980 has been surveyed by Caraculacu,[71] and his review provides a good general account of the level of activity and the results obtained. A collaborative study by a IUPAC Working Party has been in progress since 1977 which has involved detailed chemical and spectroscopic work in various laboratories. When the findings are published they will add considerably to existing knowledge.

4.3.1 Short-chain branching

Until recently, much of the information on the level of chain branching in various types of PVC samples had been obtained by infrared spectroscopy. Unlike polyethylene, where the measurement of chain branching is a relatively simple and well-established procedure,[72,73] there are problems because, if PVC gives a characteristic band comparable to the methyl deformation mode at 1378 cm^{-1} in polyethylene, it is wholly lost in the more numerous and stronger bands in the spectrum as a whole. The method in common use, dating back to the work of Cotman,[74] is to reduce the polymer to polyethylene and then use the approach cited above. Much of this work has been done with lithium aluminium hydride as reducing agent, although this has now fallen into disfavour for reasons which will be considered below.

The results of various workers[74–80] proved to be in substantial disagreement as they varied by almost an order of magnitude, between 2 and 18 branches per thousand carbon atoms. Baker *et al.*[81] were able to resolve some of these discrepancies by showing that the extinction coefficient for the 1378 cm^{-1} band is dependent on the length of the branch and by noting that, of the previous workers, some had estimated in terms of butyl branches and others as methyl groups at the end of very long chains. Baker *et al.* estimated the branches in reduced PVC as methyl groups, on a basis of two pieces of evidence. Rigo *et al.*[82] had concluded from the large yields of methane obtained in the γ-radiolysis of reduced PVC that the branches are predominantly methyl groups. Secondly, the lateral unit cell constants of reduced samples, measured by Baker *et al.*, when interpreted in the light of the results of Preedy[83] on polyethylenes containing known branches, indicated that the branches must contain one, or at most two, carbon atoms. Baker *et al.* recalculated the results of the previous workers using the extinction coefficient appropriate for methyl branches and were able to obtain better agreement, although the spread of values is greater than might have been anticipated. They concluded that commercial polymers contain five or six such branches per thousand carbon atoms.

At this comparatively low level of branching it is necessary to apply a correction for chain-terminating methyl groups, if present.[73] This requires a knowledge of the chain-terminating groups, the molecular weight, and the extinction coefficient at 1378 cm^{-1} for chain-terminating methyl groups. It was supposed, initially, that in reduced polymers one chain end terminated in a methyl group and the other in an unsaturated group. However, as it is now well substantiated (see Section 4.3.3) that most chains terminate at one end with the —CH=CH—CH_2Cl group, it is reasonable to correct on the basis of two terminal methyl groups per chain in reduced materials. The work of Baker *et al.*[81] shows that the extinction coefficient at 1378 cm^{-1} for chain terminating methyl groups is smaller than that of methyl branches by a factor of 2·5. Hence, given the molecular weight, the

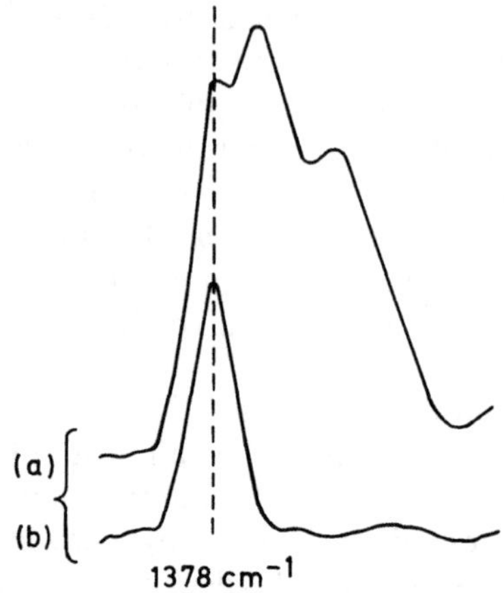

Fig. 4.4 The estimation of short-chain branching in polyvinyl chloride after reduction to polyethylene, using infrared spectroscopy (a) the spectrum of the reduced material, in which the methyl absorption at 1378 cm^{-1} is heavily overlapped by bands from methylene groups (b) the methyl band of this sample, obtained free from interference, by the use of a polymethylene wedge in the reference beam of the spectrometer to subtract the unwanted methylene bands.

necessary correction may be calculated and applied. The infrared method for estimating methyl branch concentrations requires separation of the methyl band at 1378 cm^{-1} from the stronger overlapping methylene band, as shown in Fig. 4.4, by the use of a polymethylene film in the reference beam of the spectrometer or by computer subtraction, but this is a well-established procedure for which the precision has been determined. The method is normally applied to thin, hot-pressed films but, in view of the small amount of reduced PVC often available, the use of the alkali halide disc sampling technique may be necessary. Studies on powder polyethylene specimens have shown[84] that this approach is less precise than the use of pressed films, but it yields acceptable results.

The greatest drawback of the infrared method is that it is necessary to know the nature of the branch, in order to determine the appropriate extinction coefficient at 1378 cm^{-1}, before quantitative work is possible. Furthermore, infrared spectroscopy is of a very limited value for identifying such branches.[85] Fortunately, ^{13}C-NMR spectroscopy provides the required specificity and, with the recent substantial improvements in spectrometer resolution and sensitivity, it is now the preferred approach. The sample requirement of 100–200 mg also compares favourably with that necessary for the production of hot-pressed films for infrared examination.

The method has been very well developed for the characterisation of chain branching in polyethylene[86–88] and by the examination of a series of ethylene/α-olefin copolymers, containing branches from methyl to *n*-hexyl, it has been possible to associate various resonance frequencies with specific branches. These frequencies are shown diagrammatically in Fig. 4.5 and it can be seen that a particular branch is identifiable not only by means of signals from the carbon atoms in it, but also from that of the branch-point carbon atom in the main chain. For example, the frequency of the methyl carbon atom moves as the chain length increases and signals characteristic for the additional carbon atoms also appear. Furthermore, the frequency of the branch-point carbon atom is a function of the length of the branch. Thus, in the specific case of a C_4 branch, the identity is established by the frequency of the methyl carbon signal, by the appearance of signals for carbon atoms 2, 3 and 4, and by the frequency of the signal from the branch-point carbon. By this means it is possible to characterise linear branches containing up to six carbon atoms. Those containing more than six carbon atoms are indistinguishable from *n*-hexyl.

Given a proper choice of instrument operating conditions, and these are now well established, the intensities of these signals are proportional to the concentration of a particular type of branch. Furthermore, there are two α- and β-carbon atoms in the main chain, adjacent to the branch point, and the intensities of the signals from these carbon atoms, twice those from the various carbon atoms in the chain, provide an additional parameter for quantitative work. In practice, it is usual to take the average of the concentration values obtained by these various approaches. Thus, ^{13}C-NMR spectroscopy is now well established as a sound and versatile method for the characterisation and estimation of chain branching in polyethylene.

Abbås *et al.*[89] were the first to apply the method to reduced PVC. They showed that the branches are methyl groups, thus confirming the earlier suggestions based on less definitive approaches,[81,82] and that, in the case of a suspension polymer prepared at 75 °C, the level of the branching is about four per thousand carbon atoms. They also found some evidence for branches of six or more carbon atoms, at a level of about one per thousand carbon atoms. The existence of these longer branches, at a low level, had previously been suggested by transfer constant measurements and intrinsic viscosity determinations.[90] The characterisation of long-chain branching will be considered in detail in Section 4.3.2.

The presence of chloromethyl branches in PVC prior to reduction was demonstrated conclusively by Bovey *et al.*,[91] by the use of lithium aluminium deuteride as reducing agent. This substitutes deuterium for

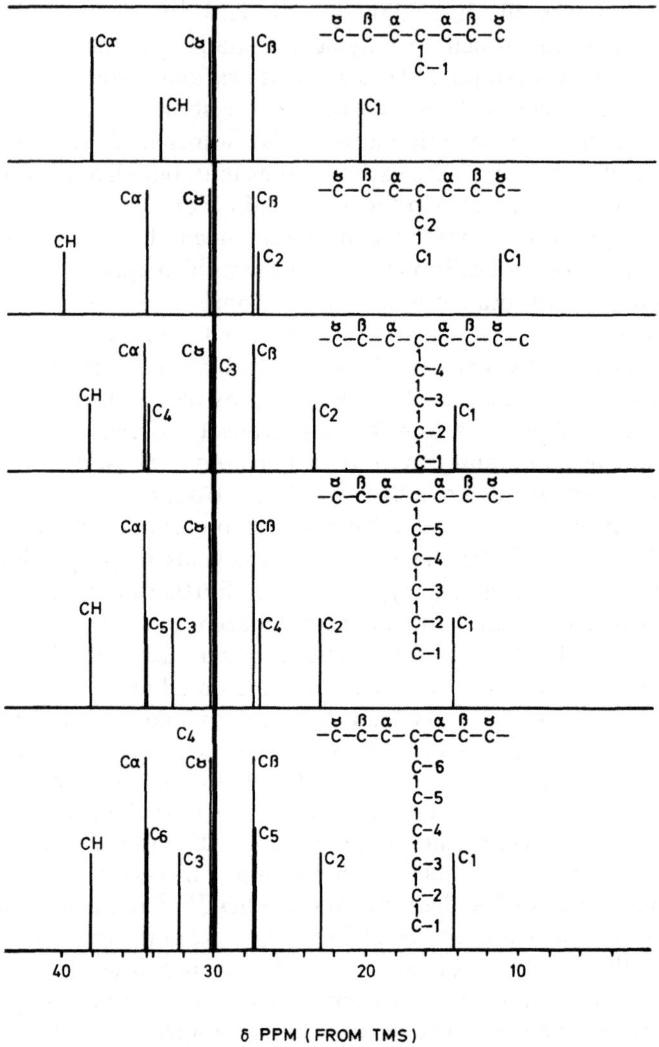

Fig. 4.5 A schematic representation of the ^{13}C-NMR spectra of polyethylene samples with branches containing one, two, four, five and six carbon atoms, respectively.

chlorine and as the deuterium nucleus has a spin of 1, and hence three spin states, the signals from those carbon atoms to which deuterium is attached are split into triplets. In the $LiAlD_4$ reduced product the methyl singlet of the product using $LiAlH_4$ appears as a triplet, providing the necessary proof. The mode of formation of these chloromethyl branches during the polymerisation process has attracted appreciable attention. As noted by Park and Saleen,[92] three mechanisms have been proposed. These involve head-to-head addition followed by chlorine atom abstraction at the end of the polymer chain,[82] hydrogen atom abstraction from an adjacent carbon atom at the end of the chain,[89] or the interaction of two growing radicals involving double hydrogen transfer in a six-membered ring system.[93] It should be possible to distinguish between the first of these possible routes on the one hand and the second and the third on the other by the preparation and subsequent reduction of the polymer prepared from α-d-vinyl chloride. The two polymer structures shown below would be formed:

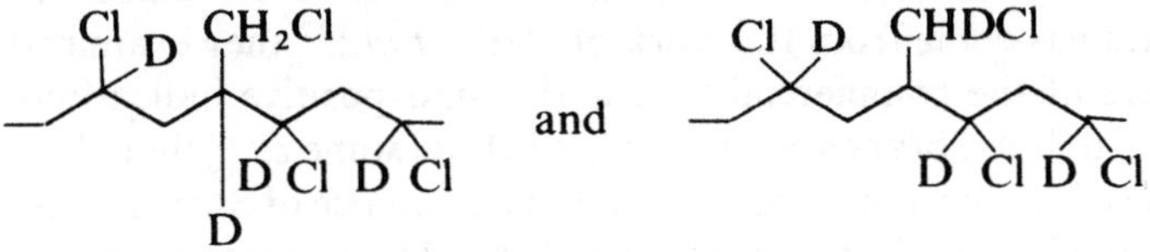

which, after the reduction with $LiAlH_4$, would give:

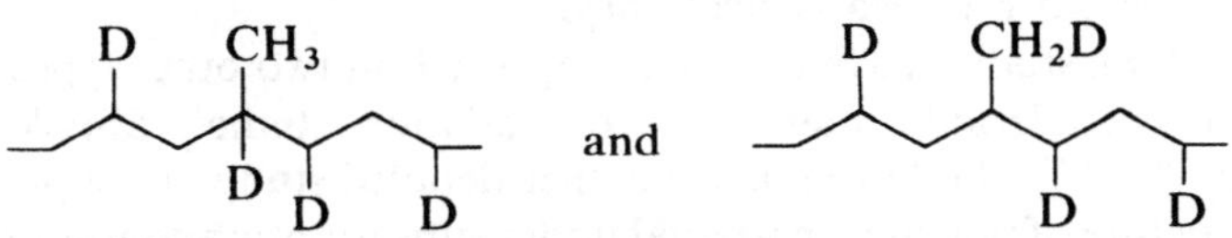

These are distinguishable, in the ^{13}C spectrum, by the presence of a single signal from the CH_3 group or a triplet from the CH_2D group. Starnes *et al.*,[94] following this route, found the methyl singlet, demonstrating that the rearrangement of head-to-head units occurs, as suggested by Rigo *et al.*[82] This conclusion is also supported by the work of Park and Saleem.[92] They measured the branch content, by reduction and infrared spectroscopy, of a series of polymers prepared using solutions of varying monomer and initiator concentrations. They found that the branch content increases with decreasing monomer concentration but appears to be independent of the initiator concentration. These results best fit the mechanism of Rigo *et al.*[82]

It has recently been shown[95] that tri-*n*-butyltin hydride is superior to lithium aluminium hydride as a reducing agent. Not only does it give a faster and more efficient reaction, judged by the residual chlorine content, but it is also free from side reactions. In the case of lithium aluminium hydride these include the generation of *trans* internal double bonds, the appearance of artefact bands that mask some of the important weaker signals characteristic of longer branches, and the preferential reduction of m diads. In this study, Starnes *et al.*[95] examined five polymers, made over the limited temperature range 43–75 °C, and found branching levels in the range 1·9 to 2·8 methyl groups per thousand atoms, with some indication of a decrease in the level as the polymerisation temperature is lowered.

Starnes and co-workers[96] have recently published studies on a polymer prepared at 100 °C after reduction with tri-*n*-butyltin hydride. In addition to methyl groups they found ethyl branches, from chloroethyl branches in the original polymer, at a level of 0·8 per thousand carbon atoms. The value dropped to 0·2 in the case of the 43 °C polymer examined in their earlier work.[95] They concluded that the reduced polymer might contain *n*-butyl branches but, if so, the concentration would not exceed about one per thousand carbon atoms. More positive proof for the occurrence of butyl branches has come from the work of Park *et al.*;[97] they examined several polymers of the commercial type and found positive indications for the presence of butyl branches, at an average level some 25 % that of the methyl branches. Furthermore, contrary to the experience of Starnes *et al.*,[96] they were not able to detect ethyl branches. Hence, there is a measure of disagreement about the nature and level of the longer branching in PVC and further work is clearly necessary.

Branching studies have also been reported on two other types of PVC, both of considerable interest for degradation studies. Sörvik and co-workers,[98–101] who have made a rather detailed study of the preparation and structure of polymers prepared under subsaturation conditions, giving a degree of monomer starvation, have examined the pattern of short-chain branching.[102] The polymers were reduced with tri-*n*-butyltin hydride and the ^{13}C spectra measured at 50 MHz. As anticipated, methyl branches were found, at a level comparable to that of commercial materials. No ethyl branches were detected but butyl branches are present at about the same concentration as their methyl counterparts. There is little doubt that they are formed by backbiting, as the result of monomer starvation during the polymerisation. Schwenk *et al.*[103] prepared two fractions, of molecular weights 800 and 1500, by methanol extraction of a bulk (mass) polymer. These were hydrogenated in the presence of Raney nickel and the products examined by ^{13}C-NMR spectroscopy. The spectra showed the presence of methyl branches at a level of five per thousand carbon atoms, and longer branches containing more than five carbon atoms, at a concentration of about two per thousand carbon atoms. With these materials, as with the subsaturation polymers, it is tempting to associate the decreased thermal stability with the greater concentration of longer branches. However, as will be seen below, other types of structural defects are present at higher levels, necessitating a careful comparison of the various parameters in order to establish possible correlations.

Not surprisingly, there has been interest in possible NMR methods for the direct determination of chloromethyl groups in PVC, to avoid the rather tedious process of reduction to polyethylene. The feasibility of such an approach, using ^1H measurements, was suggested by work described in a Firestone patent, although the materials examined contained unusually high levels of chloromethyl branches. More recently, work on this topic has figured prominently in the activities of the IUPAC PVC Working Party.[104] The CH—CH_2Cl group would be expected to give a signal in the region 3·5–4·0 ppm. Unfortunately, interference from signals from —CHCl—CH_2Cl and —CH_2—CH_2Cl chain-end groups is likely and, at present, no certain method for separating and distinguishing these is available, although there are clear indications that some solvents give preferential resolution of one or more of these groups of signals. A clear interpretation of this spectral region would also be valuable for the characterisation of saturated chain-terminating groups.

Two other techniques have been used, to a limited degree, for the characterisation of the branches in reduced PVC. The first of these, γ-radiolysis, has already been mentioned in passing, with the reference to the work of Rigo *et al.*,[82] in which the large yields of methane from the radiolysis had been taken as evidence for the presence of methyl branches. The method has subsequently been developed by Bowmer and O'Donnell[105] for polyethylene and applied by Bowmer *et al.*[106] to PVC samples reduced with tri-*n*-butyltin hydride. The results in the latter case were less clear-cut than in the former, as uncertainties arise because of the presence of unreduced PVC and combined or occluded reactant. Nevertheless, their observation that the major volatile hydrocarbon fragments are methane and butane indicates that either the branches present are methyl and butyl or that the branches are mostly, or entirely, methyl with the butane arising from residual tri-*n*-butyltin hydride reductant.

In order to distinguish between these alternative interpretations, Bowmer *et al.* examined polymers reduced with tri-*n*-propyltin hydride and found that the major products upon radiolysis are methane, propane and butane. They deduced that 95% of the propane came from the reducing agent, indicating that a comparable proportion of the butane is likely to derive from the reducing agent when tri-*n*-butyltin hydride is used. However, the fact that butane is obtained with the use of tri-*n*-propyltin hydride shows that C_4 branches are present in the original polymers. ^{13}C-NMR examination[97] showed the presence of butyl branches only when tri-*n*-propyltin hydride was used as the reducing agent, so it is clear that most of the propane and butane found when samples reduced with the appropriate tri-*n*-alkyltin hydride are examined comes from the presence of unreacted reducing agent, which produces a much larger hydrocarbon yield than the polymer. Further proof for the presence of butyl branches comes from the work of Schröder and Byrdy.[107] They used the radiolysis method to examine 12 PVC samples reduced with lithium aluminium hydride and in each case found methane and butane to be the major volatile hydrocarbon products. Hence, although the radiolysis method lacks the quantitative capability of ^{13}C-NMR spectroscopy it has yielded qualitative results very much in accord with the NMR findings.

The second non-spectroscopic technique that has been used recently in pyrolysis/gas chromatography and pyrolysis/hydrogenation/gas chromatography.[108,109] In the latter approach, 1 % palladium on Chromosorb P is used to hydrogenate the pyrolysis products at the injection port of the chromatograph. This converts the appreciable content of alkenes into alkanes, thereby simplifying the interpretation of the gas chromatograms. The results of Ahlstrom and co-workers[108] suggest that the branches are predominantly methyl groups; some ethyl branches were also indicated but, surprisingly, no butyl branches. Since these latter branches are clearly indicated by ^{13}C-NMR spectroscopy and γ-radiolysis, some suspicion must attach to the pyrolysis/hydrogenation/gas chromatography technique.

4.3.2 Long-chain branching

The information available on long-chain branching in PVC is markedly more meagre than in the case of the short branches, partly because of the more empirical nature of the methods available and, also, as a consequence of the detectably lower level of long branches. Unfortunately, ^{13}C-NMR spectroscopy is non-specific for longer branches; those with six and more carbon atoms give a signal at the same frequency. However, it is possible to distinguish such branches from saturated chain ends by the signal from the branch-point carbon atom. The results that have emerged from this approach show a measure of disagreement. Starnes and co-workers have published three papers on polymers prepared at 100 °C and although it is not clear if all their results relate to the same material, this is probably the case. In 1979[110] they reported 1·2 long-chain branches per thousand carbon atoms; a year later[96] they gave the comparable figure of 1·0, but recently[111] have reduced it to 0·3. It may be noted, incidentally, that in reducing the level of the long-chain branching they also modified their views about the presence of butyl groups, positively detecting them at a concentration of 0·9 per thousand carbon atoms. Some information on the level of long-chain branching for other types of PVC is available. Hjertberg and Sörvik[102] have found a value of 1·0 per thousand carbon atoms (with 1·7 butyl branches) for a polymer prepared under subsaturation conditions, while Schwenk *et al.*[103] have reported about 2 per thousand carbon atoms for a fraction of molecular weight 1500, obtained by methanol extraction of a bulk (mass) polymer.

Much of the work reported on long-chain branching in PVC involves measuring the hydrodynamic volume of the polymer chain. As such, it does not come within the present terms of reference but it will be considered briefly for the sake of completeness. Fuller details are to be found in the review article by Abbås.[112] These methods are based on the fact that the ratio of the intrinsic viscosity of a branched molecule to that of a linear molecule of the same molecular weight may be related through an equation of the type $[\eta]_{br}/[\eta]_{lin} = g^b$, where the value of b depends on the branching model that is used. Although a figure of 1·5 was used initially,[113] a value of 0·5 is now considered more appropriate.[114] For polydisperse samples g is related to the number of trifunctional branch points, n_w, per molecule by the equation:[114]

$$g_w = \frac{6}{n_w}\left[0{\cdot}5\left(\frac{2+n_w}{n_w}\right)^{0\cdot5} \ln\left\{\frac{(2+n_w)^{0\cdot5}+n_w^{0\cdot5}}{(2+n_w)^{0\cdot5}-n_w^{0\cdot5}}\right\}-1\right]$$

In using this equation, in conjunction with measured intrinsic viscosity values, care must be taken to break up the aggregates which are of almost universal occurrence in PVC solutions prepared at ambient temperature. Following the work of Hengstenberg and Schuch,[115] Gautron and Wippler,[116] Kobayashi[117] and Lyngaae-Jorgensen[118] found that heating at 120 °C for 3 h destroyed these aggregates in the case of commercial polymers. This procedure is probably not effective for low temperature polymers[119] but heating at 200 °C for 2 h should suffice, subject to the thermal stability being adequate. Lyngaae-Jorgensen[118] and Andersson and Sörvik[120] reported work on PVC samples, whereas De Vries *et al.*[121] examined a number of commercial polymers after reduction with lithium aluminium hydride. In both cases, long-chain branching levels of rather less than one per thousand carbon atoms were found, and these branches were probably formed during the latter stages of polymerisation, when the conversion from monomer to polymer was high.[122]

4.3.3 Unsaturation

Until recently, the qualitative and quantitative information on the nature of the unsaturation in PVC had come from the use of chemical analytical methods. These have provided a good indication of the level of the unsaturation as a whole and also that part of it not adjacent to chain ends. However, their drawback has been a marked lack of specificity for the possible types of chain-terminating unsaturation. Fortunately, the recent emergence of spectroscopic methods, particularly those based on the ^{1}H NMR spectrum, promises to remedy this shortcoming.

The estimation of unsaturation by the addition of a suitable reagent such as bromine across the double bonds has been much studied in the case of low molecular weight compounds, and the conditions necessary to avoid extraneous reactions, such as substitution, have been defined. The problems that arise in the use of such methods to estimate unsaturation in PVC are two-fold: the low level of the unsaturation, typically 1 double bond per thousand carbon atoms, and the relatively poor solubility of the polymer, which compounds the sensitivity problem. In these circumstances the blank readings obtained from the solvent system may approach the values obtained from the polymers and if the former are not constant the results can be no better than semi-quantitative.

Bromine addition is the method favoured by the majority of workers and the best approach is probably that of Morikawa,[123] using mercuric acetate as catalyst. The polymer is dissolved in dichloroethane and the bromine, somewhat in excess, is added, in solution in carbon tetrachloride. After standing for 24 h in darkness at ambient temperature the unreacted bromine is estimated by adding aqueous potassium iodide and measuring the iodine liberated by titration with sodium thiosulphate. Morikawa used the classical approach, with starch as indicator, but the colour change is difficult to follow. Hence, Boissel[124] modified the procedure by adopting potentiometric titration with platinum electrodes. The most extensive results available are those of the IUPAC Working Party, which show[104] that there are approximately 0·8 double bonds per chain. Furthermore, as the level is inversely proportional to the molecular weight, most of the unsaturation must be of the chain-terminating type.

Valuable confirmation for these findings has come from the use of an improved version of Wijs' method, in which iodine monochloride is added across the double bond. As with the bromination technique, there are several variants of this method,[125,126] of which the most recent[127] is to be preferred. A solution of iodine monochloride in trichloromethane is added at ambient temperature to the PVC sample in solution in 1,2-dichloroethane. After 15 min, when the addition reaction has reached completion, the polymer is precipitated and washed with methanol. The iodine content of the dried polymer, and hence the double bond content of the starting material, is determined by X-ray fluorescence analysis. It is also possible to assess the iodine monochloride uptake by the use of ^{125}ICl and counting techniques,[128] and comparable results are forthcoming. It should, in principle, be possible to estimate the unsaturation by catalytic microhydrogenation, using the method of Merz and Muller.[129] In practice, it is difficult to ensure complete hydrogenation without incurring hydrogenolysis of terminal allylic chlorine atoms and the method cannot compete effectively with the bromine and iodine monochloride addition techniques.

Several workers have estimated the internal unsaturation by oxidative chain scission of these double bonds followed by a molecular weight determination. Although permanganate oxidation has been advocated,[130,131] ozonolysis now finds greater favour, but there is some disagreement about the optimum conditions.[132,133] Hjertberg and Sörvik[102] have given a valuable analysis of the limitations of the method imposed by the experimental errors occurring during the molecular weight measurements after chain scission. They point out that M_n can be determined with a precision of 2–4 %, using gel permeation chromatography or viscometry; hence, double bonds in the outermost sequence of 20–40 monomer units in the chains of typical commercial polymers are not determined. Furthermore, if the internal double bonds are randomly distributed the measured values will be low by 4–8 %. There is reasonable accord between the published values,[132-134] and they show that approximately 10 % of the unsaturation is internal, except in the case of subsaturation polymers, where the proportion is appreciably greater.[102] An attempt has been made to detect conjugated double bond sequences by treatment with compounds of the type $P(OR)_3$, followed by ozonolysis; if such sequences are present there will be a reduction in the molecular weight. However, Minsker *et al.*[135] were unable to detect such unsaturation. They did not give a limit of detection but the method may well not have the required sensitivity for the low levels that are anticipated.

Although the out-of-plane deformation modes of the unsaturated groups present in polyethylene samples of various types give characteristic bands in the infrared spectrum that are of considerable diagnostic value, comparable bands are not observed in PVC because of the strong absorption from the —CHCl—CH_2— structure. Furthermore, contrary to what might have been anticipated, reduction to polyethylene followed by infrared examination, does not prove useful because, in the case of $LiAlH_4$ reductions, there are side reactions which lead to the formation of double bonds,[95] and with tri-*n*-butyltin hydride the unsaturated groups are hydrogenated.[128] It is therefore necessary to use the band arising from the C═C stretching mode, at about 1667 cm^{-1}. This has the disadvantage of having a comparatively low extinction coefficient and it is necessary to use films markedly thicker than is usual for infrared studies on polymers. The weak C═C band appears against strong background absorption but Ahrens *et al.*[127] have been able to make quantitative measurements using a Fourier transform infrared spectrometer. The measured intensities are converted into unsaturation concentrations by the use of calibration samples previously examined by the iodine monochloride method. The merit of the infrared method is that it provides the means for a comparatively rapid inter-comparison of a series of samples. The precise position of the C═C band, at 1667 cm^{-1}, is indicative of the —CH═CH—CH_2Cl unit as this frequency is also found in model low molecular weight compounds containing this structural unit.[136]

^{1}H-NMR spectroscopy has proved the most useful method for the characterisation of the chain-terminating unsaturation. The original work of Caraculacu and Bezdadea[137] involved the measurement of a group of signals arising from the hydrogen atoms attached directly to the double bond carbon atoms. They encountered sensitivity problems, obtained a poor signal-to-noise ratio and drew only tentative conclusions. As the unsaturation is predominantly terminal it is possible to increase its concentration, and so obtain a much more satisfactory spectrum, by the use of low molecular weight PVC fractions. This approach has been followed by Pétiaud and Pham,[138] and by Schwenk *et al.*,[139] and they have shown conclusively that —CH═CH—CH_2Cl is the major unsaturated structural unit.

This has also been confirmed by Hjertberg and Sörvik,[102] whose work shows that there is a clear doublet, between 4·0 and 4·1 ppm, that is more readily interpretable than the rather complex system of bands between 5·5 and 6·0 ppm (Fig. 4.6) from the hydrogen atoms attached to the double bond. The doublet is characteristic for the two hydrogen atoms of the chloromethyl terminal group, shifted downfield with respect to the signals arising from groups such as CHCl—CH_2Cl and —CH_2—CH_2Cl because

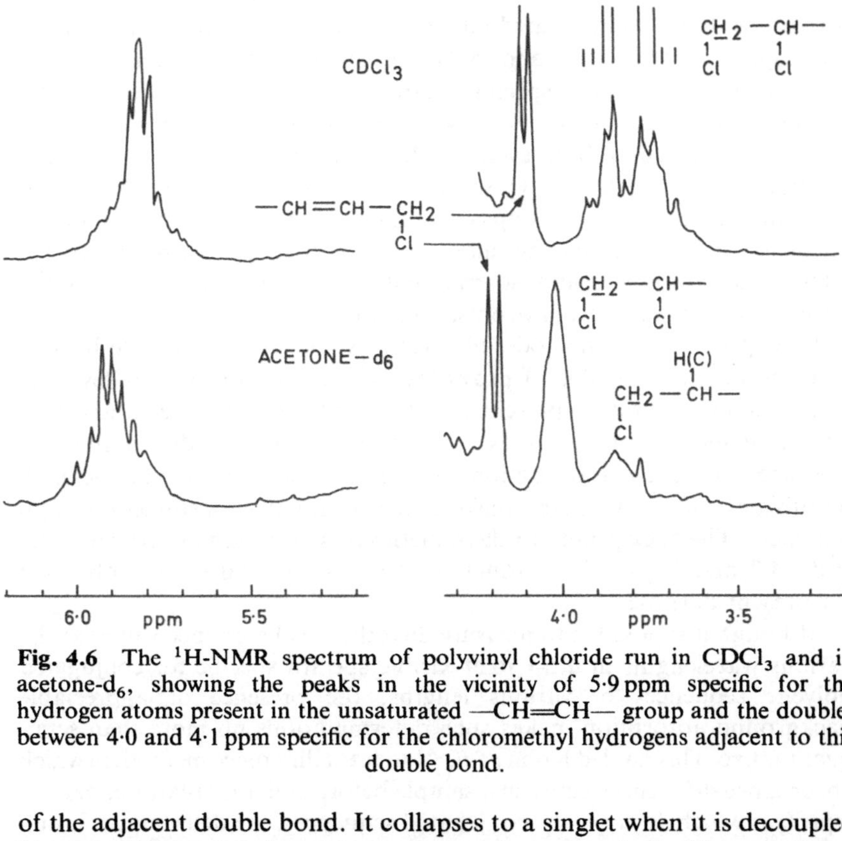

Fig. 4.6 The ^{1}H-NMR spectrum of polyvinyl chloride run in $CDCl_3$ and in acetone-d_6, showing the peaks in the vicinity of 5·9 ppm specific for the hydrogen atoms present in the unsaturated —CH=CH— group and the doublet between 4·0 and 4·1 ppm specific for the chloromethyl hydrogens adjacent to this double bond.

of the adjacent double bond. It collapses to a singlet when it is decoupled from the double bond hydrogen atoms. A very weak doublet somewhat downfield from the first doublet is also discernible in Fig. 4.6 and it has recently become clear[140] that this is also characteristic of the —CH=CH—CH_2Cl unit. In this case the main chain and the terminal chloromethyl group are disposed in the *cis* form about the double bond whereas the stronger doublet characterises the *trans* isomer. The sample examined by Hjertberg and Sörvik has a *cis*:*trans* concentration ratio of about 1:7 but until additional results become available it is not clear if this ratio varies among various commercial polymers or if it is a function of the polymerisation temperature. The concentration of *cis* units may prove to be of especial interest, because it seems probable that the *cis* isomer will split off hydrogen chloride more readily than its *trans* counterpart.

As yet, the rather complex set of signals in the interval 5·5–6·0 ppm, from the hydrogen atoms attached directly to double bonds, has not been interpreted fully. Its integrated intensity is somewhat greater than that of the signals between 4·0 and 4·2 ppm, showing that other types of unsaturation are present, especially in the case of polymers prepared under subsaturation conditions. ^{13}C-NMR spectroscopy may provide more detailed information but, even in the case of low molecular weight fractions, the signals are very weak. The use of a very high field spectrometer, coupled with long accumulation times, should surmount this problem. Such measurements may well shed light on the nature of the internal unsaturation, and so markedly supplement the rather scant information obtained from ozonisation studies.

Two spectroscopic methods, ultraviolet–visible and resonance Raman spectroscopy, are capable of providing information on the low levels of conjugated unsaturation present in virgin PVC and have been used to a limited extent. Both methods are better known in the context of characterising such unsaturation in degraded specimens, but the high sensitivities that are attainable make them very suitable for studies on virgin polymers. The background to these methods is discussed in Sections 4.4.2 and 4.4.3, and they will be considered solely in terms of the results obtained for present purposes.

Although it is possible to measure directly the absorbance values at the various wavelengths in ultraviolet–visible spectra specific for conjugated polyene sequences of various lengths, the presence of appreciable background absorption makes such determinations no more than semi-quantitative. This has led Kohn *et al.*[141] to use a difference method, in which absorbance differences between a sample before and after photooxidation, a process which destroys the conjugated sequences, are determined. Using this approach they estimated tetraene concentrations from the absorbance at 312 nm, using the extinction coefficient of 73 000 litres mol^{-1} cm^{-1} reported by Sondheimer *et al.*[142] They examined two polymers prepared by the suspension method, one in the presence of air and the other using oxygen-free conditions. The former had $4{\cdot}7 \times 10^{-5}$ mol $litre^{-1}$ of tetraene units whereas the smaller value of $2{\cdot}1 \times 10^{-5}$ mol $litre^{-1}$ was found for the latter. This observation may be of some significance for an understanding of the initiation stage of the degradation process and more detailed studies are justified. Kohn *et al.*[141] note that the determination of tetraene down to a level of 10^{-5} mol $litre^{-1}$ presents no difficulties. This corresponds to approximately 4 units per 10^7 monomer units, i.e. less than 1 unit per 2000 chains for the molecular weights encountered with commercial polymers.

Although these conjugated tetraene concentrations are low it is clear from the ultraviolet–visible spectra of most polymers that the concentrations of longer polyenes must be still smaller. Fortunately, resonance Raman spectroscopy provides the sensitivity and the specificity required to characterise these longer polyenes. It may be inferred from the work of Peitscher and Holtrup[143] that, in the case of Vestolit S 6857, a suspension polymer, the level of the conjugated polyene with $n = 20$ does not exceed one such unit per 12 000 molecules. Rather more extensive results are available from the studies of Bassez[144] and the IUPAC Working Party.[104] In some cases the level of conjugated polyenes with n in the range 10–20 was below the limit of detection and where it was detected it corresponded to total unsaturation in the range 0·0005–0·002 %, i.e. about 1 unit per 8000 or 2000 molecules, assuming a molecular weight of 35 000.

To summarise, with most types of polymer, 80–90 % of the unsaturation is present as the chain-terminating group, —CH=CH—CH_2Cl, and of the internal unsaturation, which is probably randomly distributed, the majority of it must occur as isolated double bonds. The formation of terminal double bonds is readily explicable in terms of radical transfer from polymer to monomer, and the disproportionation of polymer radicals. Various chain transfer mechanisms have been proposed,[82,91,94,138,139,145–148] the most favoured involving an initial head-to-head radical addition followed by chlorine transfer and the elimination of a chlorine radical, as shown below:

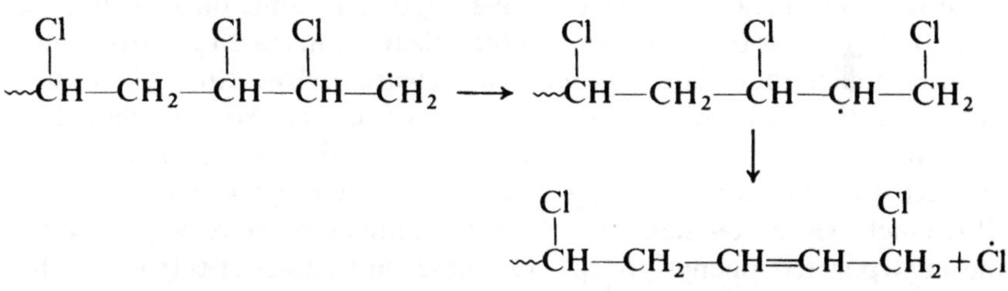

If statistical dehydrochlorination reactions occur they will result in isolated double bonds randomly distributed along the chain, as shown below:

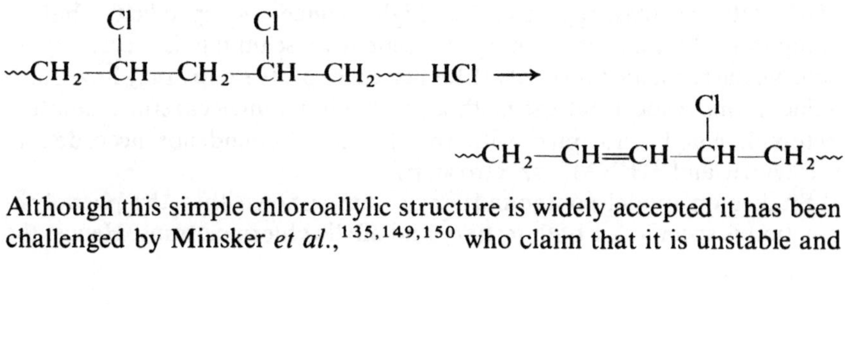

Although this simple chloroallylic structure is widely accepted it has been challenged by Minsker *et al.*,[135,149,150] who claim that it is unstable and rapidly oxidises to the ketochloroallylic group:

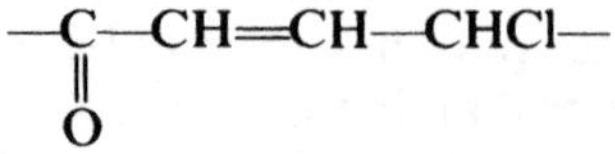

This is an $\alpha:\beta$ unsaturated ketone and will have its carbonyl stretching mode at about $1680\,cm^{-1}$ in the infrared spectrum. Although there are problems in detecting such carbonyl bands because of the presence of this functional group in initiator fragments, additives, etc., which are also present (see Section 4.3.7), it has never been detected and it is therefore reasonable to conclude that the internal unsaturation is present as isolated chloroallylic groups.

4.3.4 Labile chlorine

The term labile chlorine does not appear to have been defined precisely but, by common usage, has come to mean those chlorine atoms which are readily substituted by chemical reactions such as phenolysis, thiophenolysis and treatment with silver nitrate. Two types of chlorine atom, allylic and tertiary, are likely to undergo such reactions and it is widely assumed that because of this either or both of them are the sites for the initiation of degradation.

Of these chemical methods only phenolysis has come into general use. The conditions for the occurrence of the substitution reaction are only now properly understood. There is a tendency towards dehydrohalogenation[151] but this can be substantially eliminated by the use of an excess of phenol.[152] The mixture of PVC and excess phenol is heated at 60 °C for 100 h and unreacted phenol is removed by precipitating the polymer from solution. Three methods are available for the measurement of the concentration of the incorporated phenyl groups. Of these, ultraviolet spectroscopy has been the most widely used because of its very good sensitivity, with a limit of detection of about 0·2 incorporated groups per thousand monomer units, and the wide availability of the relevant spectroscopic equipment. ^{1}H-NMR spectroscopy gives a high degree of specificity but is comparatively time consuming, as repetitive scanning is necessary to achieve the requisite sensitivity. Fluorescence spectroscopy may have some value in this respect but the method probably requires careful evaluation before it can be accepted with the degree of confidence accorded to ultraviolet and ^{1}H-NMR spectroscopy.

Studies on model compounds[152,153] have shown that the phenolysis reaction is specific for both tertiary and allylic chlorine atoms. Hence, the fact that the results of measurements on a range of polymers correlate with the total double bond contents found by the bromination method[71] indicates that the labile chlorine atoms are largely those present in allylic chlorine units. Further evidence is provided by the results of ^{1}H NMR measurements on samples after phenolysis; they show that the chain terminating allylic chlorine unit has been substituted by phenol.

Michel *et al.*[154] have shown that thiophenol substitutes allylic chlorine atoms only. The reaction takes place at the —SH group, and no dehydrohalogenation occurs. The reaction is done in solution in dichloroethane at 60 °C for 40 h and the incorporated thiophenyl groups may be estimated by ultraviolet spectroscopy, by coulometric titration of the unreacted thiophenol or by measurement of the hydrogen chloride evolved. Good agreement between the results of such measurements and the total unsaturation, determined by bromination, was reported. Further confirmation that most of the labile chlorine atoms are present in allylic chlorine units has come from the work of Caraculacu *et al.*, using isotopic exchange of the labile chlorine with $SOCl_2^{36}$.[155] There is a good correlation of the results from this approach with those from the phenolysis method. The value of the silver nitrate method is currently open to doubt[140] because the results obtained are an order of magnitude lower than those from the bromination and phenolysis reactions. They are more in line with those from ozonolysis, suggesting that the method is estimating allylic chlorine atoms in internal double bonds. If this can be substantiated, by further work, the method could prove very useful.

Given the weight of evidence that most of the labile chlorine atoms are present in allylic chlorine units it is clearly desirable to have one or more direct methods for estimating tertiary chlorine atoms rather than to attempt to determine their concentration as the difference between the values obtained by two methods, such as phenolysis and thiophenolysis, which estimate total and allylic chlorine, respectively. Work on this topic has been reported, although, as yet, no wholly proven method is available. Caraculacu and co-workers[140] have used the phenolysis method to examine a series of samples after bromination, a reaction which converts the allylic groups into ones having chlorine atoms attached to saturated carbon atoms, so rendering the former non-labile. Hence, it should be possible to estimate the relatively low level of tertiary chlorine atoms that are believed to be associated with the point of attachment of longer branches. The results[140] show a measure of correlation with the concentration of long branches, but are not conclusive.

The second approach, and the one that is likely to provide the more decisive evidence when the method has been fully developed, is the use of ^{13}C-NMR spectroscopy to examine polymers reduced with tri-*n*-butyltin deuteride. If the branch-point carbon has a tertiary chlorine atom attached to it there will be replacement by deuterium and the ^{13}C-signal from this carbon atom will be split into a triplet. This approach has been used by Starnes *et al.*, who concluded that[96] both the branching carbon atoms of the ethyl and the long branches present in a polymer prepared at 100 °C using the bulk process had tertiary chlorine atoms attached to them. Later,[111] they modified their views slightly, concluding that tertiary chlorine atoms were present at butyl branch points but not those involved with long branches.

Further evidence for the presence of tertiary chlorines at the butyl branch points has come from the work of Saremi[156] and Hjertberg[157] and it is reasonable to conclude, semi-quantitatively in view of the sensitivity problems encountered with measurements of this type, that butyl branches and tertiary chlorine atoms occur together. The picture is less clear in the case of the long-chain branch points, partly because of the two differing sets of results given by Starnes *et al.*,[96,111] and also because of the later measurements of Hjertberg.[157] He pointed out that if both tertiary hydrogen and chlorine are associated with long chain branch points, the two structures shown below will be involved:

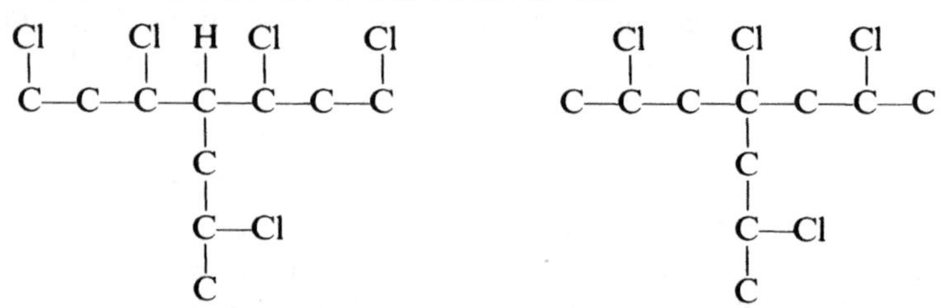

The ^{13}C spectra of the α- and β-carbon atoms, after reduction with tri-*n*-butyltin deuteride, will differ and from a detailed examination of the spectrum it was concluded that both types of tertiary atoms are present at the long-chain branch points, in the case of polymers prepared using subsaturation conditions.

It is not clear, at present, if there are fundamental differences between different types of polymer with respect to the nature of the tertiary atom, or whether the differences are apparent rather than real, and are a consequence of inadequate instrumental sensitivity, compounded by the fact that each component of the triplet from the —CD— group is one-third of the intensity of the signal from the corresponding —CH— group. Use of the very high field spectrometers now becoming available to examine a wider range of polymers should resolve this point, which is important in view of the calculations of Berens,[158] which led to the conclusion that 1 tertiary chlorine atom per thousand carbon atoms should suffice to account for the observed degradation in PVC.

4.3.5 Head-to-head structures

Reference has already been made to the role of head-to-head structures in the formation of branching and unsaturation in PVC. There is now strong circumstantial evidence that these head-to-head units are formed during the polymerisation and that, as radicals, they are markedly less stable than the conventional head-to-tail units. However, it is not clear if any of them survive the rearrangement reactions that lead to branching and unsaturation. If occasional head-to-head units are incorporated in an otherwise normal chain it is to be expected that they will lower the thermal stability because Crawley and McNeill[159] have shown that the head-to-head polymer, prepared by chlorinating *cis*-1,4-polybutadiene, is of lower thermal stability than the conventional polymer.

As yet, no certain method for the estimation of such units is available. One chemical method, that of Shimizu and Ohtsu,[160] has been used. This involves treating the sample with an excess of potassium iodide and dioxane at 100 °C for 4 days. The iodine liberated, which is a measure of the head-to-head units present, is estimated by titration with sodium thiosulphate. It is clear that this method will be subject to interference by any peroxides present and is therefore likely to yield high values in practice. It has been used by Mitani *et al.*,[161] who found 6–7 head-to-head units per thousand monomer units. More recently, much lower figures, averaging about 1 per thousand monomer units, have been obtained by the IUPAC Working Party.[140] Furthermore, a reasonably good correlation with reciprocal number average molecular weight is obtained. This lower figure is in line with the negative result obtained by Starnes *et al.*,[94] who failed to find signals in the ^{13}C spectra of various samples at the frequencies expected for the carbon atoms in and adjacent to the head-to-head unit. The limit of detection, which could probably be improved by the use of very high field instruments, was $1{\cdot}0 \pm 0{\cdot}5$ units per thousand carbon atoms and although a value somewhat below this may seem negligible it could be of significance or a contributory factor in the overall degradation process.

4.3.6 Peroxides and hydroperoxides

Although peroxide and hydroperoxide groups ought not, ideally, to occur as chemical defect structures in the PVC strain, their presence is possible because they may derive from initiator fragments or from gaseous oxygen present during the polymerisation. Furthermore, Svetly *et al.*[162] have provided cogent reasoning why the various types of chemical defect structures discussed hitherto cannot be the main initiating sites for degradation. They conclude that these activating groups must be oxygen-containing structures.

Several chemical methods are available for the estimation of peroxide/hydroperoxide groups and they have been widely used for the examination of low molecular weight compounds. The problem in the case of PVC is to find a mutually suitable solvent for the polymer and the reagents used to estimate the peroxide/hydroperoxide groups. Unfortunately, some workers have not indicated how they overcome this difficulty and they have also failed to distinguish between the two functional groups. Of the two basic methods used the one involving the oxidation of Fe^{2+} to Fe^{3+}, followed by a colorimetric determination with thiocyanate[163] or *o*-phenanthroline,[164] is the more likely to be selective for hydroperoxide groups. Iodination methods based on the Mair and Graupner method I[165] have been used on samples in solution[166] and on swollen gels;[167] in both cases it was assumed that hydroperoxide groups were being estimated.

The reactivity of peroxide groups in these conditions is difficult to determine because, although low molecular weight model compounds may prove distinctly unreactive, the reactivity of isolated peroxide groups in PVC chains may be enhanced by the presence of adjacent chlorine atoms. Nagy *et al.*[168] have used two types of iodination reaction to estimate hydroperoxides and hydroperoxides plus peroxides, respectively. In the first case iodine is liberated from added potassium iodide and in the second from hydrogen iodide. They used the method to examine the hydroperoxide/peroxide contents of oxidised PVC samples containing between 0·5 and 3 peroxide units per thousand monomer units. They indicated that the method would probably not be suitable for use with virgin polymers which, they surmised, would have peroxide levels an order of magnitude smaller. This has proved to be a sound judgement in the light of the limited number of results available from the IUPAC Working Party.[140]

4.3.7 Carbonyl groups

Carbonyl groups might be expected to occur as defect structures in PVC, for several reasons. For example, they could result from the decomposition of hydroperoxides or they could form by oxygen attack on a polyene radical during degradation.[169] The preferred approach for their characterisation is infrared spectroscopy, as the high intensity of the C=O stretching mode provides a good level of sensitivity. Furthermore, the variation of the band position with the type of carbonyl group provides a considerable degree of specificity, and it is possible to distinguish between the various major types of carbonyl-containing compounds.

Nevertheless, two important problems occur in practice. The first is the complexity and variability of the absorption found in the ν (C=O) region of the infrared spectrum. Various bands are observed which can only be indicative of artefacts. They include one at about 1770 cm^{-1}, resulting from the oxidation of tetrahydrofuran used as the solvent for casting the thin films required for this type of work, a second at 1748 cm^{-1} indicative of carbonate end-groups formed from percarbonates used as initiators, and a third at about 1722 cm^{-1}, from the phthalate often used as a solvent for the initiator. Even if these artefact bands are recognised their intensities may be such as to mask the defect carbonyl bands being sought. The second problem is a consequence of the non-availability of extinction coefficients to convert measured absorbance values into concentrations of carbonyl groups. This method is therefore most useful for comparative measurements, although it is capable of further development. A careful study of the occurrence of carbonyl defect structures is overdue.

4.4 DEGRADED POLYVINYL CHLORIDE

The problems engendered by the rather poor thermal stability of PVC have been evident from the time that this polymer became commercially important. Foremost among these is discoloration and it was shown almost 30 years ago that this is the result of the formation of sequences of conjugated double bonds. Hydrogen chloride is eliminated at an initiation point, such as a labile chlorine atom, and is then further eliminated sequentially by a 'zipper' mechanism.[170] Discoloration occurs when the conjugated sequence length exceeds about five units.[171] Consequently, a number of workers have developed methods for the characterisation of these polyene sequences in order to study the degradation reaction and also to assess the performances of the many stabilisers that have been developed. These methods fall into three categories: chemical, ultraviolet–visible spectroscopy and resonance Raman spectroscopy.

4.4.1 Chemical methods

The measurement of the hydrogen chloride evolved during the degradation will, clearly, provide an overall measure of the polyene content, although it will not give an indication of the spread of sequence lengths. Three types of method have been used for this purpose; titration, pH measurement and conductometric measurement. The first of these has been the most widely used, because of its inherent simplicity, and a number of variants have been described.[172-182] However, it suffers from a relatively poor sensitivity, the limit of detection of degradation being about 10^{-2}%. The measurement of pH provides more sensitivity and versatility, the former approaching 10^{-4}% in favourable circumstances. This approach has been used by a number of workers.[183-187] Conductometric measurements are more reliable and at least as sensitive as pH measurements,[1,2,188-192] and may well become the most widely used approach. In order to utilise fully the inherent sensitivity of the hydrogen chloride detector system it is necessary to design the degradation apparatus carefully, paying attention to the dead volume and the temperature control. These points, and others, have been discussed in detail by Abbås and Sörvik,[193] who have described a suitable reaction vessel.

The ozonolysis technique, described in Section 4.3.3, has been used to obtain values for the mean conjugated sequence length.[194] The number of scissions, x, upon ozonolysis, can be expressed by the general relationship:[185]

$$\frac{1}{\overline{DP}_n} - \frac{1}{\overline{DP}_n^0} = \frac{x - x_e}{\bar{n}} + \left(\frac{x_e}{1 - x_e} \times \frac{1}{\overline{DP}_n^0}\right)$$

where $\overline{DP}_n$ is the number-average degree of polymerisation of the polymer after ozonolysis, $\overline{DP}_n^0$ is the number-average degree of polymerisation prior to ozonolysis, $\bar{n}$ is the average polyene sequence length and x_e is the number of scissions from polyene sequences beginning at chain ends. Because it is highly probable that $x \gg x_e$, this equation may be simplified, to a good approximation, to

$$\frac{1}{\overline{DP}_n} - \frac{1}{\overline{DP}_n^0} = \frac{x}{\bar{n}}$$

Hence, the number of scissions plotted as a function of the degree of degradation should give a straight line, the inverse slope of which is equal to the average polyene sequence length as was found to be the case by Abbås.[194] However, his work also cast doubt on the validity of this approach for degradation levels lower than about 0·1%, because of the effect of the allylic chlorine initiation sites in the original polymer.

4.4.2 Ultraviolet–visible spectroscopy

The ultraviolet–visible absorption spectra of degraded PVC samples, typified by those given in Fig. 4.7, show a series of maxima superimposed upon a general background. These peaks are indicative of conjugated polyene sequences of different lengths.[195] It is possible to associate a particular peak with a specific sequence length by counting from the one at about 306 nm which is known to be characteristic for $n = 4$. However, the method is less empirical than this approach might suggest because the positions of the peaks at longer wavelengths agree well with those of known polyenes.[196] Furthermore, correlations among the spectra of such compounds show that there is a simple relation of the type $\lambda_{max} = kn^{1/2} + C$ and the use of this equation enables characteristic wavelengths to be established for n values somewhat larger than those that give discrete peaks in the spectrum.[195] The assignments for $n = 12$ and 13 in Fig. 4.7 are made on this basis.

It should also be possible to estimate the concentrations of sequences of different lengths by measuring the absorbance values at the appropriate wavelengths and the use of the Lambert–Beer law. This requires a knowledge of the extinction coefficients for the various polyene sequences in question. This information is also obtainable from measurements on simple polyenes. The results fit an equation of the type $E_{max} = a + b_n$[196,197] and, in particular, Popov and Smirnov[198] have given the specific equation $E_n = 10\,000 + 20\,000(n + 1)$, which has been used by Abbås.[193] In the work of Braun,[199] and most other workers, it has been tacitly assumed that the absorption at a wavelength specific for a particular n value is assignable to this sequence alone and that contributions from $n - 1, n + 1$, etc., are either

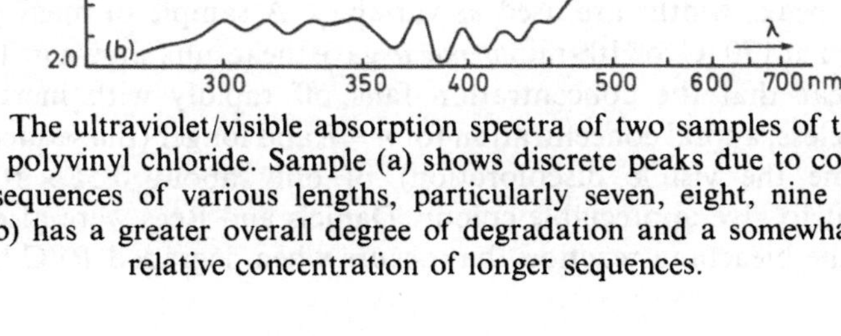

Fig. 4.7 The ultraviolet/visible absorption spectra of two samples of thermally degraded polyvinyl chloride. Sample (a) shows discrete peaks due to conjugated polyene sequences of various lengths, particularly seven, eight, nine and ten. Sample (b) has a greater overall degree of degradation and a somewhat higher relative concentration of longer sequences.

negligible or constant. This is highly implausible, not least because the increasing loss of resolution of the peaks as *n* increases suggests a steadily increasing overlap of the peaks.

The problems involved in quantitative work have been examined in greater depth by Marks *et al.*[184] and by Daniels and Rees.[200] The former workers pointed out that, in principle, for a particular *n* value, there are four types of polyene structures present. They correspond (1) to the sequence beginning at the end of the polymer chain, (2) beginning and terminating in the middle of the chain, (3) terminating at a branch point and (4) beginning and terminating at branch points:

$$CH_2{=}CH{-}(CH{=}CH)_{n-2}{-}CH{=}CHR_1 \quad (1)$$

$$R_1CH{=}CH{-}(CH{=}CH)_{n-2}{-}CH{=}CHR_2 \quad (2)$$

$$R_1CH{=}CH{-}(CH{=}CH)_{n-2}{-}CH{=}CR_2R_3 \quad (3)$$

$$R_1R_2C{=}CH{-}(CH{=}CH)_{n-2}{-}CH{=}CR_3R_4 \quad (4)$$

By comparison with a simple polyene, such as 1,3,5,7,9-decapentaene, these structures have, respectively, one, two, three and four alkyl groups attached to the conjugated sequence. This will give a bathochromic (red) shift of the absorption maxima, the magnitude of which may be calculated from an application of Woodward's rules.[201,202] On this basis, Marks *et al.* used shifts, by comparison with the parent polyene structure, of 5, 10, 15 and 20 nm and also, for simplicity, assumed that these shifts would be independent of *n* and that all four structures (1) to (4) are present in equal concentration. They were then able to obtain calculated spectra that were in reasonable agreement with their experimental data, but they did not give concentrations for the various polyenes.

Daniels and Rees[200] used computer curve fitting and, in so doing, were able to avoid some of the constraints encountered by Marks *et al.* because of their simplifying assumptions. They also took into account the fact that alkyl substitution causes both a bathochromic shift and a broadening of the peaks.[203,204] In the optimisation process, to obtain the best fit between measured and calculated spectra both the relative polyene concentrations and the peaks widths are used as variables. A sample of mass polymer, degraded at 170 °C for 100 min *in vacuo* gave the results shown in Table 4.5. It is clear that the concentration falls off rapidly with increasing *n*; nevertheless, a total concentration for $n = 8$ and longer (the sequences that determine the visible discoloration) of only about $0{\cdot}55 \times 10^{-5}$ M, is sufficient to give appreciable colour. Daniels and Rees were also able to follow the bleaching reaction that occurs when degraded PVC is treated

Table 4.5 Concentrations of polyenes of different sequence lengths in a PVC sample degraded *in vacuo* at 170 °C for 100 min

n	*Molar concentration* × 10^5
3	1·66
4	1·22
5	0·82
6	0·41
7	0·20
8	0·17
9	0·12
10	0·11
11	0·072
12	0·060

with α-hydroperoxytetrahydrofuran.[205] Their results show that the peroxide attack occurs on the first double bond in the polyene sequence.

The work of Daniels and Rees probably represents the ultimate that may be achieved in obtaining information from ultraviolet–visible spectra. The precision and accuracy of the results should be appreciably greater than those from the simple approach of assuming that the absorbance at one particular wavelength is a measure of the concentration of a specific sequence length. Even this simple approach will probably provide more realistic results than those from the ozonolysis technique. Abbås[194] has compared the two and finds that the latter gives the larger values for the mean polyene sequence length, although when various corrections are applied the difference is not excessive.

The majority of ultraviolet–visible measurements reported in the literature have been made on samples degraded in solution or as thin films. When degradation occurs in thick specimens, or with other conditions where ready dispersal of the evolved hydrogen chloride is not achieved, the intensity of absorption in the visible spectrum is greater and, frequently, the poorly resolved conjugated polyene bands are superimposed upon a broad peak centred at about 450 nm. It is not clear if this is the result of a very abnormal distribution of sequence lengths, in which *n* values in the range $n = 10$–20 dominate, or if there is a fundamentally different explanation. Schlimper has shown[206,207] that hydrogen chloride will add to conjugated polyenes to give strongly coloured carbonium ions having the general structure $-\!\!\!\!\![CH{-}\overset{+}{C}H{-}CH{=}CH{-}CH{=}CH]_n\!\!-\!Cl^-$. This complex formation is reversible; the complexes can be destroyed by treatment with ammonia and regenerated by the addition of HCl. Alternatively, Onozuka and Asahini[208] have suggested that charge transfer complexes are formed; however, the resonance Raman spectra of degraded PVC samples show no evidence for such complexes.[209] Until this fundamental point has been resolved considerable caution should be exercised in the quantitative interpretation of the spectra of samples degraded using conditions which give the broad band at about 450 nm.

4.4.3 Resonance Raman spectroscopy

In the measurement of Raman spectra, as the laser exciting frequency approaches that of an allowed electronic transition in the molecule being examined those normal modes that are vibronically active in the electronic transition exhibit a pronounced enhancement of their intensities, by a factor as great as 10^4 or 10^5. This is known as the resonance Raman effect and most, but not all, of the known cases that have been studied in detail involve totally symmetric vibrational modes and intense electronic transitions. The conjugated polyenes provide a very good example as they give intense bands arising from the C—C (ν_1) and C=C (ν_2) stretching modes. The spectrum of β-carotene is shown in Fig. 4.8.

Liebman *et al.*[210] found bands at 1124 cm^{-1} and 1511 cm^{-1} in the Raman spectra of PVC samples subjected to various heat treatments. The intensities of the two bands were greater the higher the temperature to which the samples were taken and, for a given temperature, were more intense for slow cooling than for quenching. They concluded that the appearance of the two bands was indicative of what had occurred during degradation rather than a predegradative structural reorientation, and they suggested that conjugated polyene sequences were responsible. They did not give quantitative information about the extent to which conjugated polyene sequences were formed but it is evident that, with the conditions employed, the level must have been below about 0·01 %. In the subsequent publication,[211] they refer to work by Rimai *et al.*[212] on intense bands at 1158 and 1527 cm^{-1} for β-carotene and at 1158 and 1515 cm^{-1} for lycopene, which were shown to be of enhanced intensity because of the resonance Raman effect. They then concluded that resonance enhancement is the reason why the 1124 and 1511 cm^{-1} bands appear so strongly in degraded PVC samples containing low levels of conjugated polyene sequences.

Positive proof that these two bands arise from the resonance Raman effect was provided by Gerrard and Maddams[213] on the basis of their intensities, their dependence on the wavelength of the exciting line in

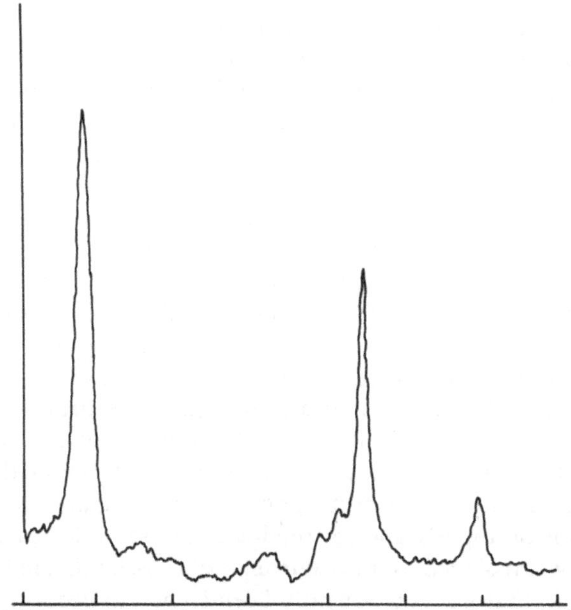

Fig. 4.8 The resonance Raman spectrum of β-carotene, a model compound for the conjugated structures present in degraded polyvinyl chloride. The peak at 1527 cm^{-1} is the C═C stretching (ν_2) vibration and the one at 1158 cm^{-1} is the C–C stretching (ν_1) mode.

relation to the visible absorption spectra of the degraded polymers, the appearance of strong harmonic and combination bands, and the systematic variation of their half-widths. By considering the published literature values for ν_2 for various low molecular weight conjugated polyenes they were able to show that there is a linear relation between ν_2 and $\log n$. This opens the way to the use of the technique, using a range of exciting wavelengths, to characterise conjugated polyene sequences of different lengths. With the range of exciting wavelengths available to them they were able to cover the range $n = 9$ to 17, with clear indications that rather longer sequences were also present. Using one of the bands of the solvent, tetrahydrofuran, in which the degraded samples were examined, they were able to obtain a comparative measure of the degradation using various conditions by taking the ratio of the intensity of ν_2 and the solvent band. By this means it was shown that for a series of samples degraded to known extents, on the basis of hydrogen chloride estimations, at 160 and 190 °C, the average polyene sequence length increases with time and temperature of degradation. Shortly after the publication of this work the rather less detailed study of Peitscher and Holtrup,[143] already discussed in Section 4.3.3, appeared. It served to confirm the findings of Gerrard and Maddams, that the technique had very high sensitivity and considerable potential.

This latter surmise was confirmed in two subsequent publications by Gerrard and Maddams, one in collaboration with Millán and colleagues at the Plastics and Rubber Institute, Madrid. The first of these[214] related to studies on two samples of different syndiotacticity degraded by γ-irradiation. The relative concentrations of various sequences over the range $n = 9$ to 17 were measured as a function of radiation dosage and syndiotacticity. The results indicate that the concentration of the longer conjugated sequences rises steadily with radiation dose in the case of the less syndiotactic polymer, a commercial material prepared by the mass route. With the more syndiotactic polymer, prepared at $-30\,°C$, the rate of increase falls off rapidly and for the highest doses a decrease is observed. This effect was assigned to the occurrence of crosslinking in the polymer, a phenomenon that has subsequently been shown to occur as the result of photodegradation and to be associated with the deterioration in the mechanical properties of polymers.[215]

The second study[68] covered four polymer fractions of appreciably different tacticity and tacticity distribution, prepared by selective extraction of samples polymerised at temperatures between +60 and $-50\,°C$. They were degraded thermally at 180 °C in the solid state to conversions in the range 1·6–5 %. The results show that the type of sequence distribution depends markedly on both the syndiotacticity and its distribution. The low temperature material had the greatest syndiotacticity and was also somewhat non-Bernoullian, and both factors may influence the distribution and also the marked propensity to crosslinking that was observed.

Gerrard and Maddams[216] have recently undertaken a background study of the method. They observed that their previous linear relation between ν_2 and $\log n$ could only be valid for a limited range of n values because extrapolation to $n = \infty$ gives $\nu_2 = -\infty$, an impossible situation. By using a ν_2 for *trans*-polyacetylene, which contains long conjugated sequences, in conjunction with the previous data, they were able to obtain an experimental relation between ν_2 and n. This revised relation gives somewhat greater n values for a given ν_2. With a ν_2 value of 1498 cm^{-1} the value of n increases from 17·0 to 19·3 and with $\nu_2 = 1491$ cm^{-1} the n value changes from 19·0 to 23·0.

The fractional n values that are obtained are a consequence of the finite resolution of the method. This was assessed in two ways, by experiment and by calculation. With the former, ν_2 was measured as a function of exciting wavelength at frequent intervals in the range 420–625 nm, using a tunable dye laser. There were no discrete steps in the curve, indicating that there is overlap between the ν_2 bands from sequences of n and $n + 1$, and possibly others. This overlap was examined in some detail by calculations based on the excitation profile of the model compound β-carotene. They show that with $n = 9$, contributions from $n = 10$ and 11 exceed 10% of that of $n = 9$. With $n = 16$, similar contributions occur from $n = 15$ and 17 to 26 and with $n = 20$ the specificity decreases still further, such that $n = 18$, 19 and 21–45 are involved. The net effect is to cause an asymmetry of the ν_2 profile, whose magnitude depends on the n value and the relative concentrations of the sequence lengths excited by a particular incident wavelength. In general, the measured n value is somewhat in excess of the true value, but the error only becomes serious for n values around 20 and then only if the concentration of still longer sequences is comparable. In practice, this is unlikely because of the crosslinking reaction.

As yet, there has been little progress towards developing the method for absolute quantitative measurements. This requires calibration data, and as low molecular weight model polyenes with the requisite conjugated sequence lengths are not available an indirect approach will be necessary. There are some indications that this will prove possible. Tonks and Page[217] have given a very simple relationship between absorption and resonance Raman intensities, using expressions derived by Hizhnyakov and Tehver,[218] and although Hassing and Mortensen[219] have subsequently suggested that it is strictly applicable at zero temperature only, it ought to provide the basis for a quantitative approach. As the molar absorptivity in the electronic spectrum is proportional to the number of conjugated double bonds[196,197] it should be possible to derive these for larger n values and then to relate them to the resonance Raman intensities. It is therefore reasonable to anticipate that the resonance Raman method, which has already proved valuable for characterising longer conjugated sequences than those amenable to examination by electronic spectroscopy, is capable of further development, particularly for quantitative work.

REFERENCES

1. Millán, J., Madruga, E. L., Bert, M. and Guyot, A. *J. Polym. Sci., Polym. Chem. Ed.*, **11** (1973), 3299.

2. Millán, J. and Carranza, M. *Rev. Plast. Mod.*, **30** (1975), 67.
3. Millán, J., Madruga, E. L. and Martinez, G. *Angew. Makromol. Chem.*, **45** (1975), 172.
4. Millán, J. and Guzman, J. *Eur. Polym. J.*, **12** (1976), 299.
5. Fuller, C. S. *Chem. Rev.*, **26** (1940), 143.
6. Krimm, S. and Liang, C. Y. *J. Polym. Sci.*, **22** (1956), 95.
7. Mizushima, S., Shimanouchi, T., Nakamura, K., Hayashi, M. and Tsuchiya, S. *J. Chem. Phys.*, **26** (1957), 970.
8. Grisenthwaite, R. J. and Hunter, R. F. *Chem. Ind.* (1958), 719.
9. Grisenthwaite, R. J. and Hunter, R. F. *Chem. Ind.* (1958), 1513.
10. Krimm, S., Berens, A. R., Folt, V. L. and Shipman, J. J. *Chem. Ind.* (1958), 1512.
11. Shimanouchi, T., Tsuchiya, S. and Mizushima, S. *J. Chem. Phys.*, **30** (1959), 1365.
12. Shimanouchi, T. and Tasumi, M. *Bull. Chem. Soc. Japan*, **34** (1961), 359.
13. Shimanouchi, T. and Tasumi, M. *Spectrochim. Acta*, **17** (1961), 755.
14. Shipman, J. J., Folt, V. L. and Krimm, S. *Spectrochim. Acta*, **18** (1962), 1603.
15. Krimm, S., Folt, V. L., Shipman, J. J. and Berens, A. R. *J. Polym. Sci.*, **A1, 1** (1963), 2621.
16. Shimanouchi, T., Tsuchiya, S. and Mizushima, S. *Kobunshi*, **8** (1959), 202.
17. Krimm, S., Folt, V. L., Shipman, J. J. and Berens, A. R. *J. Polym. Sci.*, **B2** (1964), 1009.
18. Pohl, H. U. and Hummel, D. O. *Makromol. Chem.*, **113** (1968), 190.
19. Pohl, H. U. and Hummel, D. O. *Makromol. Chem.*, **113** (1968), 203.
20. Baruya, A., Booth, A. D., Maddams, W. F., Grasselli, J. G. and Hazle, M. A. S. *J. Polym. Sci., Polym. Lett. Ed.*, **14** (1976), 329.
21. Robinson, M. E. R., Bower, D. I. and Maddams, W. F. *Polymer*, **19** (1978), 773.
22. Maddams, W. F. and Tooke, P. B. *J. Macromol. Sci., Chem.*, **A17** (1982), 951.
23. Maddams, W. F. and Southon, M. J. *Spectrochim. Acta*, **A38** (1982), 459.
24. Shimanouchi, T., Tasumi, M. and Abe, Y. *Makromol. Chem.*, **86** (1965), 43.
25. Doskocilova, D., Stokr, J., Schneider, B., Pivcova, H., Kolinsky, M., Petranek, J. and Lim, D. *J. Polym. Sci.*, **C16** (1967), 215.
26. Schneider, B., Stokr, J., Doskocilova, D., Kolinsky, M., Sykora, S. and Lim, D. *J. Polym. Sci.*, **C16** (1968), 3891.
27. Opaskar, C. G. and Krimm, S. *J. Polym. Sci.*, **A-2, 7** (1969), 57.
28. Rubčič, A. and Zerbi, G. *Macromolecules*, **7** (1974), 754.
29. Rubčič, A. and Zerbi, G. *Macromolecules*, **7** (1974), 759.
30. Germar, H., Hellwege, K.-H. and Johnsen, U. *Makromol. Chem.*, **60** (1963), 106.
31. Nakajima, A., Hamada, H. and Hayashi, S. *Makromol. Chem.*, **95** (1966), 40.
32. Krimm, S. and Enomoto, S. *J. Polym. Sci.*, **A1, 2** (1964), 669.
33. Krimm, S., Shipman, J. J., Folt, V. L. and Berens, A. R. *J. Polym. Sci.*, **B3** (1965), 275.
34. Koenig, J. L. and Druesedow, D. *J. Polym. Sci.*, **A2, 7** (1969), 1075.
35. Robinson, M. E. R., Bower, D. I. and Maddams, W. F. *Polymer*, **17** (1976), 355.

36. Moore, W. H. and Krimm, S. *Makromol. Chem. Suppl. 1* (June 1975), 491.
37. Morrow, B. A. and Cody, I. A. *J. Phys. Chem.*, **77** (1973), 1465.
38. Vandeginste, B. G. M. and De Galan, L. *Anal. Chem.*, **47** (1975), 2124.
39. Maddams, W. F. *Appl. Spectrosc.*, **34** (1980), 245.
40. Baker, C., Cockerill, I. P., Kelsey, J. E. and Maddams, W. F. *Spectrochim. Acta*, **34A** (1978), 673.
41. Baker, C., Johnson, P. S. and Maddams, W. F. *Spectrochim. Acta*, **34A** (1978), 683.
42. Baker, C., Maddams, W. F., Grasselli, J. G. and Hazle, M. A. S. *Spectrochim. Acta*, **34A** (1978), 761.
43. Goodfield, J. E., George, W. O. and Maddams, W. F. Unpublished results.
44. Bovey, F. A. and Tiers, G. V. D. *J. Polym. Sci.*, **44** (1960), 173.
45. Sewell, P. R. The nuclear magnetic resonance spectra of polymers. In: *Annual Review of NMR Spectroscopy*, Vol. 1, Ed. E. F. Mooney, Academic Press, New York, 1968.
46. Shimanouchi, T. *Pure Appl. Chem.*, **12** (1966), 287.
47. Johnsen, U. *J. Polym. Sci.*, **54** (1961), 56.
48. Tincher, W. C. *J. Polym. Sci.*, **62** (1962), S148.
49. Yoshino, T. and Komiyama, J. *J. Polym. Sci.*, **B3** (1965), 311.
50. Bovey, F. A., Anderson, E. W., Douglass, D. C. and Manson, J. A. *J. Chem. Phys.*, **39** (1963), 1199.
51. Doskocilova, D. *J. Polym. Sci.*, **B2** (1964), 421.
52. Heatley, F. and Bovey, F. A. *Macromolecules*, **2** (1969), 241.
53. Abdel-Alim, A. H. *J. Appl. Polym. Sci.*, **19** (1975), 1227.
54. Robinson, M. E. R., Bower, D. I., Maddams, W. F. and Pyszora, H. *Makromol. Chem.*, **179** (1978), 2895.
55. Carman, C. J., Tarpley, A. R. and Goldstein, J. H. *Macromolecules*, **4** (1971), 445.
56. Inoue, Y., Ando, I. and Nishioka, A. *Polym. J.*, **3** (1972), 246.
57. Carman, C. J., Tarpley, A. R. and Goldstein, J. H. *J. Am. Chem. Soc.*, **93** (1971), 2864.
58. Carman, C. J. *Macromolecules*, **6** (1973), 725.
59. Schaefer, J. *Macromolecules*, **4** (1971), 110.
60. Liu, N.-I., Tong, S. N. and Koenig, J. L. *J. Appl. Polym. Sci.*, **25** (1980), 2205.
61. Elgert, K.-F. and Kosfeld, R. Paper presented at the Third International Symposium on Poly(vinyl chloride), Cleveland, Ohio, USA, 10–15 August, 1980.
62. King, J., Bower, D. I., Maddams, W. F. and Pyszora, H. *Makromol. Chem.*, **184** (1983), 879.
63. Pham, Q. T., Millán, J. and Madruga, E. L. *Makromol. Chem.*, **175** (1974), 945.
64. Cais, R. E. and Brown, W. L. *Macromolecules*, **13** (1980), 801.
65. Coleman, B. D. and Fox, T. G. *J. Chem. Phys.*, **38** (1963), 1065.
66. Coleman, B. D. and Fox, T. G. *J. Polym. Sci.*, **A1, 1** (1963), 3183.
67. Bovey, F. A. Chapter VIII in: *High Resolution NMR of Macromolecules*, Academic Press, New York, 1972.
68. Martinez, G., Mijangos, C., Millán, J., Gerrard, D. L. and Maddams, W. F. *Makromol. Chem.*, **180** (1979), 2937.

69. Fordham, J. W. L. *J. Polym. Sci.*, **39** (1959), 321.
70. Schneider, B., Stokr, J., Doskocilova, D., Sykora, S., Jakes, J. and Kolinsky, M. *J. Polym. Sci.*, **C22** (1969), 1073.
71. Caraculacu, A. A. *Pure and Appl. Chem.*, **53** (1981), 385.
72. Haslam, J., Willis, H. A. and Squirrel, D. C. M. *Identification and Analysis of Plastics*, 2nd ed., Illiffe Books, London, 1972, 374.
73. Baker, C. and Maddams, W. F. *Makromol. Chem.*, **177** (1976), 437.
74. Cotman, J. *Ann. N.Y. Acad. Sci.*, **57** (1953), 417.
75. Nakajima, A., Hamada, H. and Hayashi, S. *Makromol. Chem.*, **95** (1966), 40.
76. Binder, L. Doctoral Thesis, Techn. Hochschule Vienna, 1962.
77. Braun, D. and Scharek, W. *Angew. Makromol. Chem.*, **7** (1969), 121.
78. Barnett, G. M., Ross, F. L. and Hay, J. N. *J. Polym. Sci.*, **5** (1967), 1467.
79. Baijal, M. D., Wang, T. S. and Dillen, R. M. *J. Macromol. Sci., Chem.*, **A4** (1970), 965.
80. Boccato, G., Rigo, A., Talamini, G. and Zilio-Grandi, F. *Makromol. Chem.*, **108** (1967), 218.
81. Baker, C., Maddams, W. F., Park, G. S. and Robertson, B. *Makromol. Chem.*, **165** (1973), 321.
82. Rigo, A., Palma, G. and Talamini, G. *Makromol. Chem.*, **153** (1972), 219.
83. Preedy, J. E. *Br. Polym. J.*, **5** (1973), 13.
84. Baker, C., David, P. and Maddams, W. F. *Makromol. Chem.*, **180** (1979), 975.
85. McRae, M. A. and Maddams, W. F. *Makromol. Chem.*, **177** (1976), 449.
86. Dorman, D. E., Otocka, E. P. and Bovey, F. A. *Macromolecules*, **5** (1972), 574.
87. Randall, J. C. *J. Polym. Sci., Polym. Phys. Ed.*, **11** (1973), 275.
88. Axelson, D. E., Mandelkern, L. and Levy, G. C. *Macromolecules*, **10** (1977), 557.
89. Abbås, K. B., Bovey, F. A. and Schilling, F. C. *Makromol. Chem. Suppl.*, **1** (June 1975), 227.
90. Carrega, M. E. *Pure and Appl. Chem.*, **49** (1977), 569.
91. Bovey, F. A., Abbås, K. B., Schilling, F. C. and Starnes, W. H. *Macromolecules*, **8** (1975), 437.
92. Park, G. S. and Saleen, M. *Polymer Bull.*, **1** (1979), 409.
93. Park, G. S. *J. Macromol. Sci., Phys.*, **B14** (1977), 151.
94. Starnes, W. H., Schilling, F. C., Abbås, K. B., Cais, R. E. and Bovey, F. A. *Macromolecules*, **12** (1979), 556.
95. Starnes, W. H., Schilling, F. C., Abbås, K. B., Plitz, I. M., Hartless, R. L. and Bovey, F. A. *Macromolecules*, **12** (1979), 13.
96. Starnes, W. H., Schilling, F. C., Plitz, I. M., Cais, R. E., Freed, D. J. and Bovey, F. A. Paper presented at the Third International Symposium on Poly(vinyl chloride), Cleveland, Ohio, USA, 10–15 August, 1980.
97. Park, G. S., Maddams, W. F. and Pyszora, H. Unpublished results.
98. Sörvik, E. M. *J. Polym. Sci., Polym. Lett. Ed.*, **14** (1976), 735.
99. Sörvik, E. M. and Hjertberg, T. *J. Macromol. Sci., Chem.*, **A11** (1977), 1349.
100. Uuskyla, L. and Sörvik, E. M. *J. Polym. Sci., Polym. Lett. Ed.*, **15** (1977), 387.
101. Hjertberg, T. and Sörvik, E. M. *J. Polym. Sci., Polym. Chem. Ed.*, **16** (1978), 645.

102. Hjertberg, T. and Sörvik, E. M. Chemical methods for characterising defect structures in PVC, Report of the IUPAC PVC Working Party, *Pure and Appl. Chem.*, in press.
103. Schwenk, U., Cavagna, F., Lömker, F., König, I. and Streitberger, H. *J. Appl. Polym. Sci.*, **23** (1979), 1589.
104. Michel, A. Report to the Third International Symposium on Poly(vinyl chloride), Cleveland, Ohio, USA, 10–15 August, 1980.
105. Bowmer, T. N. and O'Donnell, J. H. *Polymer*, **18** (1977), 1032.
106. Bowmer, T. N., O'Donnell, J. H., Park, G. S. and Saleen, M. Paper presented at the Third International Symposium on Poly(vinyl chloride), Cleveland, Ohio, USA, 10–15 August, 1980.
107. Schröder, E. and Byrdy, M. *Plaste Kautschuk*, **14** (1977), 757.
108. Ahlstrom, D. A., Liebman, S. A. and Abbås, K. B. *J. Polym. Sci., Polym. Chem. Ed.*, **14** (1976), 2479.
109. Liebman, S. A., Ahlstrom, D. H. and Starnes, W. H. Paper presented at the Third International Symposium on Poly(vinyl chloride), Cleveland, Ohio, USA, 10–15 August, 1980.
110. Bovey, F. A., Schilling, F. C. and Starnes, W. H. *Polym. Prepr., Am. Chem. Soc. Div. Polym. Chem.*, **20** (1979), 160.
111. Starnes, W. H., Schilling, F. C., Plitz, I. M., Cais, R. E. and Bovey, F. A. *Polym. Bull.*, **4** (1981), 555.
112. Abbås, K. B. *Pure and Appl. Chem.*, **53** (1981), 411.
113. Thurmond, C. D. and Zimm, B. H. *J. Polym. Sci.*, **8** (1952), 477.
114. Zimm, B. H. and Kulb, R. W. *J. Polym. Sci.*, **37** (1959), 19.
115. Hengstenberg, J. and Schuch, E. *Makromol. Chem.*, **74** (1964), 55.
116. Gautron, R. and Wippler, C. *J. Chem. Phys.*, **58** (1961), 754.
117. Kobayashi, T. *Bull. Chem. Soc. Japan*, **35** (1962), 637.
118. Lyngaae-Jorgensen, J. *J. Chromatogr. Sci.*, **9** (1971), 331.
119. Crugnola, A. and Danusso, F. *J. Polym. Sci.*, **B6** (1968), 535.
120. Andersson, K. B. and Sörvik, E. M. *J. Polym. Sci.*, **C33** (1971), 247.
121. De Vries, A. J., Bonnebat, C. and Carrega, M. *Pure Appl. Chem.*, **25** (1971), 209.
122. Hlousek, M. and Lanikova, J. *J. Polym. Sci.*, **C16** (1967), 935.
123. Morikawa, T. *Kagaku to Kogyo (Osaka)*, **41** (1967), 169.
124. Boissel, J. *J. Appl. Polym. Sci.*, **21** (1977), 855.
125. Lee, T. S., Kolthoff, I. M. and Mairs, M. A. *J. Polym. Sci.*, **3** (1948), 66.
126. Müller, K. *Functional Group Determination of Olefinic and Acetylenic Unsaturation*, Academic Press, New York, 1975, 132.
127. Ahrens, W., Brandstetter, F., Hildenbrand, P. and Simak, P. Paper presented at the Third International Symposium on Poly(vinyl chloride), Cleveland, Ohio, USA, 10–15 August, 1980.
128. Park, G. S. and Saremi, A. A. Unpublished results.
129. Merz, W. and Muller, K. *Microchim. Acta.*, **6** (1974), 60.
130. Braun, D. and Quarg, W. *Angew. Makromol. Chem.*, **29/30** (1973), 163.
131. Lindenschmidt, G. *Angew. Makromol. Chem.*, **47** (1975), 79.
132. Michel, A., Schmidt, G., Castaneda, E. and Guyot, A. *Angew. Makromol. Chem.*, **47** (1975), 61.
133. Abbås, K. B. and Sörvik, E. M. *J. Appl. Polym. Sci.*, **19** (1975), 2991.

134. Minsker, K. S., Lisitskii, V. V., Gataullin, R. F., Saveliev, A. P., Gagarina, L. P. and Gorbatsevskaia, I. I. *Plast. Massy*, **2** (1980), 6.
135. Minsker, K. S., Berlin, A. A., Lisitskii, V. S. and Kolesov, S. V. *Visokomol. Soed.*, **A19** (1977), 32.
136. Caraculacu, A. A. and Bezdadea, E. C. *Revue Roum. Chim.*, **14** (1969), 1171.
137. Caraculacu, A. A. and Bezdadea, E. C. *J. Polym. Sci., Polym. Chem. Ed.*, **15** (1977), 611.
138. Pétiaud, R. and Pham, Q. T. *Makromol. Chem.*, **178** (1977), 741.
139. Schwenk, U., König, I., Cavagna, F. and Wrackmeyer, B. *Angew. Makromol. Chem.*, **83** (1979), 183.
140. IUPAC Working Party on the Characterisation of Defects in the Molecular Structure of Poly(vinyl chloride), unpublished results.
141. Kohn, P., Marechal, C. and Verdu, J. *Anal. Chem.*, **51** (1979), 1000.
142. Sondheimer, F., Ben Efrain, D. A. and Wolovsky, R. *J. Am. Chem. Soc.*, **83** (1961), 1675.
143. Peitscher, G. and Holtrup, W. *Angew. Makromol. Chem.*, **47** (1975), 111.
144. Bassez, C. Thesis, University of Lille, France, 1978.
145. Bier, G. and Krämer, H. *Kunststoffe*, **46** (1956), 468.
146. Enomoto, S. *J. Polym. Sci.*, **A1, 7** (1969), 1255.
147. Caraculacu, A. A., Buruiana, E. C. and Robila, G. *J. Polym. Sci., Polym. Chem. Ed.*, **16** (1978), 2741.
148. Hansen, K. and Ugelstad, J. *Makromol. Chem.*, **180** (1979), 2423.
149. Minsker, K. S., Lisitskii, V. V., Kolinsky, M., Vizmazal, Z., Bort, D. H., Lebedev, V. P., Petzena, A. I. and Il'kaeva, E. M. *Plast. Massy*, **9** (1977), 44.
150. Minsker, K. S., Lisitskii, V. V., Gataullin, R. F. and Zaikov, G. E. *Visokomol. Soed.*, **B20** (1978), 571.
151. Burille, P. Thesis, University Claude Bernard, Lyons, 1979.
152. Robila, G., Buruiana, E. C. and Caraculacu, A. A. *Eur. Polym. J.*, **13** (1977), 21.
153. Caraculacu, A. A. *J. Polym. Sci.*, **A1, 4** (1966), 1839.
154. Michel, A., Burille, P. and Guyot, A. Paper presented at the IUPAC 26th International Symposium, Mainz, 1979.
155. Buruiana, E. C., Barbinta, V. T. and Caraculacu, A. A. *Eur. Polym. J.*, **13** (1977), 311.
156. Saremi, A. A. Ph.D. Thesis, University of Wales Institute of Science and Technology, Cardiff, 1981.
157. Hjertberg, T. Thesis, Chalmers University of Technology, Gothenburg, 1982.
158. Berens, A. R. *Polym. Eng. Sci.*, **14** (1974), 318.
159. Crawley, S. and McNeill, I. C. *J. Polym. Sci., Polym. Chem. Ed.*, **16** (1978), 2593.
160. Shimizu, A. and Ohtsu, T. *Kogyo Kagaku Zasshi*, **67** (1964), 966.
161. Mitani, K., Ogata, T., Awaya, H. and Tomari, Y. *J. Polym. Sci.*, **A1, 13** (1975), 2813.
162. Svetly, J., Lukas, R. and Kolinsky, M. *Makromol. Chem.*, **180** (1979), 1363.
163. Zeppenfeld, G. *Makromol. Chem.*, **90** (1966), 109.
164. Getmanenko, E. N. and Perepletchikova, E. *Fiz. Khim. Metody Anal.*, **1** (1976), 75.
165. Mair, R. D. and Graupner, A. J. *Anal. Chem.*, **36** (1964), 194.

166. Decker, C. *J. Appl. Polym. Sci.*, **20** (1976), 3321.
167. Scott, G., Tahan, M. and Vyvoda, J. C. *Chem. Ind. (London)*, **20** (1976), 903.
168. Nagy, T. T., Kelen, T., Turcsanyi, B. and Tüdos, F. *Plaste u. Kautschuk*, **23** (1976), 894.
169. Abbås, K. B. and Sörvik, E. M. *J. Appl. Polym. Sci.*, **17** (1973), 3577.
170. Arlman, A. J. *J. Polym. Sci.*, **12** (1954), 547.
171. Geddes, W. C. *Rubber Chem. Technol.*, **40** (1967), 177.
172. Druesedow, D. and Gibbs, C. F. *Nat. Bur. Stand. Circular*, **525** (1953), 69.
173. Havens, C. B. *Nat. Bur. Stand. Circular*, **525** (1953), 107.
174. Imoto, M. and Otsu, T. *J. Inst. Polytech., Osaka City Univ., Ser. C*, **4** (1953), 1.
175. Rieche, A., Grimm, A. and Mücke, H. *Kunststoffe*, **52** (1962), 265.
176. Bengough, W. I. and Sharpe, H. M. *Makromol. Chem.*, **66** (1963), 31.
177. Stepek, J., Vymazal, Z. and Dolezel, B. *Mod. Plast.*, **40** (1963), 146.
178. Talamini, G., Cinque, G. and Palma, G. *Mat. Plast. Mod.*, **32** (1966), 50.
179. Guyot, A. and Benevise, J. P. *J. Appl. Polym. Sci.*, **6** (1962), 98.
180. ASTM Method D793-49 (1965).
181. Vymazal, Z., Dolezel, B. and Stepek, J. *Kunststoffe*, **56** (1966), 86.
182. Minsker, K. S., Krats, E. O. and Pakhomova, I. K. *Vysokomol. Soedin*, **A12** (1970), 483.
183. Cittadini, A. and Palillo, R. *Chim. Ind. (Milan)*, **41** (1966), 980.
184. Marks, G. C., Benton, J. L. and Thomas, C. M. *S.C.I. Monograph*, **26** (1967), 204.
185. Geddes, W. C. *Eur. Polym. J.*, **3** (1967), 267.
186. Laczkó, M. *Plast. Kaut.*, **17** (1970), 25.
187. Stapfer, C. H. and Granick, J. D. *Proc. 161st A.C.S. Meeting, Organic Coatings and Plastic Chemistry*, **31** (1971), 204.
188. Arlman, E. J. *J. Polym. Sci.*, **12** (1954), 547.
189. Baum, B. and Wartman, L. H. *J. Polym. Sci.*, **28** (1958), 537.
190. Talamini, G. and Pezzin, G. *Makromol. Chem.*, **39** (1960), 26.
191. Braun, D. and Thallmaier, M. *Kunststoffe*, **56** (1960), 80.
192. Guyot, A. and Bert, M. *Polymer Preprints*, **12** (1971), 303.
193. Abbås, K. B. and Sörvik, E. M. *J. Appl. Polym. Sci.*, **17** (1973), 3567.
194. Abbås, K. B. *J. Macromol. Sci. Chem.*, **A12** (1978), 479.
195. Braun, D. and Thallmaier, M. *Makromol. Chem.*, **99** (1966), 59.
196. Scott, A. I. *Interpretation of the Ultraviolet Spectra of Natural Products, International Series of Monographs on Organic Chemistry*, Vol. 7, Pergamon Press, New York, 1964.
197. Braude, E. A. *J. Chem. Soc.* (1950), 379.
198. Popov, K. P. and Smirnov, L. V. *Optika Spektroskopiya*, **14** (1963), 787.
199. Braun, D. *Pure Appl. Chem.*, **26** (1971), 173.
200. Daniels, V. D. and Rees, N. H. *J. Polym. Sci., Polym. Chem. Ed.*, **12** (1974), 2115.
201. Woodward, R. B. *J. Amer. Chem. Soc.*, **64** (1942), 72.
202. Hirayama, K. *J. Amer. Chem. Soc.*, **77** (1955), 373, 382.
203. Naylor, P. and Whiting, M. C. *J. Chem. Soc.* (1955), 3037.
204. Ferguson, L. N. *Chem. Rev.*, **43** (1948), 385.
205. Matsumoto, T., Mune, I. and Watatani, S. *J. Polym. Sci.*, **A1, 7** (1969), 1609.
206. Schlimper, R. *Plaste Kautschuk*, **13** (1966), 196.

207. Schlimper, R. *Plaste Kautschuk*, **14** (1967), 657.
208. Onozuka, M. and Asahini, M. *J. Macromol. Sci.*, **C3** (1969), 235.
209. Gerrard, D. L. and Maddams, W. F. To be published.
210. Liebman, S. A., Foltz, C. R., Reuwer, J. F. and Obremski, R. J. *Macromolecules*, **4** (1971), 134.
211. Liebman, S. A., Ahlstrom, D. H., Quinn, E. J., Geigley, A. G. and Meluskey, J. T. *J. Polym. Sci.*, **A1, 9** (1971), 1921.
212. Rimai, L., Kilponen, E. G. and Gill, D. *J. Amer. Chem. Soc.*, **92** (1970), 3824.
213. Gerrard, D. L. and Maddams, W. F. *Macromolecules*, **8** (1975), 54.
214. Gerrard, D. L. and Maddams, W. F. *Macromolecules*, **10** (1977), 1221.
215. Biggin, I. S., Gerrard, D. L. and Williams, G. E. *J. Vinyl Tech.*, **4** (1982), 150.
216. Gerrard, D. L. and Maddams, W. F. *Macromolecules*, **14** (1981), 1356.
217. Tonks, D. L. and Page, J. B. *Chem. Phys. Lett.*, **66** (1979), 449.
218. Hizhnyakov, V. and Tehver, I. *Phys. Status Solidi*, **21** (1967), 755.
219. Hassing, S. and Mortensen, O. S. *J. Chem. Phys.*, **73** (1980), 1078.

Chapter 5

CHEMICAL ASPECTS OF PVC STABILISATION

E. D. Owen

Department of Chemistry, University College, Cardiff, Wales, UK

5.1 INTRODUCTION

The large-scale and world-wide production of polymers which has taken place over the last few decades has increasingly drawn attention to the relatively low stability of these materials and the consequent problem of protecting them during the processing stage as well as from the longer term photochemical ageing processes which lead to a deterioration of their physical properties. In the case of PVC, both processes take the form of a dehydrochlorination reaction accompanied in oxidative environments by chain scission and crosslinking, the mechanisms of which have been discussed in detail in the preceding chapters. Stabilisation of PVC is an attempt to prevent or at least reduce and minimise the extent of these processes and the resulting disadvantageous changes in physical properties which they bring in their wake.

A glance at any current patent application concerned with PVC stabilisation almost invariably reveals a multi-component formulation and it needs hardly be said therefore that the mechanisms by which such formulations operate are extremely complex. Many have been developed over long periods of time in mostly empirical ways and it is only in relatively recent years that attempts have been made to clarify the mechanistic details. A further consideration is that changes in formulations which bring about improved performance can result in considerable commercial advantage and so it is hardly surprising that only a small proportion of the background of information which has been built up by PVC processors is readily available in the chemical literature.

Stabilisers fall into well-defined chemical types but are almost invariably used as combinations, the compositions of which depend in a precise way on the eventual product application and the projected length of time for which stability is required. Each type has been exhaustively investigated and many excellent review articles have appeared in the literature.[1-6] This chapter will attempt to give an overall picture of the most used classes and current developments without experimental fine detail but with sufficient reference material to enable a particular point to be pursued.

Stabilisation of PVC against thermal or photochemical degradation represent aspects of the problem which are in one sense quite separate but in another are closely related since final products intended for outdoor application and for which photoprotection is therefore of paramount importance must first be processed at temperatures at which thermal degradation also presents problems. There are examples of systems in which effective thermal stabilisers act as photosensitisers and in that sense the requirements for the two types of protection are contradictory. In the early days of PVC application the emphasis was firmly on the thermal processing phase but in more recent years the rapid growth of outdoor applications of PVC has led to increasingly stringent requirements for photostabilisers. For these reasons thermal and photostability will not be separated in this chapter but discussed in parallel, with emphasis on one or the other as seems appropriate.

Several other considerations in addition to the obvious one of efficiency may influence the suitability of a stabiliser recipe for a particular application. These include some less tangible factors such as processability, recycling potential, acceptability in terms of contact with food or water supplies (i.e. toxicity) and last but by no means least—cost! Samples may be subjected to a considerable range of preparative and processing procedures and conditions before reaching the stage where the efficiency of a particular stabiliser blend can be estimated. In addition, a variety of test procedures can be applied, each appropriate to some aspect of the final application of the sample under test. The resulting large number of procedural options added to the often incomprehensible information given in the patent literature means that it is normally impossible to assess the relative merits of different stabiliser systems in even a semi-quantitative way and therefore no such attempt will be made in this chapter.

It is intended that the preceding remarks will be sufficient to emphasise the complexity of the task as well as the impossibility of covering all aspects of the subject in even a peripheral way. Even a moderate coverage would require several substantial volumes. Instead, the intention is to mention briefly the chemical and mechanistic aspects which have been established with reasonable certainty and which form a firm basis for further progress. Invariably the choice of material to include from the vast amount which is available will reflect to a large extent the interests and prejudices of the author.

5.1.1 Stabiliser evaluation

There are several reasons why it is virtually impossible to assess the relative merits of all the stabiliser systems which have been investigated. Firstly, the deliberate ambiguity of the patent literature ensures that the large majority of compounds tested and reported could not be included in any comparison. Secondly, methods of sample preparation, i.e. polymerisation, purification, mixing, calendering, extruding, pressing, moulding temperature and so on differ so widely. The very obvious reason for this is the large and increasing range of applications of PVC and the consequent variety of particular physical properties towards which the stabiliser and hence the test may be directed. In a particular application, colour constancy may be very desirable but changes in tensile properties less stringent while in a different application the reverse may be the case. This diversity of applications means that a number of test methods have evolved. They include static[7] and moving[8] oven testing followed by colour assessment,[9] roll milling[10] and torque measurement[11,12] each in association with all the usual spectroscopic and material testing techniques. The wide diversity makes it extremely unlikely that any generally applicable scale could be constructed or would be useful and therefore no such attempt is made here.

5.2 ORGANOTIN STABILISERS

Ever since the appearance of organotin compounds in the patent literature around 1936[13] their importance as highly efficient stabilisers against PVC degradation has become increasingly apparent. In the past, emphasis has centred mainly on their function as thermal stabilisers during the high temperature processing stage and affecting photodegradation only in the sense that they prevent the formation of thermal degradation products such as carbonyl compounds which may then function as photosensitisers. More recently the possibility that some may carry out a dual role has been suggested.[14] Their relatively high cost is offset to some extent by processing and other advantages which they bestow and they are now generally regarded as the first choice in applications where 'glass clear' or highly coloured PVC is required.

Most of the stabilisers in general use have structures of the type $R_nSnX_{(4-n)}$ where R is normally an alkyl group, usually methyl, butyl or octyl, and X is one of a large number of saturated or unsaturated carboxylates or mercaptide derivatives. The carboxylates are frequently esters or half esters of maleic acid and the thiol residues often derived from alkyl thioglycolates $HSCH_2COOR$. Probably the most widely used stabiliser which falls into this category is dibutyltin bis(isooctylthioglycolate), $(C_4H_9)_2Sn(SCH_2CO_2C_8H_{17})_2$. Although frequently described as 'isooctyl', the C_8H_{17} residue is derived from a commercially convenient mixture of C_8 alcohols in which 2-ethylhexanol predominates. Dialkyl tin compounds of this type are by far the most effective stabilisers when used as individual components, and tetraalkyl tins in contrast are almost completely inactive. The relative order of efficiency of the range of possible compounds is as follows:

$$R_2SnX_2 \geq RSnX_3 > R_3SnX \gg R_4Sn$$

In practice, compounds which have been shown to have low efficiency when used as a single organotin component often appear to have a greatly improved performance when compounded as part of a synergistic mixture. Monoalkyl tin compounds seem particularly effective in such mixtures with mono or tris alkyl partners. For example, a formulation composed of $BuSn(SCH_2CO_2C_8H_{17})_3$ or BuSn(laurylmercaptide)$_3$ as primary stabiliser with $Bu_3Sn(SCH_2CO_2C_8H_{17})$, $(C_6H_5)_3Sn$ (laurylmercaptide) or $(C_6H_5)_3Sn$ (thiophenate) as secondary stabiliser confers effective stabilisation of PVC. The same workers have claimed that $\leq 67\%$ of $(Bu)_2Sn(SCH_2CO_2C_8H_{17})_2$ in a particular recipe could be replaced by a non-sulphur compound, $(Bu)_2Sn(C_6H_5)_2$ or $(Bu)_3Sn$(octyl), without adversely affecting heat stabilisation or colour rating. In some cases the mono and dialkyl components may be synthesised into one complex molecule. Thus $R^1_2XSnSSnX_2R^2$ has been shown to be effective (where R^1, R^2 = phenyl, octyl or butyl, $X = C_7H_{15}CH_2O_2CCH_2S$ or $C_7H_{15}CH_2O_2C(CH_2)_2S$).

It has been generally confirmed by several workers that varying the structure of the R group in dialkyl tin compounds, R_2SnX_2, has relatively little effect on the stabiliser efficiency but that the nature of X is much more important. For dibutyl tins, for example, diacetates or dilaurates are poor stabilisers, the dilauryl mercaptide much better and the di(isooctylmercaptoacetate) extremely good. In general, mercaptoacetates or propionates confer good thermal protection but no photoprotection. Maleate esters on the other hand do confer a degree of photostability in addition to the other property of being sulphur free, which is an advantage in some applications.

The synthesis, reactions and structure of a large range of organotin compounds have been expertly and exhaustively reviewed[15] and their application in the stabilisation of PVC documented.[2]

5.2.1 Stabilisation mechanisms

The complexity of the degradation processes which have been described in previous chapters means inevitably that the stabilisation process must be at least equally complex and that effective stabilisation almost certainly depends on the efficient operation of more than one process. At least six such processes are now clearly definable:

1. Reaction with HCl and its removal from the system is a principal requirement since HCl is not only a primary degradation product but has been shown to have a catalytic effect on the rate of reaction as well as an adverse effect on the colouration through the formation of coloured polyene–HCl complexes.
2. Reaction with and stabilisation of the 'abnormal' structural features which may be present in the polymer and which introduce instability is necessary to minimise the number of initiation sites.
3. Interference with the chain mechanism, e.g. by scavenging reactive chain propagators, usually free radicals, will reduce the overall degradation rate.
4. Non-chain decomposition of peroxy and hydroperoxy compounds which have been shown to be thermally and photochemically unstable is an important requirement.
5. Reaction with conjugated polyene sequences in a way which will effectively reduce their lengths will contribute to a reduction in one aspect of the degradation, namely colouration of the polymer.
6. Rendering harmless of prodegradant substances, i.e. reaction products, which are potential sources of instability is important in some cases. Class 1 above when the prodegradant substance is HCl is a special case of this general requirement.

Although the mechanism of PVC degradation and stabilisation has usually been considered in terms of the non-oxidative conditions which have been assumed to prevail in a normal processing operation, the recent work of Scott has indicated that this may be an incorrect emphasis. He has shown[16] that mechanochemical formation of free radicals during the very early stages before melting is complete and their reactions with the small amounts of oxygen present may be very relevant to the correct interpretation of the mechanism of stabiliser function. The primary products which result from these reactions are isolated double bonds with consequent allylic chlorine, HCl and hydroperoxide groups (Scheme 1). All three are formed rapidly at high torque and have a dominant influence on the subsequent thermal and light-ageing behaviour of PVC (Fig. 5.1). Successful solutions of the problem clearly require accurate identification

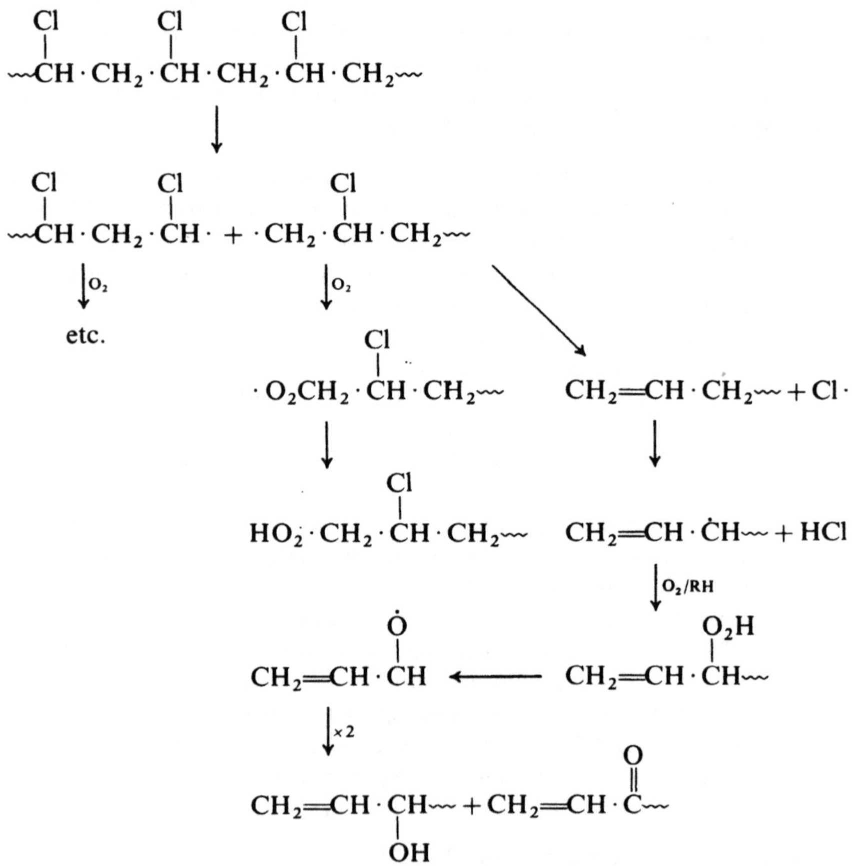

Scheme 1

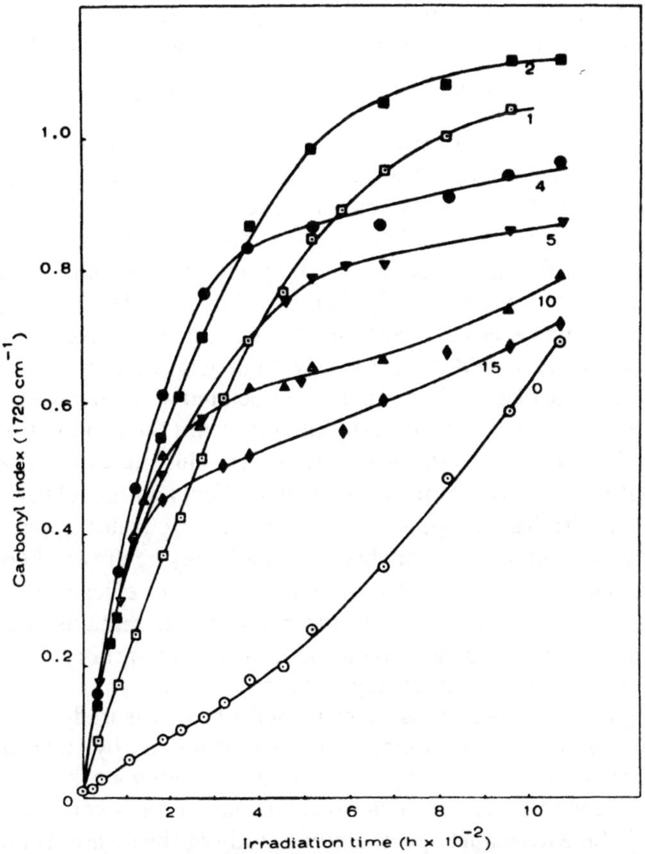

Fig. 5.1 Effect of ultraviolet light ($\lambda > 285$ nm) on unstabilised PVC processed for various times at 210 °C (numbers on curves are processing times in minutes). (Reproduced from Scott, G., Tahan, M. and Vyvoda, J., *Eur. Pol. J.*, **14** (1978), 1022, by permission of Pergamon Press Ltd.)

of the relative importance of these various processes and selection of a stabiliser formulation which will deal most adequately with them. They will be discussed in turn here with reference to organotin compounds but as will gradually become apparent they are equally relevant to the other stabiliser systems which will be discussed in due course.

(*a*) *Reactions of organotin compounds with HCl*
Solutions of dibutyl tin stabilisers heated at 179 °C in the presence of HCl

reacted without any Sn–C bond fission (eqn. (1)) and there is every reason to suppose that they would behave in a similar way in PVC.[17]

$$Bu_2SnX_2 + 2HCl \rightarrow Bu_2SnCl_2 + 2HX \quad (1)$$

The reaction product, Bu_2SnCl_2, has the advantage of being a relatively weak Lewis acid and therefore unable to catalyse further degradation although it appears[18] to be sufficiently reactive to catalyse the reaction of Bu_2SnX_2 with unstable ('labile') chlorine atoms in PVC. This possibility will be discussed further in Section 5.2.1(b). The stability of the C–Sn bonds is a crucial factor since any further reaction with HCl leading to the formation of appreciable quantities of the much stronger Lewis acids $RSnCl_3$ or $SnCl_4$ would lead to very high degradation rates and the appearance of the black colour which is associated with catastrophic failure of the polymer. Factors affecting this all important Sn–C bond fission have been examined in detail[19] and some general conclusions can be drawn. The chain length of *n*-alkyl groups has very little effect and the relative electron donating or withdrawing power of substituents does not appear to be a decisive factor (Table 5.1). Conclusions which may be drawn therefore are that facile cleavage of the Sn–X bond is a requisite for efficient reaction but that it is equally important that the Sn–C bond remains intact. Any substituent or other influence which can induce either process in the wrong direction will reduce the efficiency of stabilisation.

Although these conclusions form a useful basis for understanding the behaviour of these compounds the situation is by no means as straightforward as has been suggested so far and other workers have come to different conclusions. For example, Brooks and co-workers[20,21] have made a ^{119m}Sn Mössbauer spectroscopic study of the reactions undergone by $R_2Sn(SCH_2CO_2C_8H_{17})_2$, $R_2Sn(IOTG)_2$ (where R = butyl or octyl), Bu_2Sn (isooctyl maleate)$_2$ and ($Bu_2Sn(IOM)_2$) during the thermal degradation of PVC. They used 4% stabiliser mixtures prepared by conventional hot milling techniques which were then degraded at 185 °C for 1 h. In each case the stabiliser was converted into the corresponding dialkylmonochlorotin ester and not dialkyltin dichloride. These results generally support those of Parker and Carman[22] who have shown that dialkyltin dimercaptides readily undergo a ligand exchange reaction with dialkyltin dichlorides to form mixed halomercapto derivatives, $R_2SnClSR$, but they are contrary to those of Harrison.[23] In that case, $Bu_2Sn(IOTG)_2$ was found to undergo partial chlorine for IOTG exchange at the milling stage but complete conversion to Bu_2SnCl_2 during thermal degradation.

In a parallel Mössbauer study of the related photodegradation process[24]

Table 5.1 Approximate time for complete discoloration of PVC containing 2 % of stabiliser $R_2Sn(SCH_2CO_2C_8H_{17}\text{-i})_2$ heated at 195 °C

R	*Time (min)*
No stabiliser	10
CH_3	90
CH_3CH_2	90
n-C_3H_7	90
n-C_4H_9 (reference std)	70
t-C_4H_9	30[a]
n-C_8H_{17}	70
$CH_3CH_2OCOCH_2CH_2$	40[a,b]
$NCCH_2CH_2$	30[a]
$(CH_3)_3SiCH_2$	60
$C_6H_5CH_2$	30[a,b]
$C_6H_5CH_2CH_2$	50[a]
$BrCH_2$	10[a]
C_6H_5	50[a,b]
p-CH_3O—C_6H_4	10[a,b]
o-CH_3O—C_6H_4	20[a,b]
p-C_6H_5O—C_6H_4	10[a]
o-C_6H_5O—C_6H_4	30[a]
p-Cl—C_6H_4	30[a]
p-CF_3—C_6H_4	30[a]
o-F—C_6H_4	30[a]

[a] Black colour associated with catastrophic failure.
[b] Undergo Sn—C bond cleavage with hydrogen chloride (see text).
Reproduced from ref. 2.

the maleate stabiliser was very much more effective than the thioglycolate. The former remained virtually unchanged after 1000 h irradiation in artificial sunlight whereas the latter was rapidly converted into the monochloro ester and subsequently into stannic oxychloride. The maleate which may function by accepting energy from PVC therefore retains its stabilising function longer than the thioglycolate. The formation of the monochloro ester from the thioglycolate is further evidence that dehydrochlorination of the polymer is the important first stage in the degradation and formation of the $SnOCl_2$ is an indication of the further complexity of the processes occurring at more advanced stages of polymer and stabiliser decomposition.

(b) Reactions of organotin compounds with PVC

Since the original classic work of Frye *et al.*[25] which first showed that the X component of an organotin stabiliser becomes chemically bound to the polymer during the thermal ageing of stabilised PVC, the mechanism has subsequently been confirmed many times by different workers using a variety of experimental techniques. In nearly all cases allylic chlorine atoms formed by random elimination of HCl have been implicated as the most reactive site in the polymer. Considerable evidence in support of this idea has been accumulated from work on model compounds. Klemchuk's[26] studies in the 80–100 °C temperature range using 4-chloropent-2-ene and 2-chloro-2-methylpropane as models for allylic and tertiary chlorines, respectively, showed conclusively that the stabiliser functioned not only as an acceptor of HCl but replaced the allylic chlorine formed by the dehydrochlorination with an X group thus interrupting the 'zipper' elimination of further HCl which normally follows allylic activation. Qualitative agreement was obtained between the stabilising effect of organotin compounds on PVC stabilisation and their reactivity towards model compounds for allylic chlorine thus confirming that allylic chlorine was the most unstable site. Early model compound studies[27] showed that both 2-chlorobutane and 2,4,6-trichloroheptane, which are models of normal PVC structure, were unreactive towards organotin stabilisers for periods of 45 min at 180 °C. More recent studies,[28] however, showed that dialkyltin bis(thioglycolates) or bis(alkylmaleates) do react, though slowly, with secondary chlorine and that the reaction may be subject to catalysis. Other stabilisers, notably monoalkyltin tris(alkylmaleates) and tris(thioglycolates) react more rapidly and therefore would be expected to react appreciably and therefore wastefully with normal PVC.

Organotin carboxylates react readily with tertiary chlorine to give *t*-butyl esters in high yields (eqn. (2)) but much less readily with allylic chlorine (eqn. (3)):[18,27]

$$Bu_2Sn(O_2CR)_2 + 2Bu^tCl \longrightarrow Bu_2SnCl_2 + 2RCO_2Bu^t \qquad (2)$$

$$Bu_2Sn(O_2CR)_2 + CH_2{=}CH\cdot\underset{\underset{\displaystyle Cl}{|}}{CH}\cdot CH_3 \longrightarrow$$

$$Bu_2Sn(O_2CR)Cl + CH_2{=}CH\cdot\underset{\underset{\displaystyle O_2CR}{|}}{C}\cdot CH_3 + \underset{\underset{\displaystyle O_2CR}{|}}{CH_2}\cdot CH{=}CH\cdot CH_3 \qquad (3)$$

Dibutyltin thioglycolates, however, reacted readily with 3-chlorobut-1-ene, 1-chlorobut-2-ene or 2-chloro-2-methylpropane to give substitution products.

In general, the agreement between the effective stabilising ability of a compound and its reactivity towards model compounds leaves little room for doubt that replacement of an allylic chlorine by an X ester fragment is an important aspect of PVC stabilisation by organotin stabilisers. Allylic rearrangement which often seems to accompany the substitution may be an added stabilising influence:

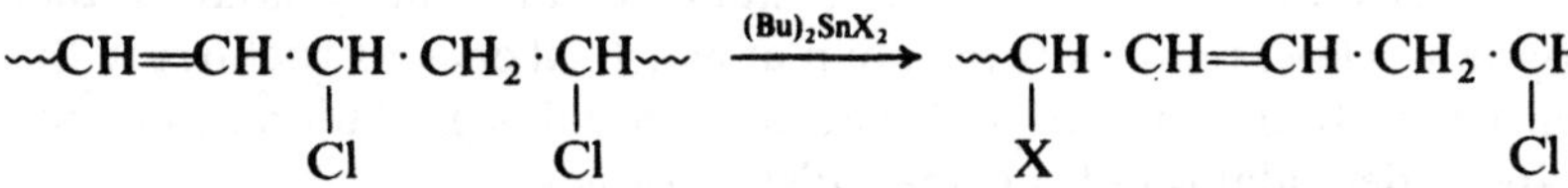
(4)

Results obtained from reactions of stabilisers with model compounds are useful only if they can be extrapolated securely to PVC and in this context experiments to measure the extent of incorporation of isotopically labelled compounds into PVC provide an invaluable link with the model compound work. It should be said though that although experimental measurement of the extent of reaction of polymer with stabiliser is not too difficult, interpretation of the results is not always straightforward. Thus ^{35}S-labelled dibutyltin dithiolate heated with PVC at 180 °C reacted in two stages.[29] An initial slow uptake over a 2–3-h period was followed by a more rapid rate of reaction which was accompanied by colouration of the polymer. At the end of the induction period the extent of incorporation of radioactivity corresponded to about 0·2 allylic chlorine atoms per 1000 monomer units. When the process was carried out at lower temperature (120 °C) at which thermal degradation was less important, even less incorporation resulted.[30] On the other hand when the labelled stabiliser reacted at 180 °C with polymer which had been degraded previously for 50 min and which was highly coloured, the colour was discharged and the rate of incorporation of radioactivity significantly increased. The uptake of ^{35}S-labelled thiol groups in the reaction following the induction period, which was also associated with strong retardation of colour formation, was shown to be due to the addition of thiol groups (formed by reaction of stabiliser with liberated HCl) to the double bonds of the degraded polymer.

Confirmatory evidence that removal of allylic chlorine leads to an improvement in stability has been obtained by treating PVC with a variety of reagents.[31–34] In addition, Starnes and Plitz[35] have shown that treatment of PVC with a relatively large quantity of stabiliser, in some cases with a requisite amount of catalyst, gave a polymer whose thermal stability was considerably improved and under standard conditions evolved HCl at a rate which was 6–9 times slower than for the untreated polymer. The amount of tin remaining in the polymer after such treatment was negligible but the amount of sulphur did appear to be an influencing factor although no obvious correlation existed between the numbers of bound mercaptide groups and the number of allylic chlorines in the original polymer. Difficulties also remain in explaining the large differences between the values obtained by Starnes from pretreated samples of PVC and those of other workers using more realistic levels of stabiliser. The optimum concentration of stabiliser appears to be about 0·5 %, further increases leading to no obvious further improvement in stability. It is usually assumed that this optimum level represents the point at which all defect structures in the polymer have been removed and that the residual degradation represents the rate of degradation of 'normal' PVC residues.

In conclusion the following broad generalisations about reactions with PVC may be made:

1. Stabilisers react with and remove certain destabilising structures which are almost certainly allylic chlorine although the levels involved are still uncertain.
2. A limiting stabilising effect is reached which does not correspond to absolute stability of the polymer and may reflect some inherent instability of the normal PVC structures.
3. Whether destabilising structures exist in the virgin polymer or are introduced during processing, the contribution of this mechanism to the efficacy of the stabiliser is considerable and allows greatly increased flexibility in processing conditions.
4. The reactions concerned are almost certainly catalysed to an extent which depends on the conditions.

(*c*) *Organotin compounds as radical scavengers and peroxide decomposers*

The feeling that some organotin compounds can contribute to the stabilisation process by interfering with free radical reactions has persisted since the idea was first put forward and it still seems justifiable although the evidence is indirect. The well known ability of tin to expand its covalency above four makes reaction with an electronegative free radical like $\dot{C}l$ a distinct possibility and if this is so it should have a profound effect on the 'zipper' free radical chain mechanism in which clorine atoms are the main chain propagators. Evidence that this is indeed the case can be inferred from the UV photolysis of CCl_4 solutions of dibutyltin diacetate, dipropionate or diethylmercaptide.[36] The volatile products formed can be satisfactorily accounted for by a reaction scheme which will not be described in full here but includes the reactions (5) and (6) below:

$$\dot{C}l + Bu_2Sn(O_2CCH_3)_2 \longrightarrow Bu_2SnCl(O_2CCH_3) + \dot{O}_2CCH_3 \quad (5)$$

$$\dot{C}l + Bu_2SnCl(O_2CCH_3) \longrightarrow Bu_2SnCl_2 + \dot{O}_2CCH_3 \quad (6)$$

Recent studies of the Mössbauer spectra of PVC stabilised with Bu_2SnX_2 (where X = isooctylthioglycolate or isooctylmaleate) or $(octyl)_2Sn(iso$-$octylthioglycolate)_2$ however, showed that in each case the stabiliser was converted into R_2SnXCl but not R_2SnCl_2.[37] ESR studies of the inhibiting effect of dibutyl tin stabilisers on the radiation-induced grafting of styrene or acrylonitrile on to PVC suggested that they operated by affecting free radical reactions.[38]

Many sulphur compounds are able to interfere with the propagation of autooxidation processes by decomposing peroxides or hydroperoxides and it is probable that organotin sulphur compounds, or the thiols liberated from them by reaction with HCl, react in this way. One report[6] which describes the reaction of dimethyltin bis(isooctylthioglycolate) with *t*-butyl hydroperoxide as a secondary antioxidant effect, lists the products as dimethyltin oxide, *t*-butanol and a disulphide (eqn. (7)):

$$(CH_3)_2Sn(CH_2CO_2\textit{i}\text{-}C_8H_{17})_2 + Bu^tOOH \xrightarrow{50\,^\circ C} (CH_3)_2Sn{=}O + (\textit{i}\text{-}C_8H_{17}CO_2CH_2S)_2 + Bu^tOH \qquad (7)$$

Another report[39] mentions the same reaction but with dibutyltin sulphate, the sulphoxide and isobutene (derived from the butyl radical) as products. Both reports agree however that the model peroxide is decomposed and an extrapolation to the decomposition of macro peroxides or hydroperoxides in PVC can safely be made. As examples of the kind of indirect evidence which may be produced, the addition to PVC of dibultytin bis(cyclohexylmaleate) or dibutyltin bis(butylthioglycolate), both of which are regarded as active decomposers of peroxides and hydroperoxides, substantially decreased the rate of dehydrochlorination under oxygen at 180 °C.[40] In addition, the report that sulphoxides and sulphones result from the reaction of dioctyltin bis(2-ethylhexylthioglycolate) may also be indirect evidence for peroxide decomposition.[41]

The role played by HCl, which has been mentioned already, has been the subject of debate for many years.[42] It now seems generally agreed that it does act as a catalyst but there are still uncertainties about the extent and detailed nature of its involvement. One such aspect which has emerged recently is its ability to catalyse the formation or destruction of hydroperoxides depending on the relative molar concentrations of the two components. When the molar ratio [ROOH]/[HCl] is less than unity, HCl is able to catalyse the formation of hydroperoxide whereas, when [ROOH]/[HCl] is greater than one the hydroperoxide concentration is reduced rapidly to zero in a pseudo first-order catalytic process.[43] The

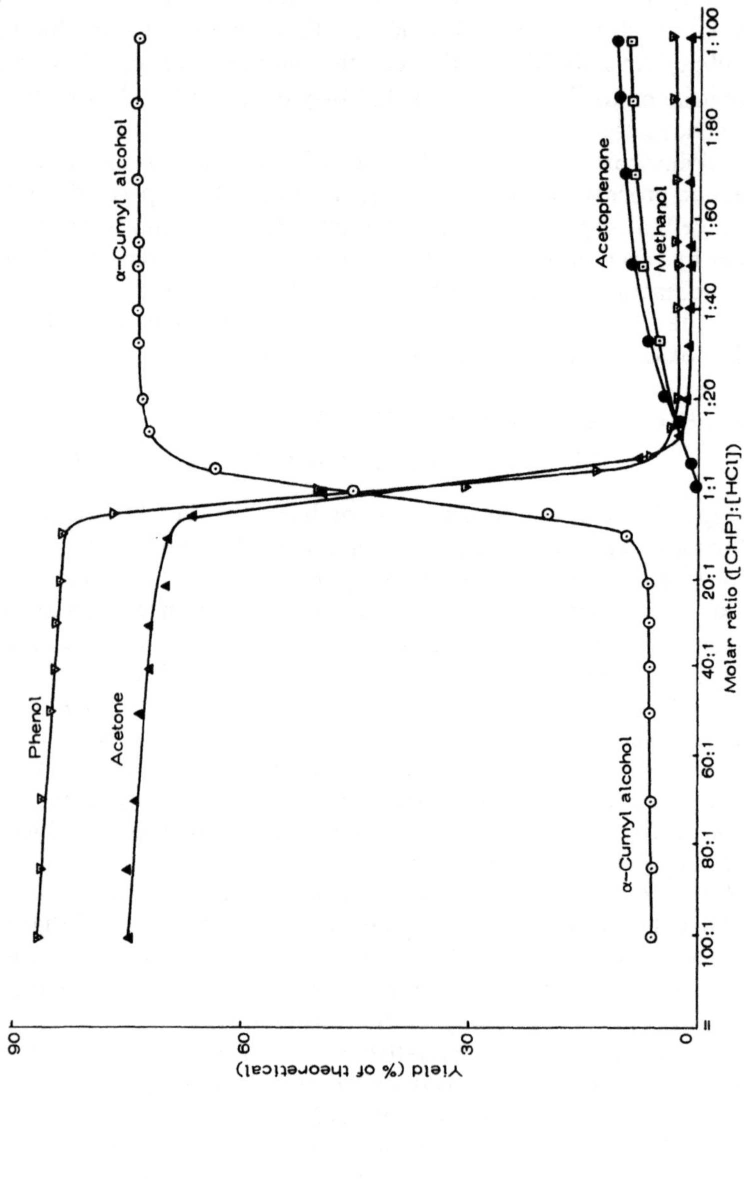

Fig. 5.2 Products formed from the decomposition of cumene hydroperoxide (CHP) (9×10^{-3} M) at varying [CHP]/[HCl] molar ratios. (After Cooray and Scott.[3])

change from pro-oxidant to anti-oxidant behaviour over a critical stoichiometric region is illustrated dramatically in Fig. 5.2 for a model system.[44] In the case of PVC the stoichiometry will obviously depend critically on the prevailing conditions and on the prehistory of the sample and this behaviour may explain some of the apparently completely contradictory results which have appeared in the literature. What is more important perhaps is that this synergistic behaviour of hydroperoxides with HCl emphasises the importance of stabilisers with a dual function which enables them to reduce the levels of both potentially destructive agents.

(*d*) *Reactions of organotin compounds with double bonds*

Addition of organotin mercaptides or maleates to partially degraded (coloured) PVC can bring about decolouration which must be due to reaction with conjugated polyene sequences.[32,45] A detailed comparison of the effects of dibutyltin maleate (DBTM) and dioctyltin thioglycolate (DOTG) in this have been made by Scott[46] from which some significant conclusions can be drawn. Figure 5.3 shows that both stabilisers have a

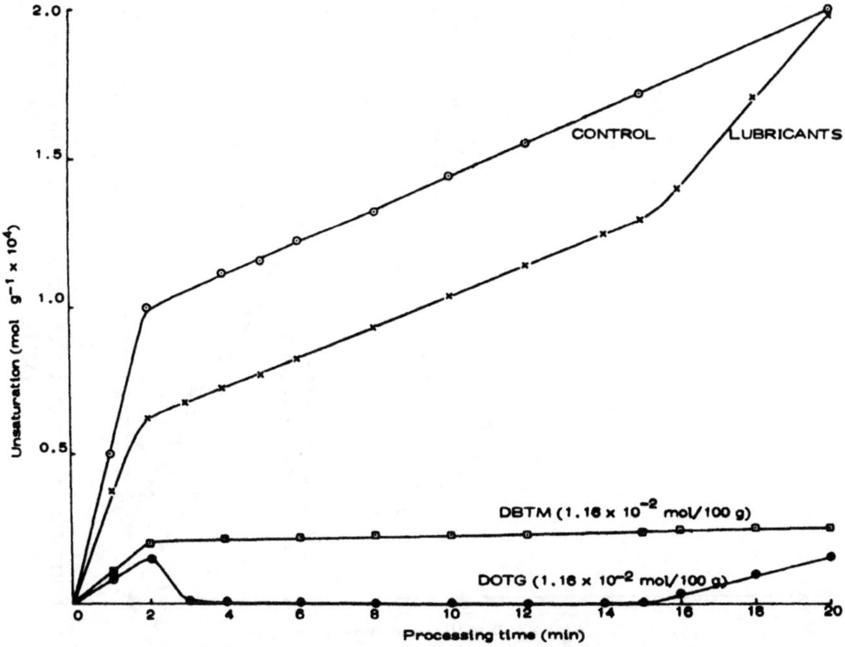

Fig. 5.3 Effect of DBTM and DOTG on the formation of unsaturation during processing of PVC at 170 °C. (After Cooray and Scott.[3])

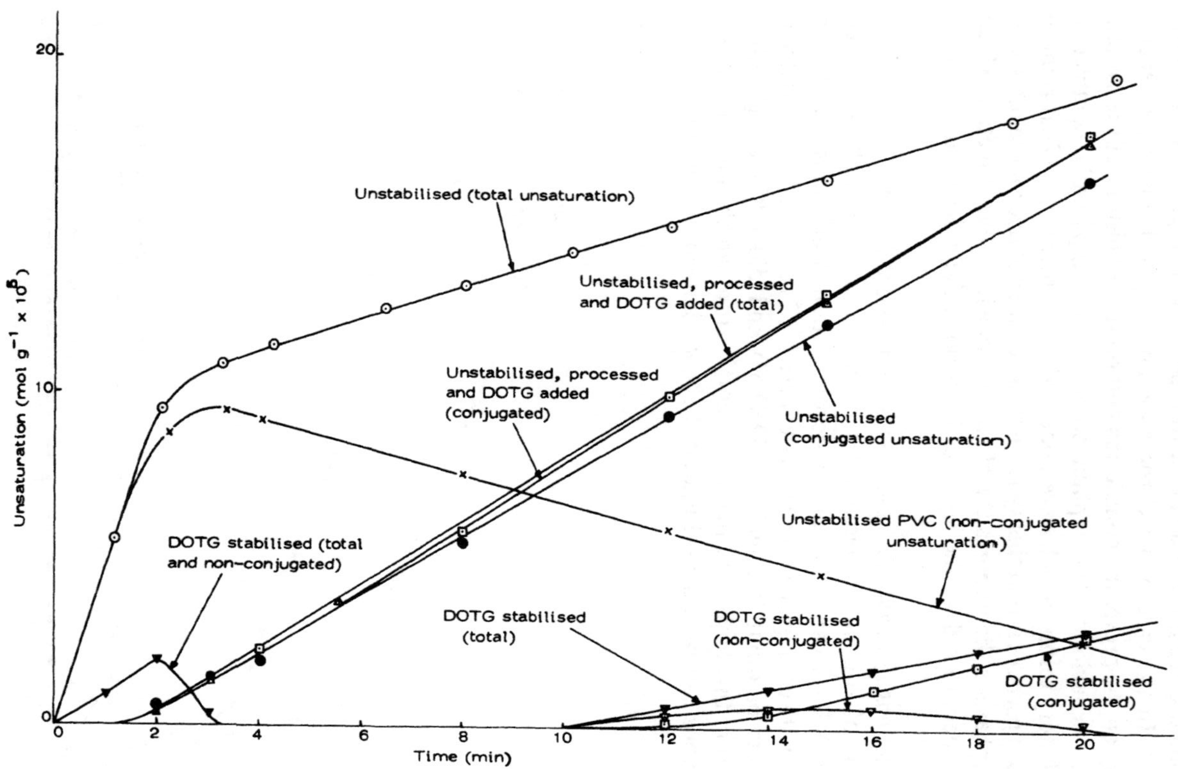

Fig. 5.4 Formation of isolated and conjugated unsaturation in PVC during processing at 170°C in the presence of DOTG ($1{\cdot}45 \times 10^{-3}$ mol/100 g). (After Cooray and Scott.[3])

profound effect in reducing the rate of formation of unsaturation but unlike DBTM, DOTG is able to destroy the unsaturation formed during the early stage. During the secondary stage DBTM continues to react with both allylic chlorine and with liberated HCl giving rise to dialkyltin dichlorides and maleic anhydride. In addition there is progressive removal of conjugated unsaturation as a result of a Diels–Alder reaction. The reason that the initial unsaturation is not removed is because maleic anhydride is unable to react with the isolated double bonds which account for the initial unsaturation. DOTG on the other hand is able to react with and remove the isolated double bonds completely. A similar pattern was observed when DOTG was replaced by a simple thiol from which Cooray and Scott[46] concluded that DOTG is a reservoir for vinyl reactive thiols produced by reaction with HCl liberated from PVC by mechanochemical macroradical formation (Scheme 1). Total unsaturation was estimated by catalysed iodination,[16] isolated double bonds by mercuric acetate[44] and conjugated unsaturation by difference. A detailed analysis of the various types is given in Fig. 5.4 from which it is clear that initially the unsaturation in the absence of additives is entirely due to isolated double bonds. Conjugated unsaturation follows after an initial induction period and increases at the expense of isolated double bonds.

(*e*) *Organotin compounds as photostabilisers*

Since pure PVC contains no chromophores capable of absorbing sunlight, its photosusceptibility is due to the presence of initiator residues or the introduction of potential photosensitisers during the processing stage. The most important of these are unsaturated sequences, peroxy and hydroperoxy groups (which have been dealt with already) and carbonyl groups, so clearly the primary function of a good photostabiliser is to prevent their formation. The relationship between a good thermal and photostabiliser emerges in an obvious way from this requirement. In this sense organotin compounds are good photostabilisers as long as the quantity which remains after processing does not undergo photolysis during subsequent outdoor use.

In general, dibutyl- or dioctyltin bis (benzyl fumarates) are good thermal stabilisers but are also effective photosensitisers,[47] isooctylthioglycolates behave similarly but are less active[6,48] and good thermal and photostability is achieved using maleate ester derivatives or synergistic mixtures of organotin maleates with mercapto esters.[49] A recent study showed[14] that this system is neither synergistic nor antagonistic with regard to thermal degradation but is synergistic with respect to photocolouration and antagonistic with respect to photoinduced carbonyl growth. In nearly all commercial situations, the photostability of polymers containing organotin stabilisers alone is not considered adequate and an additional ultraviolet absorber is incorporated. When some 'genuine' photoprotection is afforded by organotin compounds the mechanism by which they operate is not well understood. An ESR study[50] of the ultraviolet irradiation of dialkyltin bis(isooctylthioglycolate), dilaurates and di-2-ethylhexylmaleates *in vacuo* showed that a number of free radicals could be detected and a complex situation undoubtedly exists. The photostability of the compounds was in the order laurate > maleate > thioglycolate but this did not reflect their polymer photostabilising ability.

The mechanisms by which hydroperoxides, peroxides and unsaturation affect the stability of the polymer have already been discussed but the photosensitising function of carbonyl groups which initiates chain scission[51] and therefore affects the lifetime to embrittlement of the polymer is an additional factor. The effect of maleate on carbonyl formation is that of concentration dependent retardation[52] whereas thioglycolate is an accelerator.[46] In addition, the stabilities towards ultraviolet light are considerably affected by the presence of peroxides. In the absence of peroxides the stabilities are almost the same whereas when peroxides are present the thioglycolates are considerably less stable.

5.3 METAL CARBOXYLATE STABILISERS

The metal carboxylates which have played such an important part in the development of PVC are molecules having the general formula $(RCO_2)_2M$ where M is Cd, Zn, Ba or Ca and R is a linear or branched alkyl group, often a stearate ($C_{18}H_{35}$). Stabiliser formulations based on two component carboxylate mixtures (Ca/Zn or Ba/Cd) have held a prominent place in PVC stabilisation for many years but in more recent times have been extended to include Ba/Zn, Cd/Zn and Ba/Cd/Zn combinations. The continual introduction of new and improved formulations is an attempt to keep pace with or even ahead of the ever-changing and increasingly stringent demands on the quality and diversity of the product as well as the toxicological requirements imposed by the proliferation of health and safety regulations.

The early literature which has been expertly and extensively reviewed[5,53,54] indicates that the present understanding of the mechanism of operation of these stabilisers has arisen as a result of work by many groups of workers. In this context the names of Bengough and Onozuka,[55] Shimura and Sakai,[56] Anderson and McKenzie,[57] Braun and Hepp,[58] Hoang *et al.*[59] and Frye and Horst[60] deserve special mention.

The use of synergistic combinations of metal carboxylates seems to have arisen from the observation that although the carboxylates of cadmium or zinc alone provide thermally processed PVC with a good initial colour they confer poor long term stability. The carboxylates of barium or calcium on the other hand act in the reverse way and so the use of Ca/Zn or Cd/Ba mixtures became established at an early stage.

The HCl acceptor abilities of the metal carboxylates led initially to the idea that, in common with other similar substances, this property was the basis of their stabilising function. When Frye and Horst,[60] using infrared and radiolabelling techniques, showed that esterification of the polymer to form PVC–carboxylate esters occurred, it became clear that this reaction could be at least equally as important as HCl absorption since it was shown that the stability of the ester group is appreciable and can be further improved by addition of a base. The explanation for the synergistic effect began with the realisation that metal chlorides, which are by-products of the esterification reaction, are active catalysts of the dehydrochlorination. $ZnCl_2$ is most active in this respect, closely followed by $CdCl_2$, but it appears that $BaCl_2$ and $CaCl_2$ are much less active. These facts led Klemchuk[61] and independently Onozuka[62] to propose that, at least in model systems, the exchange reaction of eqn. (8) was the basis of the synergistic effect, an explanation which has since been amply substantiated:

$$ZnCl_2 + Ca(O_2CR)_2 \rightleftharpoons Zn(O_2CR)_2 + CaCl_2 \tag{8}$$

Thus, inactive (towards stabilisation/esterification) $Ca(O_2CR)_2$ reacts with reactive (towards dehydrochlorination) $ZnCl_2$ to form active $Zn(O_2CR)_2$ and inactive $CaCl_2$. The observation that substitution is accompanied by some isomerisation led Anderson and McKenzie[63] to suggest that the mechanism is ionic in nature, with both substitution and isomerisation proceeding via an intermediate delocalised carbonium ion. Although this mechanism represented a considerable improvement in understanding, some puzzling features remained unexplained. There was some suggestion[62] that although acting synergistically with the zinc, the calcium stearate depressed the stabilising efficiency of its primary partner possibly by occupying a coordinating position of the zinc and hence suppressing useful coordination with the labile chlorine of PVC. By contrast, the addition of barium stearate to cadmium stearate causes an enhancement of the induction period before loss of HCl can be detected but does not affect the rate of dehydrochlorination.[64] Thus it appears that the only interaction between cadmium and barium is the one leading to the exchange reaction (8) and no deactivation is observed.

5.3.1 Recent model compound studies

In recent years more precise details about the mechanism of the stabilisation reaction and the kinetic features which control it have been sought from studies of low molecular weight compounds which are models for the labile portions of PVC. Although such data are obtained using modern analytical techniques of great sensitivity and selectivity, the extent to which they can usefully be extrapolated to real polymer systems is always a matter of debate. In real systems physical factors such as particle size and hence rate of HCl diffusion may be rate-controlling. The extent to which the formulation is mixed and melted and hence the details of the local viscosity and rheology during the processing stage is also of crucial importance. Nevertheless there is no doubt that considerable progress has been made from such studies particularly in establishing the details of the synergistic effects mentioned previously.

One model compound which has been used with effect by Guyot and co-workers[64–66] to simulate the behaviour of chlorine made allylic by dehydrochlorination of an adjacent monomer unit is 4-chlorohex-2-ene ($4CH_2$).

Dichloromethane (DCM) solutions of $4CH_2$ in the temperature range 40–80 °C undergo reversible dehydrochlorination to form hexadiene (HD) (reaction (9)):

$$CH_3\cdot CH{=}CH\cdot \underset{\displaystyle Cl}{\underset{|}{CH}}\cdot CH_2\cdot CH_3 \xrightarrow{DCM} CH_3\cdot CH{=}CH\cdot CH{=}CH\cdot CH_3 + HCl \qquad (9)$$

Activation energies for the forward and reverse reactions are 37·8 and 8·4 kJ mol^{-1}, respectively. The strong purple colour which develops as the reaction proceeds has been attributed to a charge transfer complex between HCl and HD which acts as a catalyst for the formation of the allylic carbocation (**1**) which is an intermediate in forward and reverse directions.

$$CH_3\cdot \underbrace{CH{\cdots}CH{\cdots}CH}_{+}\cdot CH_2\cdot CH_3 \quad Cl^-$$

(**1**)

The ionic character of the reaction mechanism is inferred partly from the increase in rate which is observed as the dielectric constant of the solvent is increased and the absence of any reaction in THF (the most effective solvent for PVC) is attributed to the inability of the charge transfer complex to form in the Lewis base solvent which competes effectively for the released HCl. By comparison, $ZnCl_2$ is a sufficiently powerful dehydrochlorination catalyst to overcome this disadvantage and in the presence of $ZnCl_2$ the dehydrochlorination proceeds in THF as well as in DCM, the allylic cation in this case being associated with the $ZnCl_3^-$ gegenion. In addition to the dehydrochlorination described, some polymerisation of the hexadiene to a higher molecular weight product occurs, a reaction which is analogous to the crosslinking reaction which occurs in PVC formulations containing zinc soap stabilisers.

5.3.2 Zinc or cadmium carboxylates as stabilisers

The reaction suggested by Frye and Horst[60] and which has since been amply confirmed by other workers is described for the model compound $4CH_2$ by eqn. (10):

$$2CH_3\cdot CH{=}CH\cdot \underset{\displaystyle Cl}{\underset{|}{C}H}\cdot CH_2\cdot CH_3 + M(O_2CR)_2 \longrightarrow$$

$$2CH_3\cdot CH{=}CH\cdot \underset{\displaystyle O_2CR}{\underset{|}{C}H}\cdot CH_2\cdot CH_3 + MCl_2 \qquad (10)$$

where M = Zn or Cd and R = carboxylate, e.g. stearate ($C_{18}H_{35}$). The reaction proceeds equally well in DCM or THF solvents and like the elimination reaction is catalysed by the strong Lewis acid $ZnCl_2$. The allylic cation (**1**) has again been implicated and therefore appears to be a common intermediate for both catalysed elimination and substitution reactions:

$$2CH_3\cdot CH{=}CH\cdot \underset{\displaystyle Cl}{\underset{|}{C}H}\cdot CH_2\cdot CH_3 + 2ZnCl_2 \rightleftharpoons$$

$$2CH_3\cdot \underbrace{CH{=}CH{=}CH}_{+}\cdot CH_2\cdot CH_3 \; ZnCl_3^- \xrightarrow{M(O_2CR)_2} 2CH_3\cdot CH{=}CH\cdot \underset{\displaystyle O_2CR}{\underset{|}{C}H}\cdot CH_2CH_3 + 2ZnCl_2 + MCl_2 \qquad (11a)$$

$$\searrow \; 2CH_3\cdot (CH{=}CH)_2\cdot CH_3 + HCl + 2ZnCl_2 \qquad (11b)$$

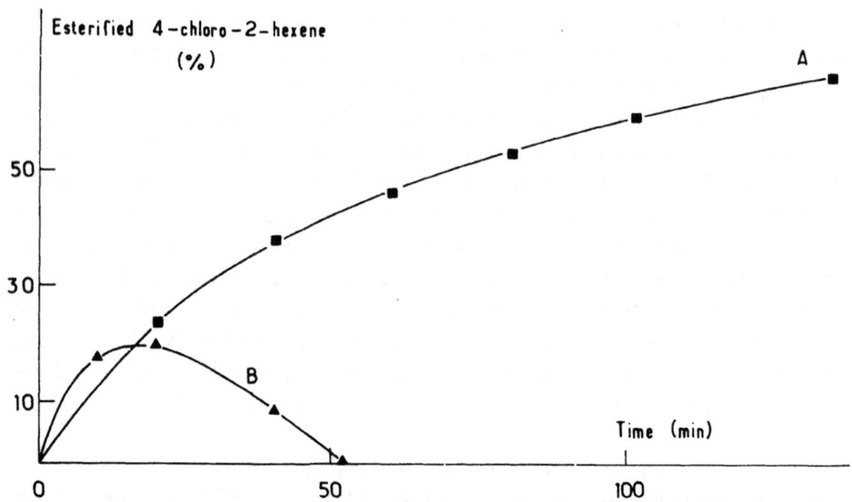

Fig. 5.5 Esterification of 4-chlorohex-2-ene in the presence of zinc stearate (0·11 M) in THF at 60 °C. [4-Chlorohex-2-ene]$_0$ = (A) 0·21 m; (B) 0·56 M. (After Guyot and Michel.[5])

Addition of bases such as pyridine or triphenylphosphine reduces the rates of both elimination and substitution reactions perhaps by interacting with the acidic catalyst $ZnCl_2$. An important role played by evolved HCl, particularly when the concentration of zinc stearate is less than the stoichiometric amount required to complete reaction (10), has also emerged from model compound studies. In such situations the concentration of hexenyl stearate ester reaches a maximum value and then decreases (Figs. 5.5 and 5.6) presumably because HCl can react with the ester to reform the unstable initial allylic compound and stearic acid (eqn. (12)). The labile sites so produced then become initiators of further degradation. Obvious consequences of these factors are seen in both model and real polymer systems. In the former case, consumption of the stabiliser coincides with the appearance of the purple colour of the hexadiene–HCl charge transfer complex and the onset of an oligomerisation reaction which is analogous to the crosslinking of PVC. In the real

$$CH_3 \cdot CH{=}CH \cdot \underset{\displaystyle O_2CR}{\underset{|}{CH}} \cdot CH_3 + HCl \rightleftharpoons CH_3 \cdot CH{=}CH \cdot \underset{\displaystyle Cl}{\underset{|}{CH}} \cdot CH_3 + HO_2CR \qquad (12)$$

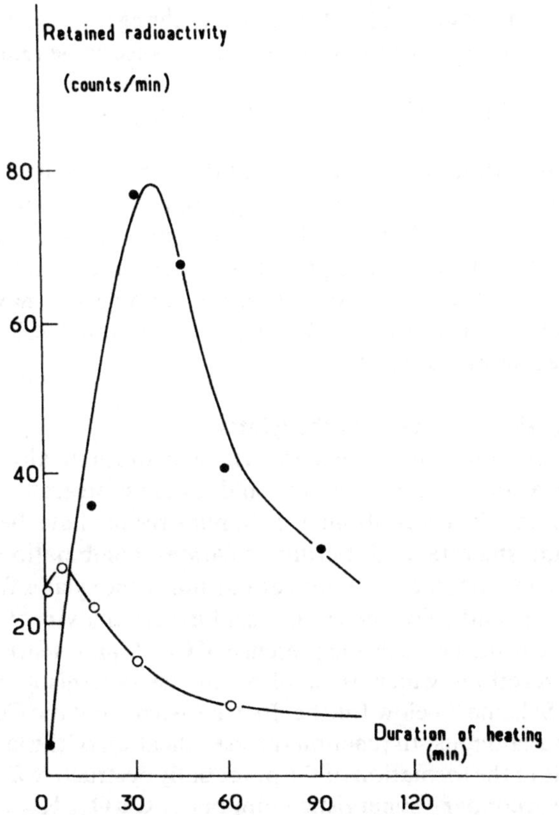

Fig. 5.6 Retention of radioactivity by PVC subjected to various heat treatments on a roll mill and in a thermostatically controlled forced-draught oven then dissolved and re-precipitated. ●, Cd 2-ethylhexanoate-1-C^{14} (milled at 160 °C and heat treated at 180 °C). ○, Zn 2-ethylhexanoate-1-C^{14} (milled at 176 °C and heat treated at 140 °C). (After Guyot and Michel.[5])

polymer case it corresponds to the onset of the catastrophic blackening which occurs when processing conditions (real or simulated) are prolonged beyond the point when the stabiliser has been consumed.

5.3.3 Cadmium and barium carboxylate stabilisers

Calcium stearate reacts with THF solutions of the allylic model compound only in the presence of catalytic quantities of $ZnCl_2$. Kinetic studies indicate that the mechanism is the same as for zinc stearate but that the rate is about 16-fold slower. The reversible exchange reaction (13), first suggested by Onozuka[62] would be expected to cause some reduction in the

$$CdCl_2 + Ba(O_2CR)_2 \rightleftharpoons Cd(O_2CR)_2 + BaCl_2 \qquad (13)$$

effective concentration of the catalyst and therefore to reduce the rate but this alone does not account for the large difference in rate for the two similar systems. Infrared studies based on estimations of the carbonyl groups of the stearate molecules provide some evidence that the formation of inactive zinc stearate–calcium stearate complexes may be partly responsible and other complexes like $ZnCl_2$–$CaCl_2$ and $ZnCl_2$–Ca stearate have also been shown to exist.

5.3.4 Synergistic mixtures of carboxylates

The rate of the elimination reaction is most dramatically reduced by synergistic mixtures of zinc stearate and calcium stearate, particularly when the ratio of Zn:Ca is about 1:5. Similar results have been obtained with cadmium stearate and barium stearates combinations.[63] In the absence of calcium or zinc, cadmium or barium carboxylates function only as HCl acceptors and colorimetric and acidimetric analysis of the polymer matrix after treatment reveals the presence of $CaCl_2$ and stearic acid.[65] The sequence of reactions which is involved in the synergistic behaviour is illustrated in Scheme 2 below for the Zn–Ca system, and the Cd–Ba system is completely analogous. In reaction (a), esterification of the labile chlorine of PVC results in the formation of the potentially destructive $ZnCl_2$ as a by-product. The secondary synergistic component $Ca(O_2CR)_2$ converts this back to $Zn(O_2CR)_2$ and in so doing is itself converted into inactive $CaCl_2$ (reaction (b)). Allylic chlorines which are continually being created by the random dehydrochlorination which occurs during heat treatment are dealt with in this way until the primary component ($Zn(O_2CR)_2$) is consumed. At

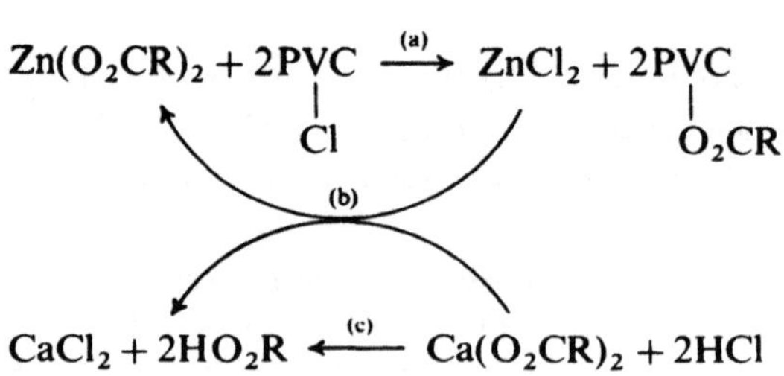

Scheme 2

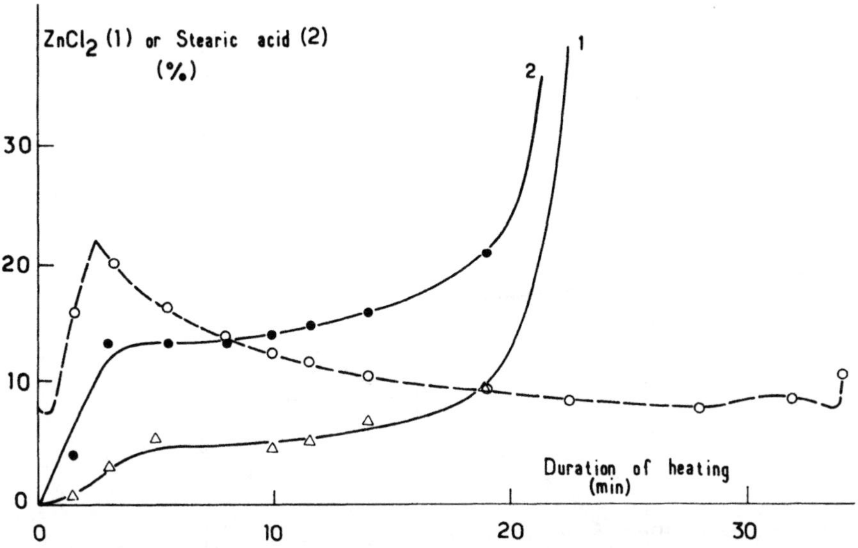

Fig. 5.7 Zinc chloride (1) and stearic acid (2) per cent accumulation in PVC during heating in the Brabender plasticorder at 190 °C. PVC (100 phr), DOP (30 phr), zinc stearate (3·3 phr), calcium stearate (3·3 phr). The broken line represents the plastogram. (After Guyot and Michel.[5])

this stage the concentration of $ZnCl_2$ begins to increase rapidly and is accompanied by the appearance of stearic acid which is formed by reaction (c) (Fig. 5.7). Similar behaviour is observed with Cd–Ba mixtures,[67] in this case the substitution reaction is observed only in the presence of catalytic quantities of $CdCl_2$ and the elimination reaction becomes totally suppressed when the ratio of Cd:Ba is about 1:3.

Mechanical consequences of the synergistic effect are illustrated for the Ca/Zn system in Fig. 5.8 which shows the effect of increasing mole fraction of calcium stearate on the action time (T_A) and crosslinking rate (V_R) of PVC processed in a Brabender plasticorder at 190 °C. T_A is the time between the appearance of the first and second maxima in the torque v. processing time curve. The former is usually indicative that the gelation point has been reached and the second the moment when degradative chain scission reactions overcome the rapidly increasing rate of crosslinking which occurs when the stabiliser is consumed. T_A is therefore an indication of the length of processing time for which the stabiliser system remains effective.

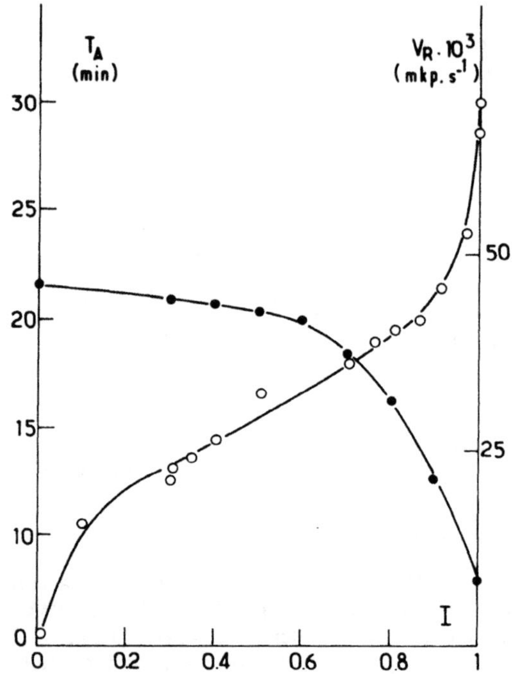

Fig. 5.8 Influence of the relative molar amounts of calcium stearate in the presence of zinc stearate on the action time, T_A (○) of the stabiliser and the crosslinking rate, V_R (●) of PVC in the Brabender plasticorder at 190 °C. PVC (100 phr), DOP (30 phr), total stearate (6·6 phr). (After Guyot and Michel.[5])

5.3.5 Stabilisers in real polymer systems

Much of the work just described and the conclusions drawn relate to model compound studies and the problem of deciding how relevant this is to real polymer systems has been mentioned already. The multi-component nature of the formulations which have evolved for satisfactory thermal processing of the polymer adds an extra dimension of complexity when the behaviour of individual components is being examined. In addition, the methods used are invariably indirect and the results frequently capable of more than one interpretation. The methods generally in use have been summarised already and in more recent studies attention has been directed towards the goal of developing sensitive and selective analytical methods which are capable of measuring the consumption or formation of individual compounds. Infrared spectroscopy is one such technique which has been used[68] to detect remaining metal carboxylates by monitoring the absorption of the $CO\bar{O}$ stretching vibrations at $1510\,cm^{-1}$ and $1550\,cm^{-1}$ and the formation of $ZnCl_2$ has been measured[62] utilising the absorption (592 nm) of its complex with dithizone. Atomic absorption, chromatographic and the coulometric titration methods mentioned previously are being developed in several laboratories and the next major advances may be expected from this direction.

5.4 ORGANIC PHOSPHITE STABILISERS

Organic phosphites have been used to stabilise PVC as well as polyolefins, natural rubbers and other heterochain polymers, probably since these materials assumed commercial significance. As well as having an overall stabilising effect they have usually been assumed to contribute towards preserving the original colour of the polymer during processing and use and in the case of PVC are most often used as synergistic combinations with other stabiliser components, antioxidants, etc. Their good polymer compatibility, low volatility and low reactivity towards molecular oxygen are added advantages in this respect. Although their extensive use in such a wide variety of polymer systems might lead to the expectation that the mechanisms by which they operate would by now be well understood this is certainly not the case. The main reason for this uncertainty is undoubtedly their high reactivity towards many distinctly different reagents in a wide variety of situations and a gradual realisation has developed that they participate in a number of processes, some operating synergistically and each making a contribution to the overall stabilising effect. General conclusions about the relative importance of the processes which are applicable to several systems are almost impossible to draw. They undoubtedly vary from one system to another and even within one system during the induction period created by the stabilising system before degradation begins.

Although these uncertainties make it difficult to characterise and quantify unambiguously the roles played by phosphites in every case, nevertheless many of their basic properties have been established with reasonable confidence and some firm ideas about the roles they play have emerged in recent years. It now seems clear that as well as acting as terminators of chain processes by reacting with peroxy radicals and suppressing degenerate chain branching by decomposing peroxides, processes with which they have long been associated, phosphites are also able to react in other ways. These include reacting with labile structural sites in the polymer, suppressing the catalytic action of HCl and metal halides and possibly using these substances to operate in a catalysed termination of kinetic chains. These various roles which have been mentioned are listed below (1–5) and will first be discussed briefly in turn, then some general observations made about their relative importance.

1. Reactions with proton donors, in particular HCl.
2. Reactions with hydroperoxides.
3. Reactions with free radicals.
4. Reactions with ketoallychloride groups.
5. Reactions with metal salts.

5.4.1 Reactions of phosphites with HCl

Phosphites share with the other classes of stabilisers discussed so far the ability to react with and remove from the system the potentially catalytic HCl without producing equally or even more active prodegradant products. What follows may well apply to a greater or lesser extent to other electrophilic reagents which catalyse the degradation, but HCl is undoubtedly of primary importance.

The sequence of reactions (14)–(16) below can usefully be regarded as a general description of the reactions of HCl with the various phosphorus containing species but the contribution made by each depends critically on the phosphite structure, i.e. whether alkyl, aryl or mixed alkyl/aryl.

$$(RO)_3P + HCl \longrightarrow (RO)_2P(H){=}O + RCl \qquad (14)$$

$$(RO)_2P(H){=}O + HCl \longrightarrow (RO)P(OH)(H){=}O + RCl \qquad (15)$$

$$(RO)P(OH)(H){=}O + HCl \longrightarrow H_3PO_3 + RCl \qquad (16)$$

In a detailed study of the effect of the nature of R on the reactivity and on the stabilising ability, Verizhnikov *et al.*[69] examined a series of trialkyl, dialkylaryl, alkyldiaryl and triaryl phosphites and came to the conclusions listed below which appear to be generally applicable:

1. Trialkyl phosphites, e.g. tributyl phosphite (TBP), trihexyl phosphite (THP) or trioctyl phosphite (TOP) take part in the

sequence of reactions (14)–(16) forming dialkyl phosphites, monoalkyl phosphites and H_3PO_3. Reaction (14) proceeds at an appreciable rate even at 20 °C with reaction (15) at least an order of magnitude slower.

2. Dialkylaryl phosphites such as dihexylphenyl phosphite (DHPP) lose an alkyl group forming alkylaryl phosphites (which are possible prodegradant substances) then react further to monoaryl phosphites.
3. Alkyldiaryl phosphites, e.g. hexyldiphenyl phosphite (DPHP) takes part only in reaction (14), forming inert diaryl phosphite $(ArO)_2P(H){=}O$ and this must be a major part of the reason why they represent the most effective class of mixed phosphites.
4. Triaryl phosphites are stable towards HCl.

Since HCl is a major decomposition product of the degradation of PVC the mode of action of phosphite stabilisers described above has an extra significance for PVC compared with other polymer systems where HCl is not a problem. However it must also be remembered that phosphites almost always form part of a multi-component formulation in which other components may compete effectively for HCl. Hybart and Rowley have investigated the combined effect of tridecyl phosphites and metal stearates in the stabilisation of PVC.[70] They concluded that the stabilising effect of phosphites depended on the ability of the phosphonates formed by reaction with HCl to neutralise metal ions probably by chelation. Finally, although de-alkylation of organic phosphites has been discussed entirely in terms of HCl it should be remembered that an analogous process involving carboxylic acids derived as by-products of the reaction of HCl with metal stearates (eqn. (17)) may be significant in some circumstances.[71]

$$(RO)_3P + R'COOH \longrightarrow (RO)_2\underset{\displaystyle H}{\underset{|}{P}}{=}O + R'COOR \qquad (17)$$

5.4.2 Reactions of phosphites with hydroperoxide groups

The reactions of phosphites with hydroperoxides have long been regarded as some of the most important stabilising mechanisms, particularly when operating as synergistic components in conjunction with phenolic antioxidants. Bimolecular rate constants (k_0) for the reaction, phosphite + hydroperoxide, have been measured for a wide variety of phosphites and attempts made to use Labachnik's substituent/reactivity[72] correlation to predict which are the most effective hydroperoxide decomposers. A selection of values obtained which are typical of the large number which have been documented is given in ref. 4.

Radical yields calculated from the rate of consumption of nitroxyl radicals were low ($\sim$0·02%) for aliphatic compounds and only slightly higher (up to 5%) for aromatics. These results were interpreted by Pobedimskii and Buchachenko[73] in terms of a 'concealed radical' mechanism in which a very effective solvent cage mechanism operated so that escape of radicals from the cage was inefficient and was the rate controlling factor (eqn. (18)):

$$R'OOH + (RO)_3P \rightleftharpoons [R'\dot{O} \cdots HO\dot{P}(OR)_3] \longrightarrow R'OH + O{=}P(OR)_3 \tag{18}$$

More recent results using the newer technique of CIDNP have strongly suggested that an alternative possibility, namely that radical and non-radical reactions proceed in parallel and in competition, is more realistic.[74] The relatively minor importance of the former route is the reason for the low radical yields. More detailed descriptions of the CIDNP data have been provided for the reactions between *t*-butyl hydroperoxide (TBH), and triethyl phosphite (TEP), in styrene, cumene hydroperoxide (CHP) and TEP in styrene, and CHP and pyrocatechol phosphite (PC) in benzene.[4] Evidence which also indicates that radical and non-radical reactions may proceed in parallel has been described by Humphries and Scott.[75] The rate of oxygen uptake in the AIBN-initiated oxidation of cumene in oxygen at 50 °C was rapidly reduced to zero by catechol phosphates of the type shown below (Fig. 5.9).

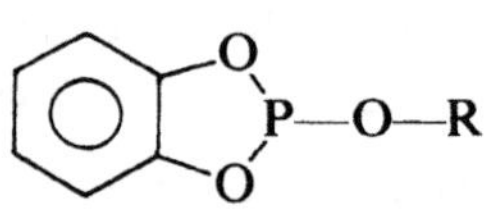

R = —CH(CH$_3$)$_2$ **(2)**

R = —C$_6$H$_5$ **(3)**

R = 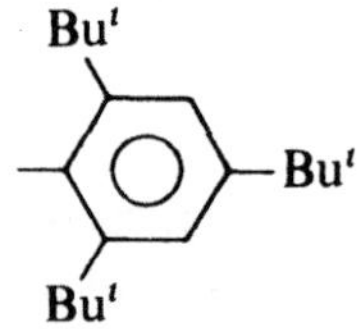

(4)

Groups **(2)** and **(3)** behaved similarly but the hindered phenol derivative **(4)** required a large initial oxygen uptake before it became effective. By contrast, triphenyl phosphite under the same conditions was auto-accelerating after a slight induction period. With cumene hydroperoxide as

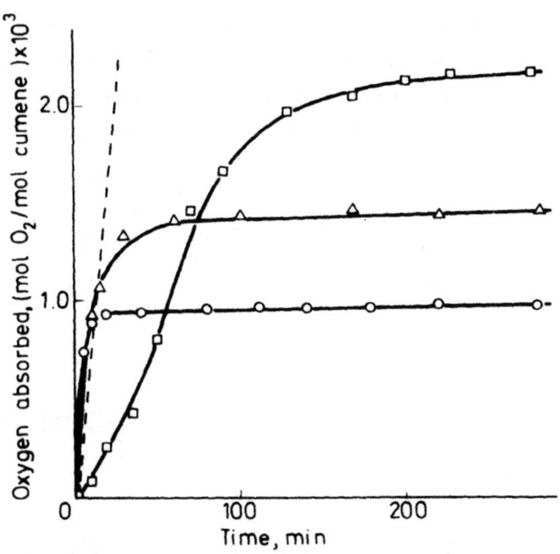

Fig. 5.9 Inhibition of AZBN-initiated oxidation of cumene in oxygen at 50 °C by 0·2 M catechol phosphites. $[AZBN]_0$, 0·02; □, hindered phenyl phosphite; △, isopropyl phosphite; ○, phenyl phosphite; ---, no phosphite. (After Humphries and Scott,[75] by permission of Butterworths.)

initiator in oxygen at 75 °C more dramatic effects were observed with (**4**) (Fig. 5.10).

It was concluded that whereas triphenyl phosphite acted by the stoichiometric destruction of hydroperoxides (eqn (19)), the hindered

$$(C_6H_5O)_3P + ROOH \rightarrow (C_6H_5O)_3P{=}O + ROH \qquad (19)$$

phenyl phosphites or possible products derived from them acted catalytically. The non-catalysed reaction has been described previously in some detail.[76] Further work on the catalysed process using hindered phenyl catechol phosphites showed that the effective catalyst is a substance derived from the phosphite. One such substance is the phosphate ester (**5**), but it is unlikely that it is the powerful antioxidant or is involved in its

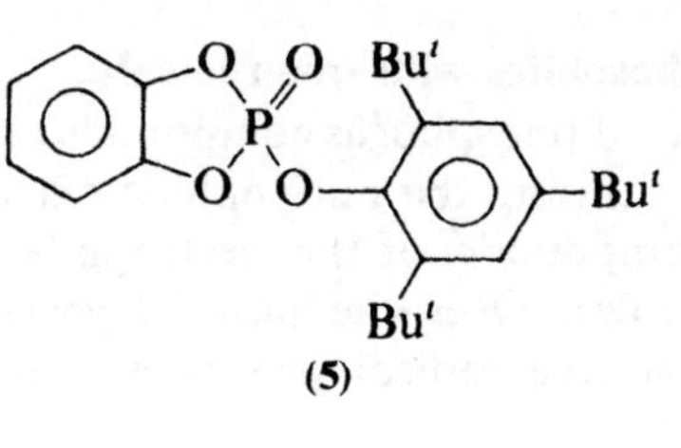

(**5**)

production. The catalyst seems to behave as a strong Lewis acid giving phenol and acetone as products in contrast to the α-methylstyrene which results from the radical mechanism (eqns. (20)–(22)) at low hydroperoxide/phosphite ratios. Measurements of the rate of polymerisation of

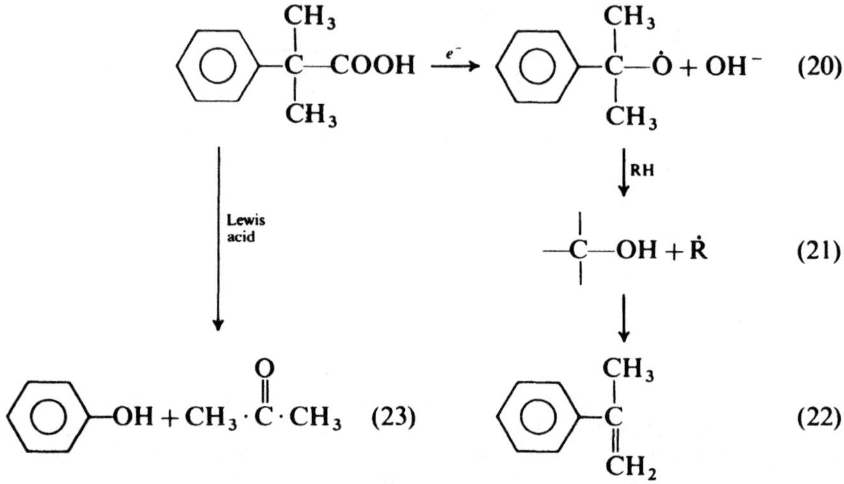

styrene initiated by cumene hydroperoxide and in the presence of catechol phosphite is consistent with a mechanism whereby initiating radicals escape with low efficiency (5–10 %) from a solvent cage. The main reaction leads to the formation of other products by a non-radical process and it is one of these which leads to the formation of the powerful Lewis acid discussed above. Its exact structure has yet to be elucidated but the hydroxy phosphate (6) and its hydrolysis product (7) have been suggested as possible candidates.

(6) (7)

5.4.3 Reactions of phosphites with free radicals

Many of the reactions of phosphorus compounds proceed by a heterolytic mechanism, thereby utilising the nucleophilic reactivity of the lone pair electrons in P(III) compounds, or the electrophilicity of the phosphorus atom in P(V) molecules. Over the past 20 years or so, however, an increasing number of free radical processes have been identified and

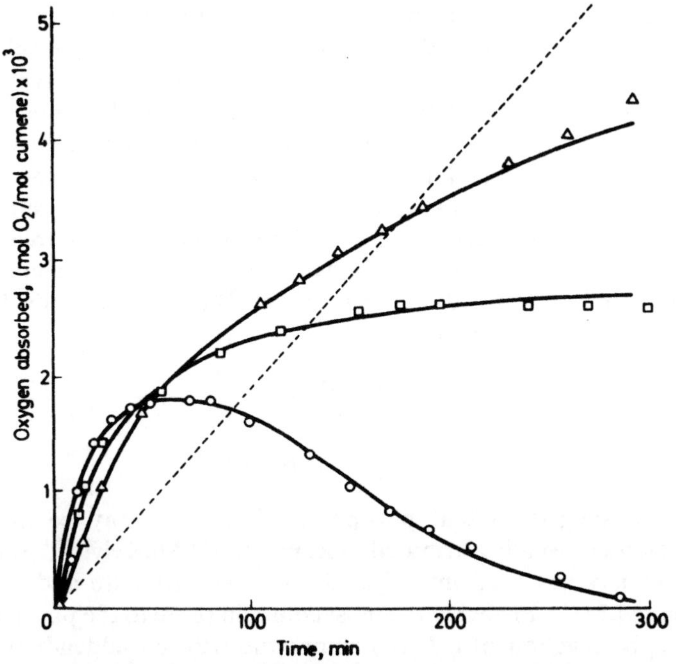

Fig. 5.10 Inhibition of cumene hydroperoxide-initiated oxidation of cumene in oxygen at 75 °C by catechol-hindered phenyl phosphite. $[CHP]_0$, 0·2 M; △, 0·002 M; □, 0·01 M; ○, 0·02 M; ---, no phosphite. (After Humphries and Scott,[75] by permission of Butterworths.)

described in comprehensive reviews.[77,78] In 1956 Hoffman reported[79] that the reaction between trialkyl phosphites and mercaptans (eqn. (24)), could

$$RSH + (C_2H_5O)_3P \rightarrow RH + S{=}P(OC_2H_5)_3 \qquad (24)$$

be brought about at elevated temperatures or at lower temperatures in the presence of light and suggested that a radical process may be involved. At about the same time an analogous reaction between trialkyl phosphites and alkyl disulphides was investigated by Walling and Rabinowitz[80] who suggested that both reactions may involve radical chain processes (eqns. (25)–(28)):

$$R'\dot{S} + P(OR)_3 \rightarrow R'{-}S{-}\dot{P}(OR)_3 \qquad (25)$$

$$R'{-}S{-}\dot{P}(OR)_3 \rightarrow \dot{R}' + S{=}P(OR)_3 \qquad (26)$$

$$\dot{R}' + R'SSR' \rightarrow R'{-}S{-}R' + R'{-}\dot{S} \qquad (27)$$

$$\dot{R}' + RSH \rightarrow R'H + R\dot{S} \qquad (28)$$

The type of experimental evidence on which the suggestion was based involved a comparison of the reaction between triethyl phosphite and isobutyl mercaptan in the presence and absence of a typical free radical chain initiator, namely AIBN. In the presence of 1·84 mole % of the initiator the reaction was complete in 2 min compared with 30 min in its absence. The average reaction chain length was estimated to be of the order of 5400. These early results which were very relevant to the subsequent understanding of the reactions of phosphite compounds with free radicals could be rationalised in terms of the available low energy *d*-orbitals which could be used to expand the valency of P(III) compounds forming phosphoranyl radicals (**8**) (eqn. (29)).

$$R{-}PX_3$$

(8)

$$\dot{R} + PX_3 \longrightarrow R{-}\dot{P}X_3 \qquad (29)$$

The first suggestion that phosphoranyl radicals may be important intermediates is usually attributed to Ramirez and McKelvie.[81] Since then a large variety has been investigated and their structure and reactivity documented.[82-84] Probably the most common route to the phosphoranyl radical is by reaction of a P(III) compound with an addendum radical produced thermally or photochemically from a peroxide or hydroperoxide. For example the ESR spectrum of the *t*-butoxytrimethyl phosphoranyl radical (**9**) was observed following the photolysis of di-*t*-butyl peroxide in the presence of trimethyl phosphine in cyclopropane.[85]

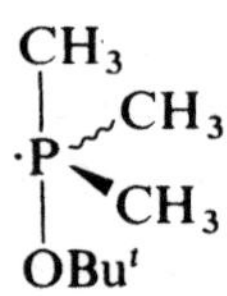

(9)

$$Bu^tOOBu^t \xrightarrow{h\nu} 2Bu^t\dot{O} \qquad (30)$$

$$Bu^t\dot{O} + P(CH_3)_3 \rightarrow (9) \qquad (31)$$

$$(9) \rightarrow (CH_3)_2POBu^t + \dot{C}H_3 \qquad (32)$$

When other phosphines with larger alkyl groups were used, no phosphoranyl radicals were detected but large signals resulted from the displaced alkyl radical. Later work by these[86] and other workers described the ESR spectra of a variety of phosphoranyl radicals formed by the attack of alkoxy radicals, derived photochemically or thermally, on trialkyl phosphites.

$$Bu^t\dot{O} + P(OAlk)_3 \rightarrow Bu^tO\dot{P}(OAlk)_3 \qquad (33)$$

The appearance of alkyl radicals derived from the addendum alkoxy radicals or from the phosphite indicates that the phosphoranyl radical may undergo α or β fission depending on the nature of the group attached to the phosphorus and on the experimental conditions (eqn. (34)).

$$Bu^tO\dot{P}X_3 \xrightarrow{\alpha} Bu^tOPX_2 + \dot{X} \qquad (34a)$$

$$Bu^tO\dot{P}X_3 \xrightarrow{\beta} \dot{B}u^t + O{=}PX_3 \qquad (34b)$$

For example, an ESR study of 10% cyclopentane solutions of several phosphites, including $(CH_3O)_3P$, $(C_2H_5O)_3P$, $(i\text{-}C_3H_7O)_3P$, $(n\text{-}C_4H_9O)_3P$ and $(t\text{-}C_4H_9O)_3P$ and containing 15% of di-*t*-butyl peroxide were irradiated (UV) at −70°C and the phosphoranyl radical detected in each case.[87] In contrast none was detected for triphenyl phosphite or trialkyl phosphite presumably due to rapid scission of the radical. As the temperature was increased the concentration of alk˙ was increased at the expense of the phosphoranyl radical. The presence of oxygen poses additional complications which are well illustrated by reference to the ultraviolet irradiation of dialkylperoxide solutions in trialkylphosphites at low temperatures. ESR spectra of these systems in the presence of oxygen indicated that radicals (**10**) and (**11**) were present. When photolyses were

$$alk{-}O{-}\dot{P}(O{-}alk')_3 \quad \textbf{(10)}$$

$$alk{-}O{-}P(O{-}alk')_3(\cdot O_2) \quad \textbf{(11)}$$

$$\textbf{(10)} + O_2 \longrightarrow \textbf{(11)} \qquad (35)$$

carried out at higher temperatures the spectrum of (**10**) persisted for only a short time before being replaced by that of (**11**) which is consistent with the removal of oxygen in a chain process (eqns. (36–39)).

$$(RO)_4\dot{P} + O_2 \rightarrow (RO)_4P\dot{O}_2 \qquad (36)$$

$$(RO)_4P\dot{O}_2 + (RO)_3P \rightarrow (RO)_4P{-}\dot{O} + (RO)_3P{=}O \qquad (37)$$

$$(RO)_4P{-}\dot{O} + (RO)_3P \rightarrow (RO)_4\dot{P} + (RO)_3P{=}O \qquad (38)$$

$$(RO)_4P{-}\dot{O} \rightarrow (RO)_3P{=}O + R\dot{O} \qquad (39)$$

These reactions are certainly of importance in the free radical initiated auto-oxidation of trialkyl phosphites particularly at low temperature where β scission of the alkoxyphosphoranyl radicals is retarded relative to its reaction with oxygen (eqns. (40a, b)).

$$(RO)_4\dot{P} \xrightarrow{\beta\ (slow)} (RO)_3P{=}O + \dot{R} \qquad (40a)$$

$$(RO)_4\dot{P} \xrightarrow{O_2\ (fast)} (RO)_4P\dot{O}_2 \qquad (40b)$$

ESR spectra of the closely related dialkylphosphonyl radicals (**13**) have been observed[88] following the photolysis of di-*t*-butyl peroxide in the presence of dialkylphosphites (dialkylphosphonates (**12**)). Identical but

$$(RO)_2P(H){=}O \quad (\mathbf{12}) \qquad\qquad (RO)_2\dot{P}{=}O \quad (\mathbf{13})$$

more intense spectra were obtained when di-*t*-butylperoxide was photolysed in the presence of tetraethylpyrosphosphite (**14**). In contrast to the reaction of $Bu^t\dot{O}$ radicals with trialkylphosphites no phosphoranyl

$$(C_2H_5O)_2POP(OC_2H_5) \quad (\mathbf{14})$$

radicals derived from (**14**) could be detected but after prolonged photolysis the spectrum of radical (**15**) was observed (formed by addition of $Bu^t\dot{O}$ to *t*-

$$(Bu^tO)_2\dot{P}(OC_2H_5)_2 \quad (\mathbf{15})$$

butoxydiethoxy phosphite) as well as that of $Bu^{t\cdot}$ derived from (**15**) by β scission (eqn. (43)).

$$Bu^t\dot{O} + (\mathbf{12}) \rightarrow Bu^tOH + (\mathbf{13}) \qquad (41)$$

$$Bu^t\dot{O} + (\mathbf{14}) \rightarrow Bu^tOP(OC_2H_5)_2 + O{=}\dot{P}(OC_2H_5)_2 \qquad (42)$$

$$(\mathbf{15}) \rightarrow \dot{B}u^t + O{=}P(OC_2H_5)_2(Bu^tO) \qquad (43)$$

More detailed information regarding the mechanisms of radical reactions of organic phosphites have been provided by Russian workers[89] using the technique of chemically induced dynamic nuclear polarisation

(CIDNP) of the ^{31}P nucleus. They showed that the thermal decomposition of dicyclohexylperoxydicarbonate (DCP) or of dilauroyl peroxide (DLP) in tetrachlorethane at 75–100 °C in the presence of dimethyl phosphite results in a negative polarisation. This indicates that the reaction proceeds via addition of a primary radical, R$\dot{O}$, to the double bond of the phosphite followed by disproportionation of the oxyphosphorane radical formed with another R$\dot{O}$, i.e.

$$R\dot{O} + H\text{—}P(=O)(OCH_3)_2 \longrightarrow RO\text{—}P(H)(\dot{O})(OCH_3)_2 \xrightarrow{R\dot{O}} RO\text{—}P(=O)(OCH_3)_2 \quad (44)$$

rather than the alternative but kinetically equivalent mechanism:

$$R\dot{O} + H\text{—}P(=O)(OCH_3)_2 \longrightarrow \dot{P}(=O)(OCH_3)_2 \xrightarrow{R\dot{O}} RO\text{—}P(=O)(OCH_3)_2 \quad (45)$$

On the other hand, when cumylhydroperoxide reacted with an excess of the triaryl phosphite (**16**) in benzene a small positive polarisation resulted which could best be accounted for by the scheme of eqn. (46).

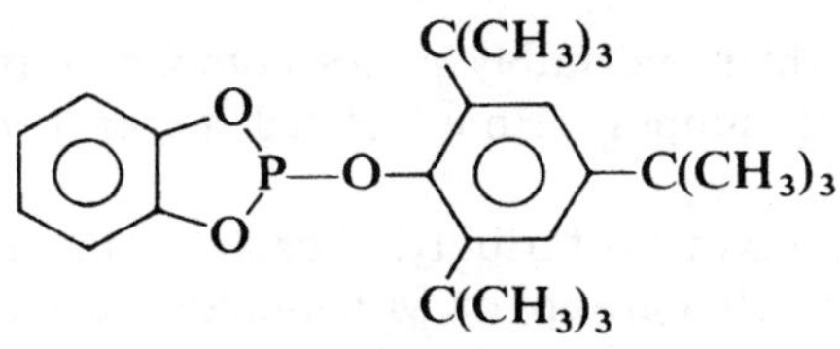

(16)

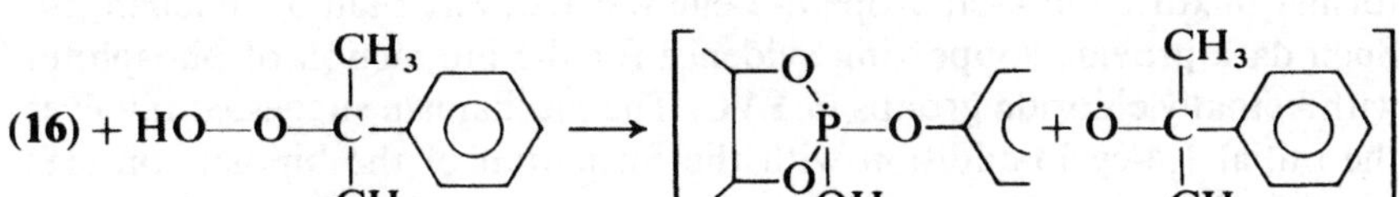

(46)

These authors have made an intensive study of the relative oxidation rates of a large number of phosphites having a wide variety of structures. They have documented the reasons for the differences in behaviour of alkyl and of aromatic phosphites, in particular those with hindered phenoxyl groups.[4]

5.4.4 Reactions of phosphites with ketoallylchloride groups

It has been suggested recently by Minsker and Fedoseyeva[90] that at relatively high rates of dehydrochlorination of PVC, the content of ketoallylchloride groups (**17**) is a predominant factor in determining the rate of initiation of polyene sequences and hence development of colour in the polymer.

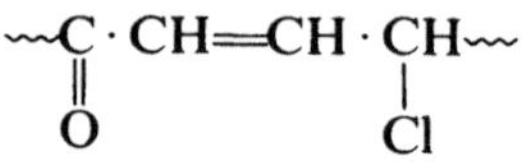

(**17**)

Further, the rate is substantially influenced by the presence of organic phosphites due to their chemical interaction with isolated double bonds in PVC. The following experimental evidence is typical of that on which this proposal is based. Heating PVC with various phosphites at 80–100 °C under conditions where the rate of HCl elimination is negligible results in quantitative removal of internal double bonds.[4] The molecular mass of the PVC measured by viscometery following ozonolysis and hydrolysis is the same as that before treatment. The presence of phosphorus in the PVC was reliably detected by emission spectroscopy after extraction of the unreacted phosphite but no quantitative determination could be made. Finally the polymer treated in this way eliminated HCl at a lower initial rate than untreated polymer.

Such evidence, which inevitably is not completely free of ambiguity, receives substantial support from related experiments using model compounds.

When equimolar mixtures of tributyl phosphite with methyl vinyl ketone (a model for the ketoallyl group) or with 4-chloropent-2-ene (a model for the chloroallyl group) were heated at 80 °C the phosphite reacted in the former mixture but 4-chloropent-2-ene was recovered almost unchanged. Such data provide supporting evidence for the interaction of phosphites with ketoallylchloride groups in PVC. The mechanism suggested involves the initial 1,4-cyclo-addition with the formation of the bipolar ion (**18**) which is then transformed either into the cyclic oxaphospholene (**19**) or in the presence of HCl into the stable ketophosphonate (**20**).

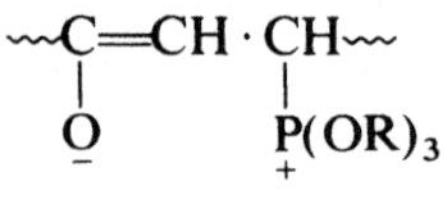

(**18**)

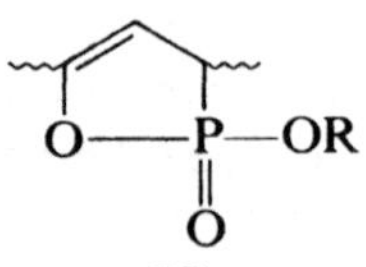

(**19**)

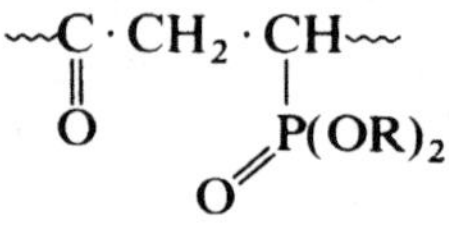

(**20**)

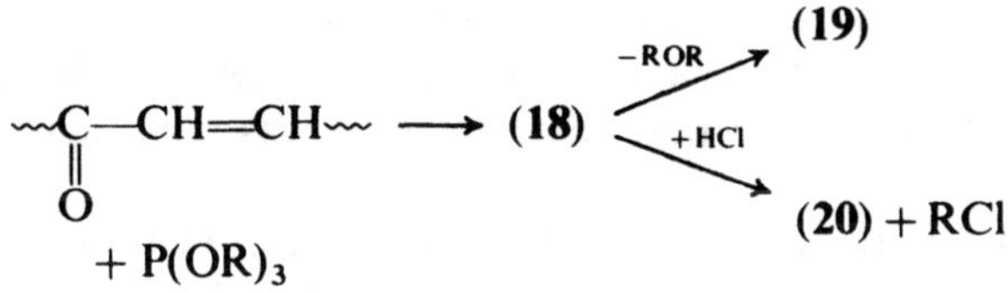

(47)

(48)

Tri-isooctyl phosphite reacts with PVC at a measurable rate even at room temperature and results in a decrease in the content of ketoallyl groups in the polymer. Although this decrease results in improved thermal stability of the polymer[90] in the initial stages under conditions of PVC thermal degradation, the rate of HCl elimination gradually increases to the level of the untreated polymer as the reaction proceeds and the initial level of ketoallylchloride groups is restored. Analysis of the kinetic data, however, shows that when the phosphite concentration is greatly in excess of that of the ketoallylchloride groups (which is always the case in practice), the ratio of the rate constants for ketophosphonate formation and decomposition is of the order of 1000.

On the basis of his experiments using PVC and low molecular weight analogues, Minsker and Fedoseyeva[90] claim to have shown unequivocally that the reaction of phosphites with ketoallylchloride chromophores is the main stabilising process which dominates all other reactions of phosphite in the polymer.

5.4.5 Reactions with metal salts

The ability of organic phosphites to complex with and thereby reduce the catalytic activity of various transition metal ions is often quoted as an important stabilisation process but direct evidence of this is not easy to find. Such complexes would be expected to be highly coloured and some confirmation of this has emerged[91] from studies of coloured model compounds involving Ti^{4+} and V^{5+}. However, such coloured complexes appear to be very sensitive to oxygen, moisture or increases in temperature and the colour stabilising effect appears to be related to efficient hydrolysis of the complexes to colourless products. Stabiliser compositions involving phenolic antioxidants present an additional complication since these substances also have an affinity for metal ions, an interaction which gives rise to colouration but whose response to moisture and oxygen may be different from that of the phosphite–metal complexes.[92] The overall effect of the formulation in suppressing colouration depends on the competition between the two systems. For example it has been shown that addition of an excess of diphenylisooctyl phosphite to a hexane solution of the complex of

2,2′-methylene-bis(4-methyl-6-*t*-butylphenol) (AO-2246) with $VOCl_3$ eliminated the bisphenol from the complex. An ESR examination of the resulting solution showed that the phosphite was coordinated with the metal ion.

Most of the work reported involving phosphite–metal ion interaction has been concerned with the stabilisation of polyethylene where the degradation mechanism is different from that of PVC and so care should be taken in making any extrapolation between one system and another. An added factor as far as PVC is concerned is that the most relevant metal ions are Zn^{2+}, Cd^{2+}, Ba^{2+}, Ca^{2+} and Sn^{4+} for which little or no information is available.

5.5 PHOTOSTABILISATION

The reasons for the photochemical instability of PVC have been dealt with in detail in Chapter 3 and the problems which this poses in the increasingly important area of outdoor application of the polymer are immediately apparent. What is more, the relationship between the thermal and photochemical processes emerges from this in a very obvious way, in that the most serious sensitisers of the photodegradation (namely unsaturation—hydroperoxy and conjugated carbonyl groups) arise during processing as products of the thermal degradation. Samples which have been thermally stabilised in an apparently adequate way as far as processing is concerned, in the sense that minimal colouration, chain scission or crosslinking occurs, may be photosensitive if the stabiliser formulation has not been chosen with photostabilisation in mind. It is for this reason that formulations intended for outdoor application contain a specific component which can survive the processing stage and thereafter afford a degree of photoprotection to the polymer by minimising the extent of the photodegradation. Many accounts of the mechanism of photostabilisation of polymers in general have appeared in the literature[42,93,94] and only some general remarks and some aspects which are particularly relevant to PVC will be mentioned briefly here.

In general terms the efficiency of any photochemical process reflects the competition between photophysical processes like fluorescence, intersystem crossing and internal conversion which do not result in any permanent chemical change and the photochemical processes like free radical production, intermolecular rearrangements, elimination or ionisation nearly all of which do cause permanent changes which are reflected in a deterioration in optical and mechanical properties. Photostabilisers attempt to minimise the photochemical effects by maximising the photophysical processes. They fall into three broad categories, namely:

1. Ultraviolet absorbers.
2. Excited state quenchers.
3. Scavengers of photolytically produced free radicals and other chain propagators.

The characteristics of group (3) have already been dealt with in relation to thermal degradation and the same general criteria apply when the transients are produced photolytically. Only groups (1) and (2) will be dealt with here.

5.5.1 Ultraviolet absorbers

The function of an ultraviolet absorber is to compete effectively for photons in the spectral region where potential photosensitisers absorb, namely about 280–350 nm, and to dissipate the absorbed energy in a harmless way by means of the photophysical processes mentioned earlier. The absorber's efficiency depends primarily on how effective this process can be made. The question of how much light is left unabsorbed and therefore potentially harmful is a critical factor and has been discussed and conveniently illustrated by Heller[93] using model systems containing acetophenone as sensitiser and 2-(2-hydroxy-5-methyl-phenyl)-benzotriazole as ultraviolet absorber. For example, at absorber levels of 0·1 % (which represents the lower end of the range used in normal practise) only 50 % of the incident (330 nm) light reaches a depth of 49 μm and 10 % reaches 154 μm. The importance of reactions occurring in a thin surface layer is therefore apparent and this aspect of the problem will be returned to later.

The amount of radiation absorbed by a sensitiser in the presence of an ultraviolet absorber as a fraction of its absorbance in the absence of the sensitiser is called the 'residual absorption (R)' and is a measure of the protective power of the absorber. The higher the numerical value of *R* the poorer is the effect of the absorber. Values of *R* depend on the relative concentrations and molar extinction coefficients of the various absorbing species present and can be calculated easily assuming that the Beer–Lambert Law is valid. For a particular absorber concentration there may be very little sensitiser absorption left in a thick sample whereas in a thinner sample the higher residual absorption necessitates higher absorber concentrations. Limits on the extent to which absorber concentrations can be increased are imposed by the shape of the absorption spectral band. If the band tails appreciably into the visible region then increases in levels bring about unacceptable yellow colouration. No ideal absorption band shape exists and some compromise must usually be reached between maximum protection and minimum colouration.

A further requirement of any ultraviolet absorber which is not concerned with its photochemical properties is that it be compatible with the polymer. This condition that a true solution is formed applies to all additives and is not always easy to fulfil. In general, low molecular weight additives tend to accumulate in the amorphous phases of the polymer and at lattice imperfections leaving crystalline regions largely unprotected. Nominal concentrations of say 0·1% therefore may be exceeded by several-fold in some regions and be correspondingly reduced in others.

Apart from these primary properties which are concerned with the competitive absorption of the absorber some further requirements are (1) that the absorber should be photochemically stable and inert, (2) that it should have no sensitising properties and (3) that its volatility should be low so that it does not diffuse out of the polymer at an appreciable rate. In general it is the balance between these various factors which determines whether the performance of the absorber is satisfactory in practice.

Large numbers of potentially useful substances have been investigated and many structures have appeared in the patent literature but relatively few have found general application. Those which have are derivatives of either salicylates, *o*-hydroxybenzophenones or benzotriazoles. The salicylates were the first light protective group to be used but have now been largely superseded in polymer applications although derivatives of (**21**) are still widely used in the cosmetic industry as ‘suncreams’.

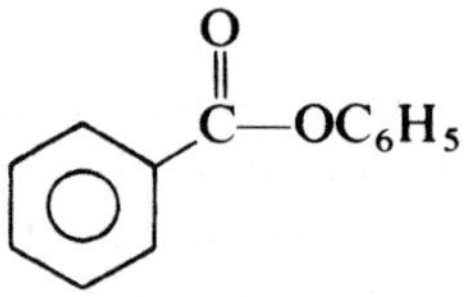

(**21**)

Hydroxybenzophenones

These compounds represent the most commercially used class of ultraviolet absorbers, partly on account of their high photostability. A large variety of structures has been prepared and investigated but most of those used successfully are related to types (**22a–c**).

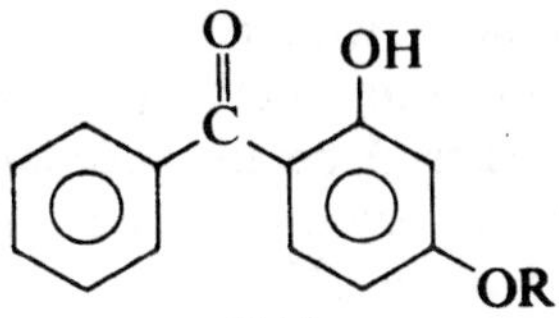

(22a)

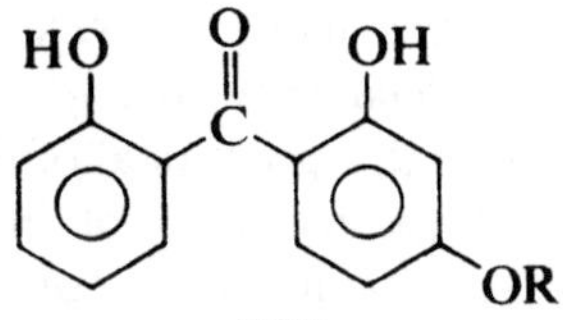

(22b)

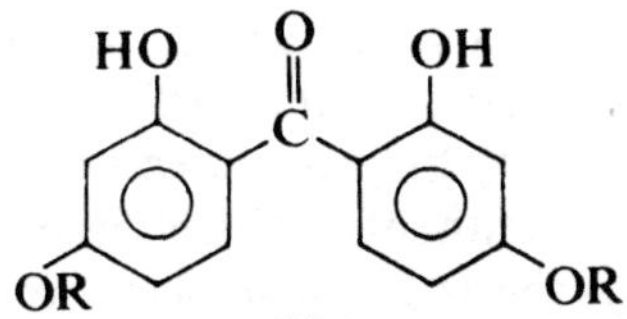

(22c)

The mechanism by which they operate is thought to involve an interplay between keto and enol forms which can be represented thus:

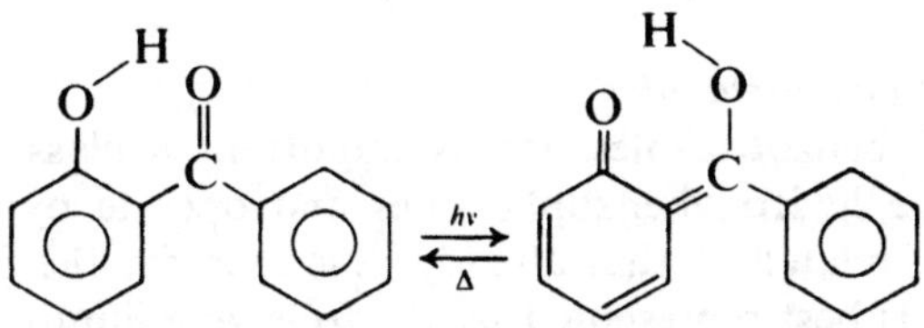

(keto form) (enol form)

In the ground state the keto form is thermodynamically more stable but in the excited singlet state the reverse is the case.

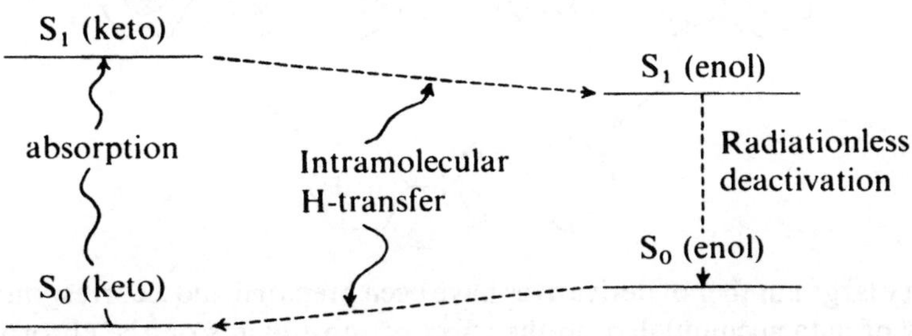

There has been and still is much speculation as to which particular excited states are involved since several possibilities exist of which the most likely are listed below, where S and T refer to singlet and triplet states:

(i) S_0(keto) → S_1(keto) → S_1(enol) → T_1(keto) → S_0(keto)

(ii) S_0(keto) → S_1(keto) → S_1(enol) → T_1(enol) → S_0(enol) → S_0(keto)

(iii) S_0(keto) → S_1(keto) → T_1(keto) → T_1(enol) → S_0(enol) → S_0(keto)

(iv) S_0(keto) → S_1(keto) → S_1(enol) → S_0(enol) → S_0(keto)

Detailed studies using fluorescence combined with nanosecond[95] and picosecond[96] flash spectroscopy appear to favour schemes (iii) and (iv) in general although a great deal of diversity exists. A detailed account of the kinetics and spectroscopy of intramolecular proton transfer in a wide variety of systems has been compiled by Klöpffer.[97]

It has already been mentioned that the long term performance of polymers is often limited by loss of stabiliser from the polymer matrix by evaporation following diffusion to the surface or by solvent extraction. Efforts to restrict the mobility have included the design of stabilisers with long aliphatic alkyl substituents which impart improved polymer compatibility and the synthesis of substances like 2,4-dihydroxy-4'-vinylbenzophenone[98] which can be copolymerised with the bulk monomer and hence immobilised or grafted only on the surface to form a thin ultraviolet protective screen.[99]

Hydroxyphenyl benzotriazoles

Hydroxyphenyl benzotriazoles are as important a class of ultraviolet absorbers as the hydroxybenzophenones and operate by a mechanism which is closely related to that already described for the former group. Their structure is best represented by the following limiting structures:

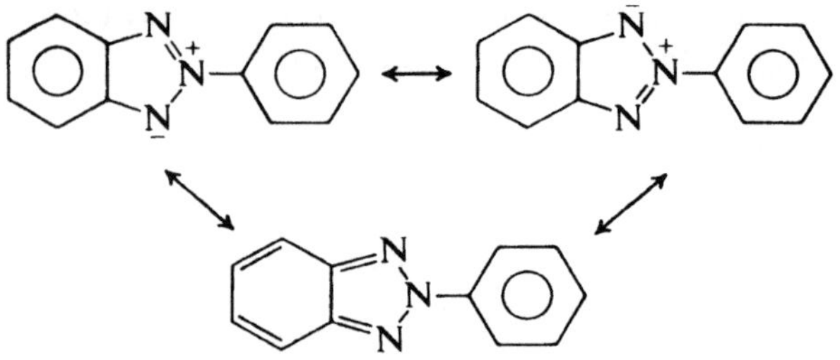

A very large number of derivatives have been prepared and a corresponding body of data accumulated on the effect of substituents on the absorption spectra, light fastness and ultraviolet absorber efficiency of these molecules. In general it appears that the parent molecule has a strong absorption maximum at about 310 nm. Blue shifts in the maximum produced by the introduction of an auxochrome into the *p*-position of the phenyl ring is cancelled by introducing a further substituent into the *o*-position relative to the benzotriazole ring. Clearly, the steric hindrance of the *o*-substituent is sufficiently large to twist the two rings far enough out of plane for effective conjugation to disappear. When the phenyl ring contains an *o*-hydroxy substituent the absorption is split into two maxima of roughly equal intensity. The shorter wavelength absorption (~300 nm) corresponds to the conformation with the planes of benzotriazole and phenyl rings at right angles to each other and the longer wavelength absorption (~340 nm) to the conformation with rings much nearer planarity and containing an intramolecular N---H bond (**23**). The two conformations are present in roughly equal amounts. It is the longer wavelength absorption which is the basis of the intramolecular hydrogen transfer process and therefore of the ultraviolet absorber properties. In general molar extinction coefficients are higher than for hydroxybenzophenones and so benzotriazoles are generally superior stabilisers.

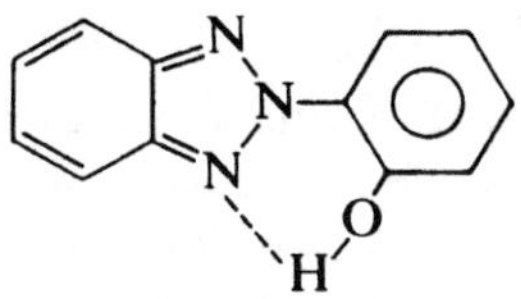

(**23**)

The spectroscopic details of the photophysical processes involved in the action of benzotriazoles have been studied in detail by Werner *et al.*[100] and others and it is clear that many similarities exist between them and hydroxybenzophenones. A complication however is that in addition to operating by the screening effect, benzotriazoles enhance their photostabilising abilities by acting as acceptors in an energy transfer process from triplet excited polymer to stabiliser. The stabiliser molecule in its excited singlet state (ultraviolet absorber) and triplet state (energy transfer acceptor) is rapidly deactivated ($\tau(S_1) = \sim 60$ ps and $\tau(T_1) = < 20$ ns) and the intramolecular hydrogen bond is the origin of the rapid deactivation. There has also been some suggestion that they may also take part in radical scavenging processes.[101]

5.5.2 Excited state quenchers

One of the photophysical processes mentioned above which may participate in the deactivation of an excited polymer chromophore is energy transfer, a process whereby the excitation energy of the polymer (whether acquired directly or in a sensitised process) is transferred to another (quencher) molecule.

$$\text{Polymer}^* + \text{Quencher} \rightarrow \text{Polymer} + \text{Quencher}^* \qquad (49)$$

An immediate contrast between this group and the absorbers discussed in the previous section is that the quenchers need not have high absorptions or indeed any absorption at all in the wavelength region where the polymer absorbs. Several types of energy transfer are possible and are characterised initially by the types of polymer and quencher excited states which participate. Restrictions which govern the efficiency of the various types may be understood in terms of the Wigner spin selection rules details of which may be found elsewhere.[102] The most commonly encountered types are singlet–singlet transfer (eqn. (50)) and triplet–triplet transfer (eqn. (51)) but many variations exist and the detailed spectroscopic and kinetic variations have been reviewed.[103]

$$\text{Polymer}(S_1)^* + Q(S_0) \longrightarrow \text{Polymer}(S_0) + Q(S_1)^* \quad (50)$$

$$\text{Polymer}(T_1)^* + Q(S_0) \longrightarrow \text{Polymer}(S_0) + Q(T_1)^* \quad (51)$$

An important requirement of course is that the excited state quencher formed by the energy transfer process is able to dissipate its energy harmlessly. One efficient quencher which does not fulfil this requirement and is therefore a powerful initiator of photooxidative degradation and hydroperoxide formation is molecular oxygen, which is formed in its singlet excited state ($^1O_2$) by energy transfer from a sensitiser triplet.

$$^3\text{Sens}^* + {}^3O_2 \longrightarrow \text{Sens}(S_0) + {}^1O_2^* \quad (52)$$

$$^1O_2 + \text{Polymer} \longrightarrow \text{Polymer hydroperoxide} \quad (53)$$

An effective quencher of the photooxidative degradation therefore may act by quenching either the excited sensitiser molecule or the singlet oxygen derived from it or both. The quenching of singlet oxygen has been investigated extensively by many groups of workers and a wide variety of mechanistic possibilities have been established,[104–106] many of which occur at diffusion controlled rates. Many of the quenchers which have found commercial application are Ni(II) chelates, the structures of some of which are given below. Some controversy still exists in some cases concerning the details of the role they play, i.e. whether their ability to quench singlet oxygen or sensitiser triplets is of prime importance.

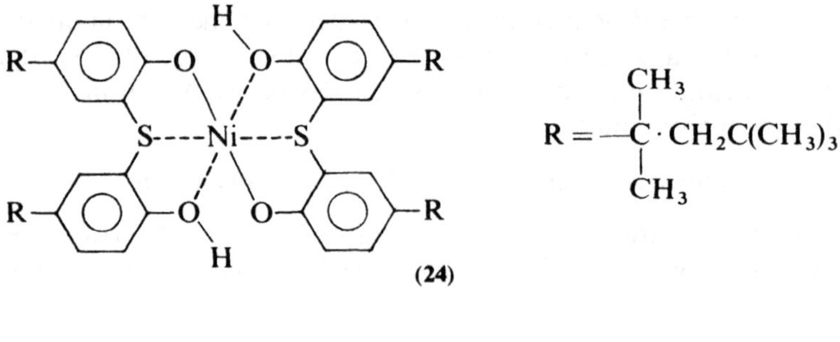

(24)

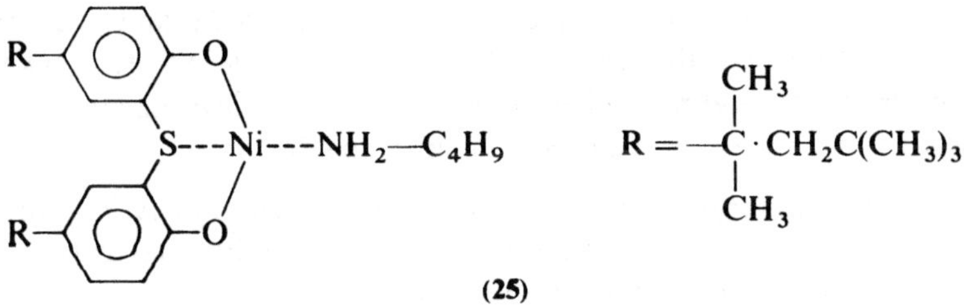

(25)

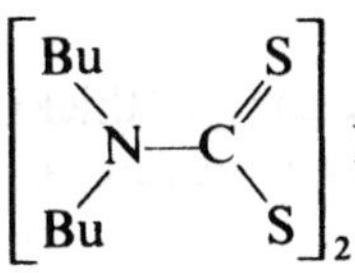

(26)

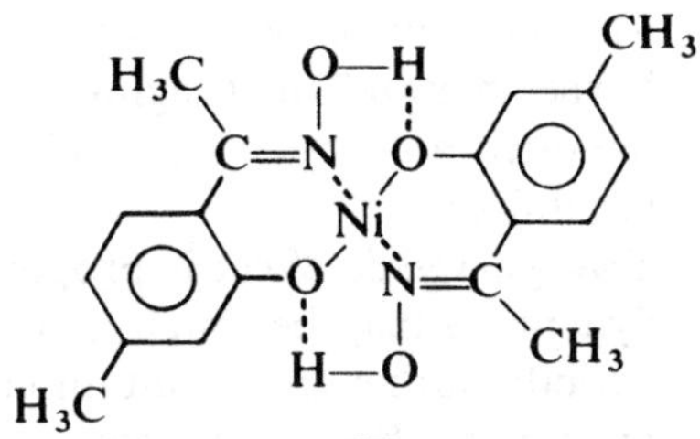

(27)

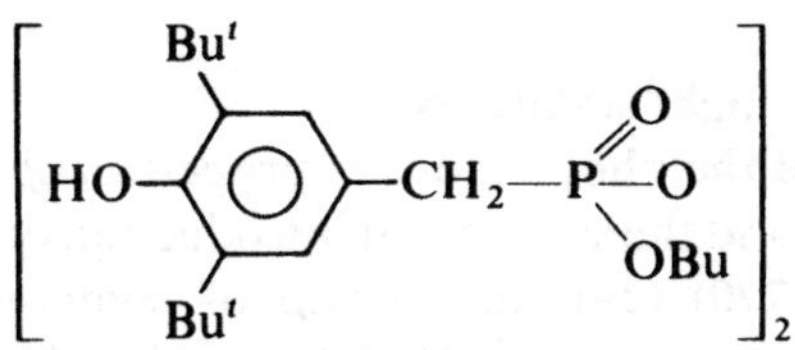

(28)

Synergistic mixtures

When ultraviolet absorbers (Abs), whether of the hydroxybenzophenone or hydroxyphenyl benzotriazole type, are used in conjunction with antioxidants (AOx) a synergism is observed in the sense that the protective action of the mixture (measured by the induction period before the onset of photooxidation) is considerably greater than the sum of either component separately. In addition it is clear that thin films are oxidised more rapidly than thick ones. Several mechanisms have been suggested to explain this phenomenon but the one most generally accepted is the diffusion mechanism which can be summarised as follows.

Introduction of Abs decreases the effective thickness of the layer in which photoprocesses take place and in deep layers of the polymer the rate of initiation and therefore of consumption of AOx will be less in the presence of Abs. In the surface layers the screening effect of Abs is less and consumption of AOx is hardly affected. For this reason AOx diffuses from the bulk to the surface and the overall synergistic effect depends on the relative magnitudes of the rate of diffusion of AOx and the rate of photooxidation of the surface layers. It is possible to make the generalisations listed below based on relatively simple diffusion theory[107] although more work is required to clarify the details:

1. Since synergism arises as a result of diffusion of AOx molecules their diffusion coefficients and the mobility/plasticity of the polymer will be important factors.
2. The light intensity will also be a critical factor which should be borne in mind in designing testing procedures where the light intensity may be several fold higher than that encountered in general use.
3. The synergistic effect is proportional to the sample thickness.[108]
4. Synergism may be much more evident when exposure to light is periodic rather than continuous.
5. Synergism has also been shown to occur when peroxide decomposers are combined with absorbers.[108]

5.5.3 Hindered amine light stabilisers

Since the discovery of the photostabiliser properties of 2,2,6,6-tetraalkylpiperidine in 1967[109] and the subsequent introduction of the closely related compound (Tinuvin 770) (**29**), this group of compounds, which have become known as hindered amine light stabilisers (HALS), have established a prominent place in the photostabilisation market. Although they are used extensively in the stabilisation of polyolefins they have not found a large application in the case of PVC up to this time and so will be mentioned only briefly here.

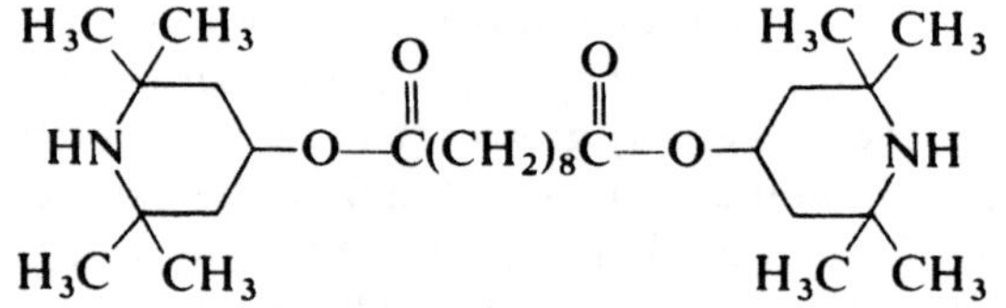

(29)

The mechanism by which HALS operate has been the object of much interest but remains incompletely understood. What is certain is that they are multifunctional stabilisers whose activity is not restricted to one particular aspect of the degradation process. The problem of establishing the relative importance of the various contributory factors is a familiar one and has proved as difficult as for other related systems. Their roles as ultraviolet absorbers, quenchers of excited states (sensitisers or singlet oxygen) or peroxide decomposers (at least at ambient temperatures) appear to be unimportant although they do form HALS–hydroperoxide associates which may well be a vital aspect of their function. Decomposition of these associates by light or by reaction with $R\dot{O}_2$ radicals leads to the formation of amine oxidation products including nitroxyl radicals which may also have a photoprotective function by acting as scavengers of alkyl radicals in a cyclic process. The formation of the HALS–ROOH associates has the added advantage of accumulating the stabilisers in the region where maximum photoprotection is required.

Although the diversity of behaviour of the systems means that as yet the details remain incompletely understood it also contributes towards making the performance of HALS superior to other classes of ultraviolet stabilisers. A greater application in the case of PVC may emerge in due course. More details may be found in the various comprehensive reviews which have appeared in the literature.[110,111]

5.6 SECONDARY STABILISATION

Substances are often included in PVC formulations the functions of which are to facilitate some aspect of the stabilisation process. These substances would not be sufficiently active to function as stabilisers in their own right and are known as secondary stabilisers. A large variety of such substances exists, many of which have been developed in a totally empirical fashion and the mechanism by which they function is often obscure. Nevertheless, many are invaluable aids to the production of a stable product and a few of the most widely used are mentioned briefly in the following sections.

5.6.1 Epoxy compounds

Epoxy compounds, of which epoxidised soya bean oil (EPO) is a typical example, are frequently used as co-stabilisers with zinc and cadmium carboxylates. During the processing stage some are consumed and some grafted onto the polymer but their presence induces a long term stability by delaying the onset of the blackening and crosslinking stage. Their main function seems to be reaction with $ZnCl_2$, thereby delaying the catalytic degradative reaction of the Lewis acid. In addition, model compound studies have shown that as well as reacting with HCl to form chlorhydrin, epoxy compounds react with allylic chlorine to form allylic chloro ethers (**30**) and (**31**) (eqn. (54)).

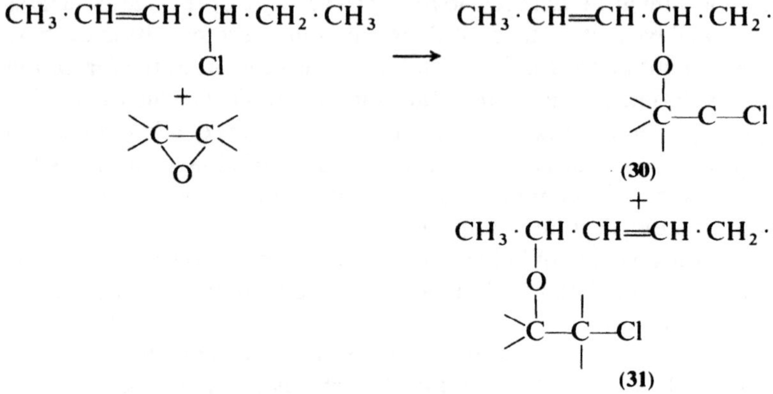

(54)

5.6.2 α-Phenylindoles (α-PI)

This secondary stabiliser (**32**) enhances the quality of the colour in rigid PVC supposedly by reacting with allylic chlorine. Model compounds indicate that it reacts first via the *β*-ethylenic group to give (**33**) then subsequently via the aromatic ring to form (**34**) and in the presence of $ZnCl_2$ a tri-substitution product (**35**).

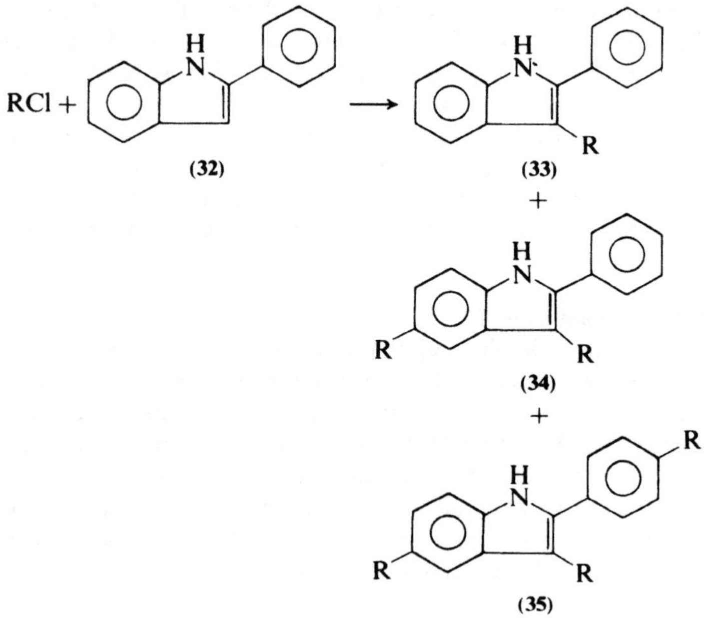

In the presence of the polymer α-PI prevents colouration by decreasing the length of conjugated polyene sequences so that the initial yellow colour observed when epoxy compounds alone are used is absent. The pink colour observed during the processing is due to an HCl–α-PI complex which is eliminated by adding an HCl acceptor such as an epoxy compound. These considerations mean that α-PI is classed as a short term secondary stabiliser which has little effect on the appearance of $ZnCl_2$ and is almost invariably used in conjunction with epoxy compounds.

5.6.3 Polyols

Polyols, mainly mannitol and sorbitol, have been used as long term secondary stabilisers which probably operate by complexing with $ZnCl_2$. Under normal processing conditions they would undergo extensive decomposition and it may be that the effects observed are due to reactions of the decomposition products than to the compounds themselves.

5.7 CONCLUSIONS

The degradation and stabilisation of PVC has been and still is a subject of intense interest which is reflected by the large number of papers which continue to appear in the literature. The extent of the subject is such that this chapter was undertaken with some misgivings and they have been amply confirmed as it has progressed. The number and variety of aspects of the problem which have emerged and which could be treated in more detail are legion and quite beyond the amount of space available here. Nevertheless it is hoped that some of the better-understood aspects of the stabilisation process have been exposed sufficiently to complement the other more detailed chapters. If the subject is difficult then a summary is even more so but some points are listed below which perhaps emphasise some of the most important aspects of the problem.

1. The wide range of application of PVC has resulted in at least an equal number of multi-component formulations being developed. The consequent complexity makes it extremely doubtful whether any rigorous conclusions could be formulated which would apply in all cases.
2. All the stabilisers discussed are multi-functional and the problem of establishing a relative order of importance for these functions would be difficult even if it was felt that such an order is a constant characteristic which is probably not the case.

3. The mechanisms by which the stabilisers operate change considerably with time, being quite different during the early and later stages of processing.
4. Relations between thermal and photodegradative processes emerge in a very obvious way once it is realised that photodegradation is a sensitised process. Effective sensitisers are substances which survive or are formed during the thermal processing stage. The choice of stabilisers is therefore particularly critical for outdoor applications.
5. Many of the conclusions drawn about stabilisation are derived from work on model compounds. The difficulties of extrapolating such results to real polymer systems should always be borne in mind and the search continued for better analytical techniques for handling such systems.
6. The extent to which synergistic and/or antagonistic interactions operate between components of a formulation has only become apparent in relatively recent times and is certainly an area where a greater understanding is required.
7. Work in the general area of polymer stability is such that a relatively small technical improvement may lead to substantial commercial reward and so it is unfortunate though hardly surprising that a good deal of information is confidential and not freely available.

REFERENCES

1. Ayrey, G., Head, B. C. and Poller, R. C. *Macromol. Rev.*, **8** (1974), 1.
2. Ayrey, G. and Poller, R. C. in: *Developments in Polymer Stabilisation—2*, Scott, G. (Ed.), Applied Science Publishers Ltd, London, 1980, 1.
3. Cooray, B. B. and Scott, G. in: *Developments in Polymer Stabilisation—2*, Scott, G. (Ed.), Applied Science Publishers Ltd, London, 1980, 53.
4. Pobedimskii, D. G., Mukmeneva, N. A. and Kerpicknikov, P. A. in *Developments in Polymer Stabilisation—2*, Scott, G. (Ed.), Applied Science Publishers Ltd, London, 1980, 125.
5. Guyot, A. and Michel, A. in: *Developments in Polymer Stabilisation—2*, Scott, G. (Ed.), Applied Science Publishers Ltd, London, 1980, 89.
6. Wirth, H. O. and Andreas, H. *Pure Appl. Chem.*, **49** (1977), 627.
7. Ives, G. C., Mead, J. A. and Riley, M. M. *Handbook of Plastics Test Methods*, The Plastics Institute, London, 1971, 303.
8. Stapfer, C. H. *Plast. Eng.*, **33** (1977), 35.

9. MacAdam, D. L. *Colour Measurement*, Springer-Verlag, Berlin, 1981.
10. *RAPRA Micromill.* Rubber and Plastics Research Association of Great Britain, Shrewsbury.
11. Malyshev, L. N. and Zavarova, T. B. *Plast. Massy*, **8** (1975), 42.
12. Scott, G., Tahan, M. and Vyvoda, J. *Europ. Polym. J.*, **14** (1978), 877.
13. Yngve, V. Brit. Pat. 497 879 (December 1938); US Pat. 2 219 463 (October 1941).
14. Bellenger, V., Verdu, J., Carette, L. B., Vymalalova, Z. and Vymazal, Z. *Polym. Deg. Stab.*, **4** (1982), 303.
15. Davies, A. G. and Smith, P. J. *Comprehensive Organometallic Chemistry*, Wilkinson, G. (Ed.), Pergamon Press, New York, 1982, 519.
16. Scott, G., Tahan, M. and Vyvoda, J. *Chem. Ind.* (1976), 903.
17. Brasington, R. D. Ph.D. Thesis, University of London, 1971.
18. Ayrey, G., Poller, R. C. and Siddiqui, I. H. *J. Polym. Sci.*, **A-1, 10** (1972), 725.
19. Man, F. P. Ph.D. Thesis, University of London, 1978.
20. Allen, D. W., Brooks, J. S., Clarkson, R. W., Mellor, M. T. J. and Williamson, A. G. *Chem. Ind.* (1979), 663.
21. Allen, D. W., Brooks, J. S., Clarkson, R. W., Mellor, M. T. J. and Williams, A. G. *J. Organometal. Chem.*, **199** (1980), 299.
22. Parker, R. G. and Carman, C. J. *Adv. Chem. Ser.*, **169** (1978), 363.
23. Harrison, P. G., King, T. J. and Healy, M. A. *J. Organometal. Chem.*, **182** (1979), 17.
24. Brooks, J. S., Clarkson, R. W., Allen, D. W., Mellor, M. T. J. and Williamson, A. G. *Polym. Deg. Stab.*, **4** (1982), 359.
25. Frye, A. H., Horst, R. W. and Paliobagis, M. A. *J. Polym. Sci.*, **A2** (1964), 1765, 1785, 1801.
26. Klemchuk, P. P. *Prog. Chem.*, **85** (1968), 1.
27. Ayrey, G., Poller, R. C. and Siddiqui, I. H. *J. Polym. Sci.*, **B-8** (1970), 1.
28. Troitskii, B. B., Troitskaya, L. S., Denisova, V. N., Novikova, M. A. and Luzinova, Z. B. *Eur. Polym. J.*, **13** (1977), 1033.
29. Alavi-Moghadam, F., Ayrey, G. and Poller, R. C. *Polymer*, **16** (1975), 833.
30. Siddiqui, I. H. Ph.D. Thesis, University of London, 1970.
31. Minsker, K. S., Kazachenko, D. V., Abdullina, R. E., Kouler, R. B. and Berlin, A. A. *Polym. Sci., U.S.S.R.*, **15** (1973), 974.
32. Suzuki, T. *Pure Appl. Chem.*, **49** (1977), 519.
33. Nichols, J. D. *J. Polym. Sci.*, **B11** (1973), 705.
34. Gaylord, N. G. and Takahashi, A. *J. Polym. Sci.*, **B8** (1970), 349.
35. Starnes, W. H. and Plitz, I. M. *Macromolecules*, **9** (1976), 633.
36. Ayrey, G., Man, F. P. and Poller, R. C. *J. Organomet. Chem.*, **173** (1979), 171.
37. Aleksandrov, A. Yu., Gol'Danskii, V. I., Zavarova, T. B. and Korytko, L. A. *Vyoskomol. Soed., Ser. B*, **13** (1971), 784.
38. Lagner, H., Zeppenfeld, G. and Barti, A. *Plast. Kautschuk*, **13** (1966), 76.
39. Stapfer, C. H. and Granick, J. D. *J. Polym. Sci.*, **A-1, 9** (1971), 2625.
40. Troitskii, B. B., Troitskaya, L. S., Lenisova, V. N., Dubova, Z. B. and Novikova, M. A. *Tr. Khim. Tekhnol.*, **2** (1974), 120.
41. Figge, K. and Findeiss, W. *Angew Makromol. Chem.*, **47** (1975), 141.
42. Owen, E. D. *ACS Symp. Ser.*, **151** (1981), 217.

43. Cooray, B. B. and Scott, G. *Chem. Ind.* (1979), 741.
44. Cooray, B. B. and Scott, G. *Eur. Polym. J.*, **16** (1980), 169.
45. Minsker, K. S., Fedoseva, G. T., Zavarova, T. B. and Krats, E. O. *Vyoskomol. Soedin., Ser. A*, **13** (1971), 2265.
46. Cooray, B. B. and Scott, G. *Polym. Deg. Stab.*, **2** (1980), 35.
47. Grover, P. N., Wirth, H. O. and Stroh, V. K. Ger. Pat. 2 516 168 (November 1975).
48. Voetter, H. D. *Pop. Plast.*, **23** (1978), 24.
49. Ito, Y. and Mori, E. Japan. Pat. 7 866 953 (June 1978).
50. Rabeck, J. F., Canback, G. and Ranby, B. *J. Appl. Polym. Sci.*, **21** (1977), 2211.
51. Scott, G. and Tahan, M. *Eur. Polym. J.*, **11** (1975), 535.
52. Scott, G., Tahan, M. and Vyvoda, J. *Eur. Polym. J.*, **15** (1979), 51.
53. Chevassus, F. and de Broutelles, R. *The Stabilisation of Polyvinylchloride*, Edward Arnold, London, 1963.
54. Onozuka, M. and Asahina, M. *J. Macromol. Sci. Rev. Macromol. Chem.*, **C3** (1969), 235.
55. Bengough, W. I. and Onozuka, M. *Polymer*, **6** (1965), 625.
56. Shimura, T. and Sakai, A. *Bull. Chem. Soc. Japan*, **40** (1967), 995.
57. Anderson, D. F. and McKenzie, D. A. *Am. Chem. Soc. Div. Org. Coatings Plast. Chem. Prepr.*, **31** (1971), 740.
58. Braun, D. and Hepp, D. *Angew Makromol. Chem.*, **44** (1975), 131.
59. Van Hoang, T., Michel, A. and Guyot, A. *Eur. Polym. J.*, **12** (1976), 357.
60. Frye, A. H. and Horst, R. W. *J. Polym. Sci.*, **40** (1959), 419; **45** (1960), 1.
61. Klemchuk, P. P. *Adv. Chem. Sci.*, **85** (1968), 1.
62. Onozuka, M. *J. Polym. Sci.*, **A-1, 5** (1967), 2229.
63. Anderson, D. F. and McKenzie, D. A. *J. Polym. Sci.*, **A-1, 8** (1970), 2905.
64. Michel, A., Van Hoang, T. and Guyot, A. *Adv. Chem. Sci.*, **169** (1978), 386.
65. Van Hoang, T. Thesis, University Claud Bernard, Lyon, France, 1975.
66. Van Hoang, T., Michel, A. and Guyot, A. *Eur. Polym. J.*, **12** (1976), 337.
67. Stepek, J., Vymazal, J. and Czako, E. *J. Makromol. Sci. (Chem.)*, **A12** (1978), 401.
68. Czako, E., Vymazol, Z., Volka, K., Stibor, I. and Stepek, J. *Eur. Polym. J.*, **15** (1979), 81.
69. Verizhnikov, L. V., Kirpichnikov, P. A., Kolyubakina, N. S., Mukmeneva, N. A. and Karpova, S. I. *Vyoskomol. Soed. Ser. A*, **13** (1971), 714; translation in *Polym. Sci. USSR*, **13** (1971), 808.
70. Hybart, F. J. and Rowley, G. N. *J. Appl. Polym. Sci.*, **16** (1972), 715.
71. Kukhtin, V. A. and Pudovik, A. N. *Uspekhi Khimii*, **38** (1969), 96.
72. Mastrykova, T. A. and Labachnik, M. I. *Uspekhi Khimii*, **38** (1969), 1751.
73. Pobedimskii, D. G. and Buchachenko, A. L. *Izv. AN. SSSR, Ser. Khim.* (1968), 1181.
74. Pershin, A. D., Pobedimskii, D. G. and Buchachenko, A. L. *Izv. AN. SSSR, Ser. Khim.* (1975), 581.
75. Humphries, K. J. and Scott, G. *Pure Appl. Chem.*, **36** (1973), 163.
76. Walling, C. and Rabinowitz, R. *J. Am. Chem. Soc.*, **81** (1959), 1243.
77. Cadogan, J. I. G. *Adv. Free Radical Chem.*, **2** (1967), 203.

78. Schipper, P., Jansen, E. H. J. M. and Buck, H. M. *Topics in Phosphorus Chemistry*, **9** (1977), 407.
79. Hoffman, F. W., Ess, R. J., Simmons, T. C. and Hanzel, R. S. *J. Am. Chem. Soc.*, **78** (1956), 6414.
80. Walling, C. and Rabinowitz, R. *J. Am. Chem. Soc.*, **79** (1957), 5326.
81. Ramirez, F. and McKelvie, J. *J. Am. Chem. Soc.*, **79** (1957), 5829.
82. Pobedimskii, D. G., Mukmeneva, N. A. and Kirpichnikov, P. A. *Russ. Chem. Rev.*, **41** (1972), 555.
83. Bentrude, W. G. *Phosphorus Sulphur*, **3** (1977), 109.
84. Roberts, B. P. *Adv. Free Rad. Chem.*, **6** (1980), 225.
85. Kochi, J. P. and Krusic, P. J. *J. Am. Chem. Soc.*, **91** (1969), 3944.
86. Krusic, P. J., Mahler, W. and Kochi, J. K. *J. Am. Chem. Soc.*, **94** (1972), 6033.
87. Davies, A. G., Griller, D. and Roberts, B. P. *J. Chem. Soc., Perkin Trans. II* (1972), 993.
88. Davies, A. G., Griller, D. and Roberts, B. P. *J. Am. Chem. Soc.*, **94** (1972), 1782.
89. Pobedimskii, D. G., Kirpichnikov, P. A., Samitov, Yu. Yu. and Goldfarb, E. I. *Org. Mag. Reson.*, **5** (1973), 503.
90. Minsker, K. S. and Fedoseyeva, G. T. *Destruktsya i stabilisatsiya polivinilhorida*, Khimiya, Moscow, 1979, 155.
91. Ahmadullina, A. G., Yarkova, E. G. and Mukmeneva, N. A. *J. Prikl. Khimii*, **46** (1973), 1291.
92. Kirpichnikov, P. A., Mukmeneva, N. A. and Ahmadullina, G. A. *Plasticheskiye Massy* (1971), 14.
93. Heller, H. J. *Eur. Polym. J.—Suppl.* (1969), 105.
94. Heller, H. J. and Blattmann, H. R. *Pure Appl. Chem.*, **30** (1972), 145.
95. Anderson, R. W., Jr., Damaschen, D. E., Scott, G. W. and Talley, L. D. *J. Chem. Phys.*, **71** (1978), 1134.
96. Werner, T. *J. Phys. Chem.*, **83** (1979), 320.
97. Klöpffer, W. *Adv. Photochem.*, **10** (1977), 311.
98. Tirrell, D. A. *ACS Symposium Series*, **151** (1981), 43.
99. Gupta, A., Scott, G. W. and Klinger, D. *ACS Symposium Series*, **151** (1981), 27.
100. Werner, T., Woessner, G. and Kramer, H. E. A. *ACS Symposium Series*, **151** (1981), 1.
101. Hodgeman, D. K. C. *J. Polym. Sci., Polym. Lett. Edn*, **16** (1978), 161.
102. Turro, N. J. *Modern Molecular Photochemistry*, Benjamin/Cummings Publishing Co., California, 1978.
103. Owen, E. D. in: *Developments in Polymer Photochemistry—1*, Allen, N. S. (Ed.), Applied Science Publishers Ltd, London, 1980, 1.
104. Zweig, A. and Henderson, W. A., Jr. *J. Polym. Sci., Polym. Chem. Edn.*, **13** (1975), 717.
105. Carlson, D. J., Suprunchuk, T. and Wiles, D. M. *J. Polym. Sci., Polym. Lett. Edn*, **11** (1973), 61.
106. Guillory, J. P. and Cook, C. F. *J. Am. Chem. Soc.*, **95** (1973), 4885.
107. Ivanov, V. B., Rosenboim, N. A., Anghert, L. G. and Schlyapintokh, V. Ya. *Dokl. Akad. Nauk. SSSR*, **241** (1978), 609.

108. Ivanov, V. B., Lozovskaya, E. L. and Schlyapintokh, V. Ya. *Polym. Photochem.*, **2** (182), 55.
109. Sankyo Co. Ltd, Brit. Pat. 1 196 224 (June 1967).
110. Sedlar, J., Marchal, J. and Petruj, J. *Polym. Photochem.*, **2** (1982), 175.
111. Felder, B., Schumacher, R. and Sitek, F. *ACS Symposium Series*, **151** (1981), 65.

Chapter 6

POLYMERIC ADDITIVES FOR POLYVINYL CHLORIDE

JOHN T. LUTZ JR

Rohm and Haas Company, Bristol, Pennsylvania, USA

6.1 INTRODUCTION

Polymers are commercially successful due to two characteristics: (1) the inherent physical and/or chemical properties are uniquely useful, and (2) the polymer will readily accept a wide array of modifying additives to alter the base properties for the better.

Acrylic polymers such as Plexiglas® are widely used for their superior optical properties, ease of fabrication and excellent resistance to the elements of outdoor weathering. Polyethylene is low cost and can be processed at relatively low temperatures to provide useful articles having good toughness and, depending on thickness, good flexibility at ambient and low temperatures. These are two examples of polymers that have inherently desirable properties and do not require extensive use of additives. In most cases, polymers of this kind are difficult to modify in order to change their inherent properties.

Polyvinyl chloride (PVC), on the other hand, is an excellent example of a polymer that cannot be commercially successful in its neat form. Although PVC has been commercially available for more than 50 years, it was not until after 1945 that PVC became a major world-wide 'plastic'. The softening point of commercial PVC is between 130 and 230 °C which is also the range of initiation and propagation of thermal degradation.[7,14–17] Although it is possible to fuse PVC in the laboratory for study without the addition of stabilisers and lubricants, such an attempt on a commercial scale would be disastrous to both the product and the processing equipment.

The stabilisation of PVC during processing and during use is covered in other chapters of this book. It is the purpose of this chapter to discuss how essentially non-reactive organic polymer additives affect the processing rheology and physical performance properties of PVC compounds.

6.2 PROCESSING AND PROCESSING AIDS

As commercially produced, PVC exists as a tangle of relatively short, stiff polymer chains. Due to stereoregularity and a high degree of polarity, the PVC chains tend to form 'ordered' or 'crystallite' areas where the chains are attracted together. For this reason, PVC is often called a pseudo-crystalline polymer. The degree of 'crystallinity' has been described variously from 5 to 20%.[2,9–13]

The first commercially successful use of PVC was for flexible sheeting. The PVC was softened under heat and shear (or heat alone if emulsion-polymerised PVC plastisols were used) in the presence of a chemical plasticiser. Early plasticisers such as diethyl phthalate, dibutyl phthalate and tricesyl phosphate were, literally, solvents for PVC or P(VC/VAc) even at room temperature. These highly compatible solvent–plasticisers quickly penetrated the 'crystalline' regions of PVC and allowed the chains to unwind. The presence of the plasticiser interspersed among these PVC chains acted as an 'internal lubricant' to allow the chains to slip over each other without breaking under stress. These polar plasticisers were able to attach themselves to PVC polymer chains via van der Waals' forces to create an innerspring configuration that gave the cooled plasticised PVC a high degree of elongation while retaining a portion of the relatively high tensile strength of the PVC polymer. Thus, temperatures could be used to 'melt' the PVC—or put it into 'solution'—that were comfortably below the decomposition point of the polymer. Nonetheless, thermal stabilisers were—and always will be—needed to prevent incipient or catastrophic decomposition during processing. Lubricants will also be needed to provide release of the melt from the hot metal of the processing equipment.

Thus, compatible plasticisers can be considered to have been the first processing aids—apart from the essential stabilisers and lubricants without which processing would be at least difficult, and probably impossible. However, rigid PVC cannot tolerate more than about 3 phr (parts per hundred of PVC resin) of plasticiser without seriously detracting from heat distortion (DTUFL) and impact properties.

The early polymeric processing aids were various copolymers of methyl

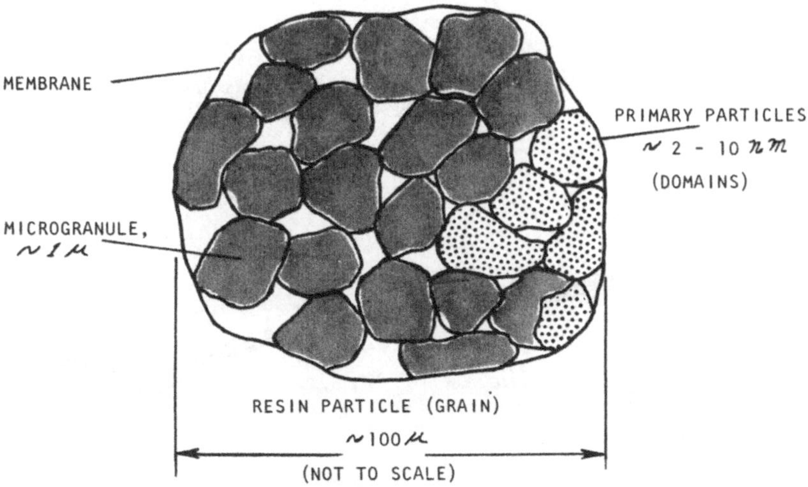

Fig. 6.1 Structure of suspension-polymerised PVC resin particle.

methacrylate, styrene, and acrylonitrile that were used proprietarily by suppliers of PVC compounds.[18,19] Today, several types of polymers similar to those mentioned above are available as polymeric additives.[20–26] The most successful are copolymers of methyl methacrylate, styrene–acrylic, methacrylate–PVC grafts and α-methyl styrene. Due to the chemical dissimilarities, the mechanisms by which they are effective must be different. Among those processing aids cited, relative efficiencies for imparting the desired effects on the PVC compound are quite different.

To understand more clearly the functions of processing aids, one must know more about PVC morphology. Relatively recent studies have been conducted to define the changes that occur as PVC powders are transformed into the molten state. PVC—especially the most common suspension-polymerised PVC—consists of the primary PVC particles clustered into PVC 'grains' that are agglomerated and encapsulated by a pericellular membrane consisting of the suspending agent (polyvinyl-(alcohol)) with some grafted PVC chains.[1,2,5–8] Pictorially, the structure as commonly accepted is shown in Fig. 6.1.

The use of more sophisticated torque rheometers and electron microscopy has assisted a variety of investigators in deducing the gradual breakdown of the cell, then the grain and, finally, the PVC particle into the melt structure. Berens and Folt[3] are generally credited with the first publication of the particulate nature of PVC melts in 1967. Shortly

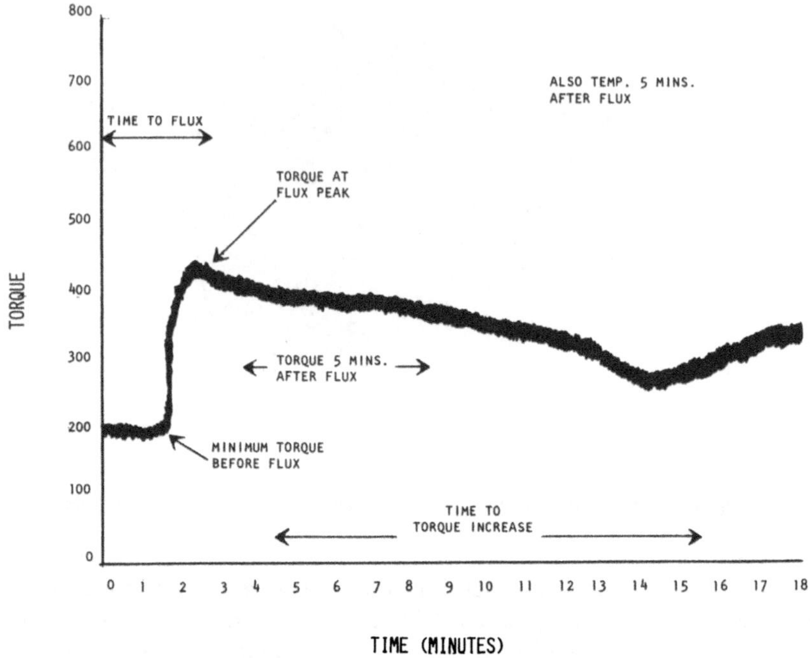

Fig. 6.2 Typical Brabender chart.

afterwards, Collins and Krier[4] showed a dual flow mechanism in PVC in the range of 160–230 °C; this was interpreted in terms of a low temperature particulate flow and a high temperature molecular flow. These landmark papers are credited with starting the widespread investigations on the particulate structure of PVC melts, the mechanism of fusion of PVC and elucidation of the effect of structure and morphology on the physical properties of PVC.

The development of the PVC melt can be followed by studying the transition of PVC powder compounds to a melt and finally to degradation—using a torque rheometer typified by the Brabender Plastigraph® or the Haake Rheocord®. These units are sometimes called 'bowls' or 'Kneader Chambers'.[31] In them, the powder is masticated by rollers as the temperature of the material increases through shear and heat transfer from the heated bowl. Typical torque rheometer curves are presented in Figs. 6.2 and 6.3. The melt viscosity goes through several transitions as the PVC grain is broken down into primary particles which in turn are broken down to the molecular level.[11] Rabinovitch[17] uses the same

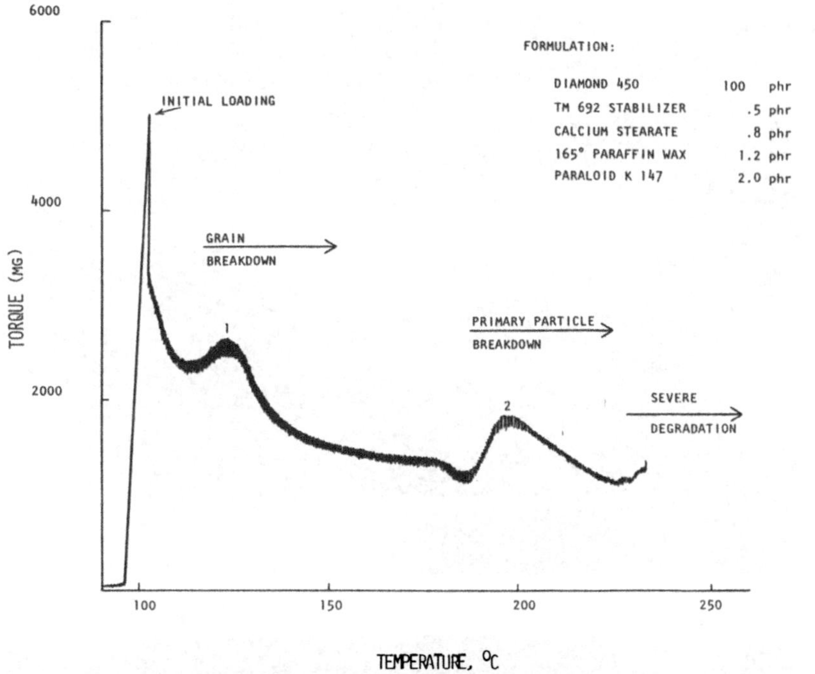

Fig. 6.3 PVC morphology changes during processing (temperature-programmed torque rheometer).

technique and identifies the peaks (1) as compaction and (2) as fusion. Scanning electron microscopy has been used to verify the morphology of the PVC samples taken from the bowl at the various peaks. One of the functions then of a polymeric processing aid is to hasten the particle breakdown and to enhance the fusion and homogenisation of the PVC melt.

Because methacrylate polymers and PVC are close in chi-values (or solubility parameters), it was assumed that these compatible systems were not unlike those of plasticisers in PVC. Studies were carried out in the early 1950s by Rohm and Haas on a two-roll plastics compounding mill. These studies clearly demonstrated that the acrylic processing aid caused or allowed the PVC to melt, or flux, faster and to form a homogeneous melt that had a high degree of extensibility while hot. At high processing aid levels (5–10%), the PVC had the consistency of hot taffy. Its counterpart, with only stabiliser and lubricant, was 'dry', or 'short', and was not homogeneous. These experiments are presented in Figs. 6.4 and 6.5. With

Fig. 6.4 (a) Unmodified PVC (b) PVC containing Paraloid K120ND.

Fig. 6.5 (a) Unmodified PVC (b) PVC containing Paraloid K120ND.

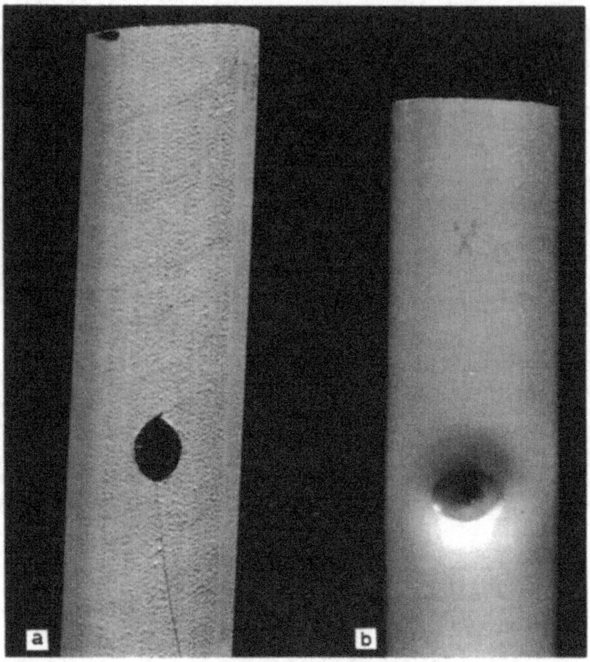

Fig. 6.6 (a) Unmodified PVC (b) PVC containing 1·5 phr Paraloid K120N.

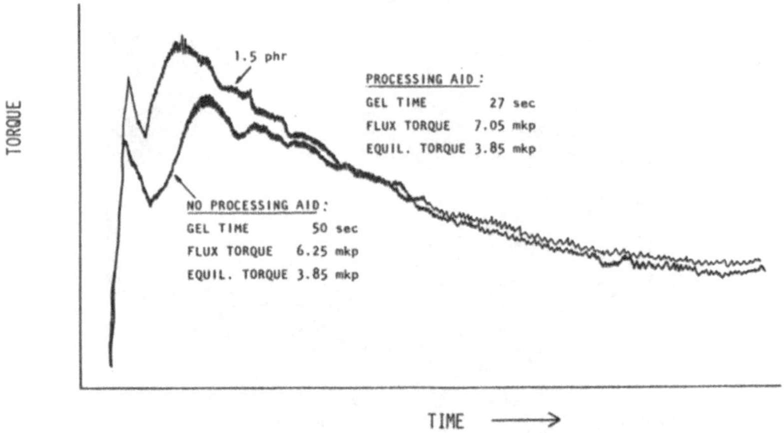

Fig. 6.7 Gottfert kneader chamber fluxing curves.

Table 6.1 Torque rheometer, 177 °C, 60 rpm, No. 6 roller head

	PVC (K = 62)	
	Unmodified	*5 % Acrylic processing aid*
Fusion time (min)	4·7	1·2
Fusion temp. (°C)	155	148
Fusion torque (m-g)	5 350	6 700
Equilibrium temp. (°C) (10 min)	165	164
Equilibrium torque (m-g) (10 min)	4 100	4 100

this 'crude-by-current-standards' understanding of the role of processing aids, it was possible to demonstrate their utility pragmatically in commercial pipe extrusion trials. (Laboratory extrusion lines of any value were virtually non-existent then.) Pipe producers found that they could extrude pipe much faster, obtain excellent gloss and produce a tougher pipe (Fig. 6.6) by incorporating processing aids.

As more sophisticated evaluation techniques became available, more was learned about their function and mechanism. The use of a torque rheometer quantifies what had originally been qualitatively observed. In Fig. 6.7, the fluxing curves of PVC compounds with and without 1·5 phr acrylic processing aid are compared. The curve for the compound containing the processing aid shows faster fusion (about one-half the time required for the control PVC) and a higher flux torque. The higher torque is due to two factors: (1) the breakdown of the PVC grains to primary particles—and then to the melt—is much more rapid and thorough, and (2) the temperature at which fusion takes place is lower, therefore giving the 'melt' a higher viscosity. Note that the equilibrium or steady melt torque of the two compounds is identical, indicating that the processing aid does not contribute to higher melt viscosity *per se*. The data in Table 6.1 further magnify these differences when a higher level (5 %) of acrylic processing aid is used, but even at this concentration, the acrylic processing aid does not contribute to higher equilibrium melt viscosity when the melt temperatures are the same. Krzewki and Collins[33] used capillary entrance pressure loss as a measure of elasticity and fusion. In their studies with 0, 2 and 5 phr acrylic processing aid, they demonstrated that increased levels of processing aid either reduced the temperature required for a given degree of fusion, or increased the degree of fusion at a given temperature with temperatures up to about 190 °C. Beyond that temperature, no advantages

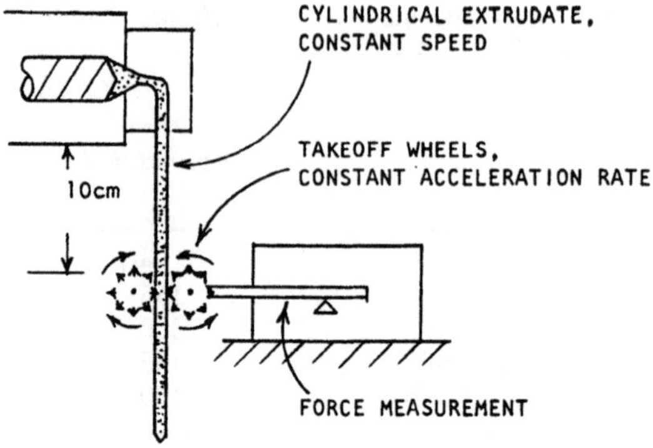

Fig. 6.8 Schematic of Rheo-Tens apparatus.

were seen for the processing aids. The ability to achieve fusion at lower temperatures lowers the demands on the PVC stabiliser and results in more-stable PVC compounds. This approach is especially important where PVC compound is first extruded into pellets and then fabricated into parts. It is also important in the use of regrind and scrap. The cumulative effects of heat history on the degradation of PVC can be quite significant.[34]

The use of the Goffert Rheo-Tens® enables us to further quantify the improvement in hot melt strength conferred by the acrylic processing aids. The Rheo-Tens is shown in Fig. 6.8. The constant-rate extrudate exiting the vertical die is pulled at a constantly increasing rate thereby stretching the hot melt. The take-up wheels are attached to a force-measuring device that records the elongation and relative tensile strength of the melt as indicated in Fig. 6.9. The effect of processing aid is presented in Fig. 6.10. The greater melt strength and elasticity of the PVC containing the acrylic processing

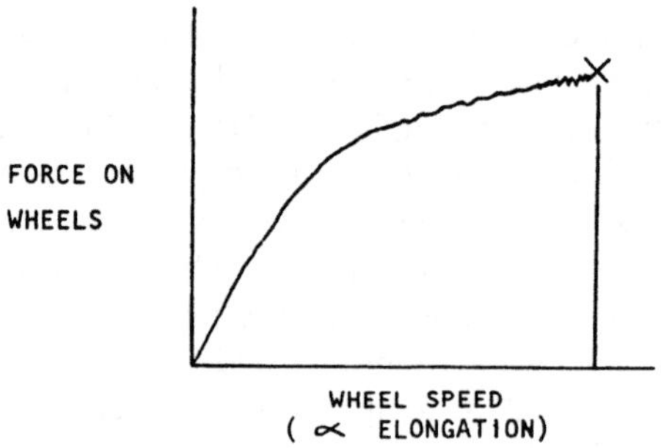

Fig. 6.9 Response of plastic to Rheo-Tens evaluation.

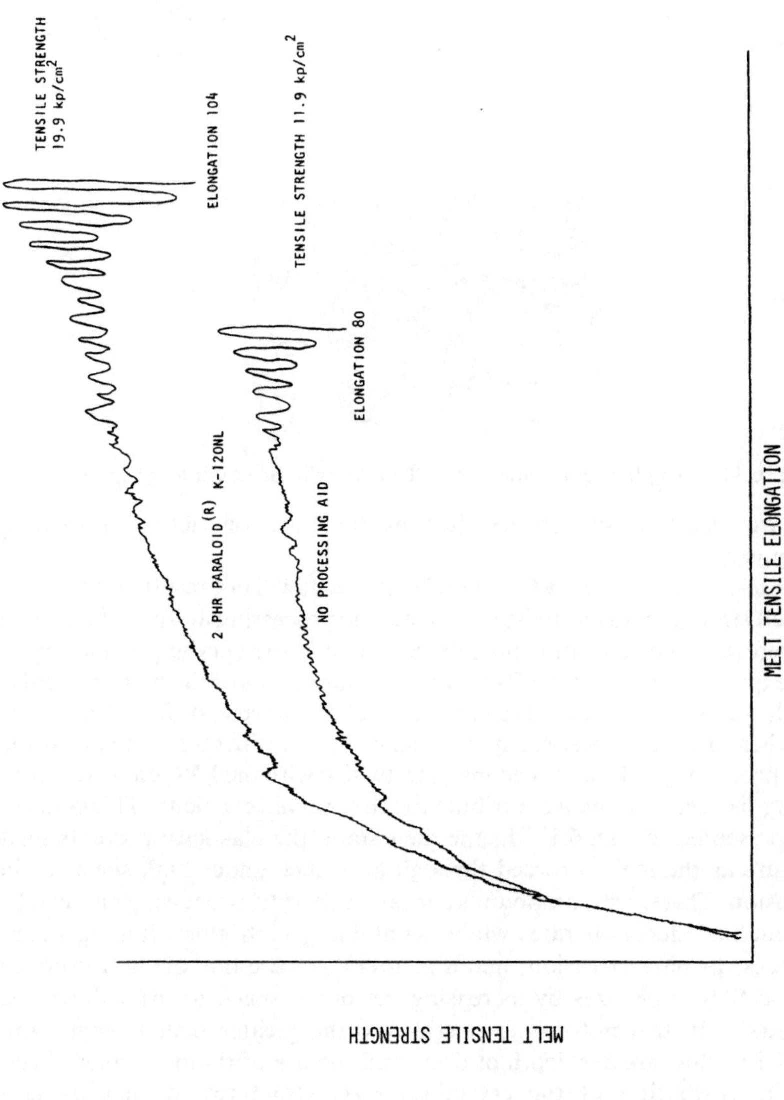

Fig. 6.10 Melt strength measurements at 182°C (Gottfert Rheo-Tens).

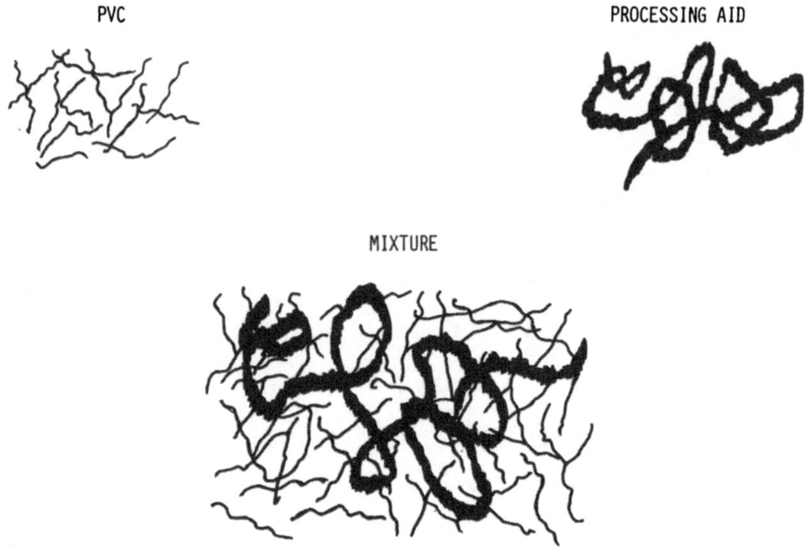

Fig. 6.11 Graphic representation of PVC/acrylic processing aid interaction.

aid are clearly demonstrated just as they are on actual processing equipment.

Merely reducing the PVC to a molecular melt will not result in the degree of melt strength and elasticity seen in acrylic processing aid-modified PVC. Therefore, a mechanism is postulated whereby the acrylic polymer melts more quickly than does the PVC, and the long chains of the high molecular weight acrylic polymer tend to bind or 'glue' the shorter, stiffer, PVC chains together causing greater shear to enhance fusion. Further, in the melt the long processing aid chains become intertwined with the PVC chains to form spring-like centres which contribute to the observed elasticity. This concept is represented in Fig. 6.11. In the melt state, the elasticity prevents melt fracture as the melt is forced through an orifice under high shear, as in extrusion. Therefore, compounds containing acrylic processing aids can be extruded at increased rates while maintaining high gloss, homogeneous surfaces. In pipe extrusion, manufacturers can use one die and produce several SDR-pipe sizes by increasing the puller speed to draw down the extrudate. In thermoforming operations, the greater melt strength and elasticity allow greater depth of draw and the use of thinner-walled sheet. The redistribution of the crystallite PVC structures in smaller-than-original domains allows the acrylic processing aid-modified sheet to be thermoformed at lower temperatures (Fig. 6.12).

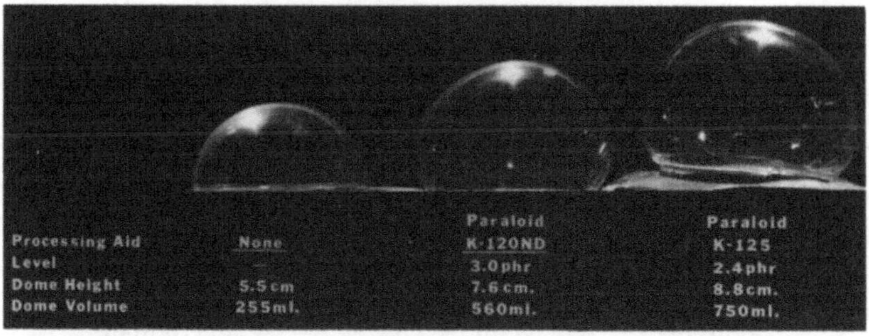

Fig. 6.12 Thermoformability as measured by blown parts.

At this point, it will be instructive to determine the effects of acrylic processing aids on the physical properties of PVC, and to compare those properties with those of plasticised PVC. Table 6.2 contains the comparisons of PVC containing either 10% dioctyl phthalate (DOP) plasticiser, or 10% acrylic processing aid. Ten per cent processing aid is unusually high for rigid PVC, but is included to accentuate performance contributions. Optical properties are not dramatically affected by either additive in the sample test formulation used here. In formulations containing impact modifiers, the processing aid can be used to adjust clarity. These effects will be discussed later. One of the major deterrents to the use of liquids in rigid PVC is the rapid deterioration of service temperature as measured by DTUFL. The use of 10% DOP causes a reduction of 22 °C in DTUFL. On the other hand, the acrylic processing aid does not cause a reduction in DTUFL. The 1 °C increase measured in the PVC compound containing the acrylic processing aid is not considered

Table 6.2 Effects of plasticiser or processing aid on physical properties of PVC ($K = 69$)

Property ($\frac{1}{8}$-in thick specimens)	*Unmodified*	*10% Dioctyl phthalate*	*10% Proc. aid*
% White light transmission	88	86	85
DTUFL at 1·81 MPa (°C)	75	53	76
Tensile impact strength (kJ/m)	53	59	181
Tensile elongation at break (%)	85	104	94
Tensile stress (max.) (MPa)	56·6	55·5	56·5
Tensile modulus (MPa)	3 080	3 024	3 092
Flexural stress (MPa)	86·8	87·5	88·9
Flexural modulus (MPa)	3 128	3 094	3 123

Table 6.3 Effect of plasticiser on the impact strength of PVC

PVC K value	*Impact modifier Acryloid® KM611 (phr)*	*Plasticiser DOP (phr)*	*Izod impact strength, $\frac{1}{8}$ in, 23°C (J/m)*
61	10	0	1 150
	10	3	760
	10	5	280 (brittle)
	10	10	123 (brittle)
55	10	0	930[a]
	10	3	370[a]
	10	5	155 (brittle)

[a] Scattered results (mixed ductile and brittle failures).

to be significant, however, this type of performance is consistent. Tensile stress is not significantly altered by either additive at this concentration, while tensile elongation is enhanced by both additives. The increased elongation provided by the plasticiser is obtained at the expense of a significant decrease in modulus which is typical when plasticisers are used, but prohibitive when PVC must retain its rigid character. The acrylic processing aid imparts increased elongation (although not equivalent to that contributed by the DOP), but without a decrease in modulus. Therefore the toughness (defined as the area under a stress–strain curve) is imparted by the acrylic processing aid under low-rate tensile stress. The toughness imparted by acrylic processing aids is more dramatically apparent under high-rate stress as in the tensile-impact test. Here, the entangled long acrylic chains and the redistributed PVC crystallites combine to give strength and ductility to PVC containing processing aid but no impact modifier. As in the tensile data, flexural data confirm the detrimental effects of plasticiser and the neutral or slightly positive effects of acrylic processing aids on the strength and modulus of rigid PVC.

The negative effect of plasticisers on the impact strength of rigid PVC is not readily apparent from the data in Table 6.2 because the impact strength of the unmodified PVC is so low. However, in a PVC formulation containing impact modifier, the effect is dramatic. In Table 6.3 the effect of as little as 3 phr plasticiser has a disastrous effect on (notched) Izod impact strength. The effect varies somewhat depending on the *K*-value (molecular weight) of the PVC used, because the impact strength of the PVC depends,

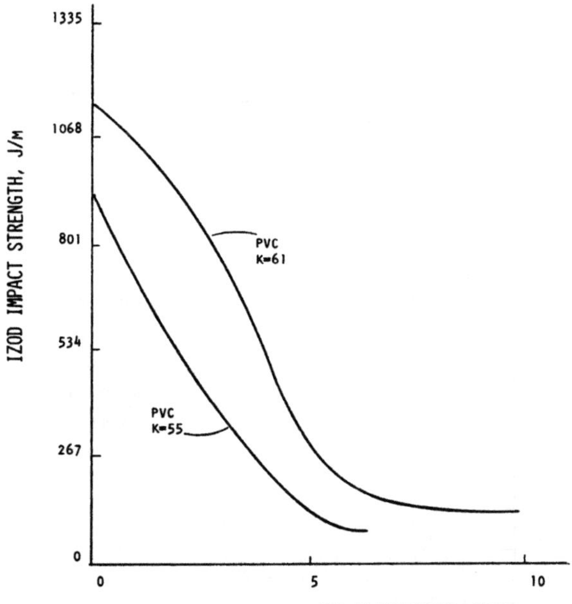

Fig. 6.13 Effect of plasticiser level on impact strength of PVC (impact modified with 10 phr MBS impact modifier).

in part, on its molecular weight. Therefore the use of truly minor amounts of plasticiser in a low *K*-value ($K = 55$) PVC compound dramatically reduces the impact strength. Between 3 and 5 phr plasticiser reduces the compound to brittle failure. In higher *K*-value ($K = 61$) PVC compounds, the absolute impact values do not drop off as rapidly as the plasticiser content is increased, but, again, 3–5 phr plasticiser induces brittle failure (Fig. 6.13).

The effects of processing aid on the impact strength and ductility of impact-modified PVC are illustrated in Fig. 6.14. The presence of an effective impact modifier at 10 phr yields tough PVC. The addition of processing aid further improves the ductility of the compound. Although an optimum improvement in impact strength is achieved between 2 and 2·5 phr, additional acrylic processing aid does not have a detrimental effect on impact strength. Therefore, acrylic processing aids can be used over a wide range of concentrations to accommodate processing as well as physical performance requirements.

The improvement in impact strength through the use of processing aids also extends to filled systems (e.g. filled PVC sewer pipe). The data in

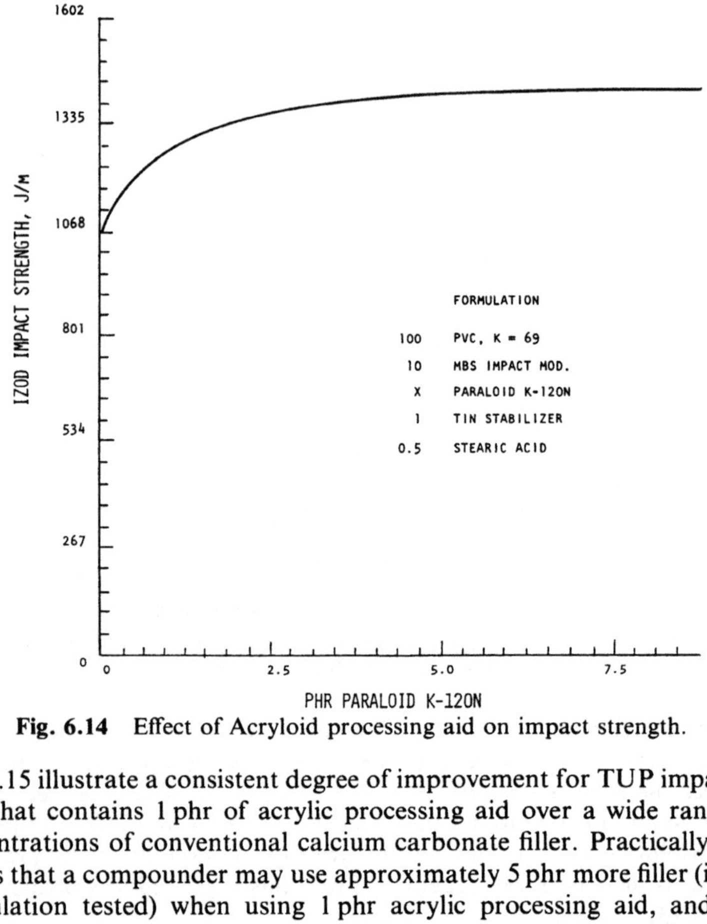

Fig. 6.14 Effect of Acryloid processing aid on impact strength.

Fig. 6.15 illustrate a consistent degree of improvement for TUP impact in pipe that contains 1 phr of acrylic processing aid over a wide range of concentrations of conventional calcium carbonate filler. Practically, this means that a compounder may use approximately 5 phr more filler (in the formulation tested) when using 1 phr acrylic processing aid, and still maintain the impact of the lower $CaCO_3$ content pipe containing no processing aid. Depending on the cost of PVC, the use of increased $CaCO_3$ can result in cost savings. Extensive commercial pipe extrusion trials[42] show that the use of acrylic processing aids in pipe formulations results in improved dimensional control of wall thickness (i.e. more in-spec. pipe and less excess weight to compensate for erratic wall thickness), higher extrusion speeds without loss of quality, improved appearance and physical properties and improvement in productivity (faster start-ups and less down-time for cleaning); so, overall, a better product at lower cost.

The terms 'all-acrylic' or 'acrylic' processing aid should be used with care when selecting a processing aid for commercial use. The chemical

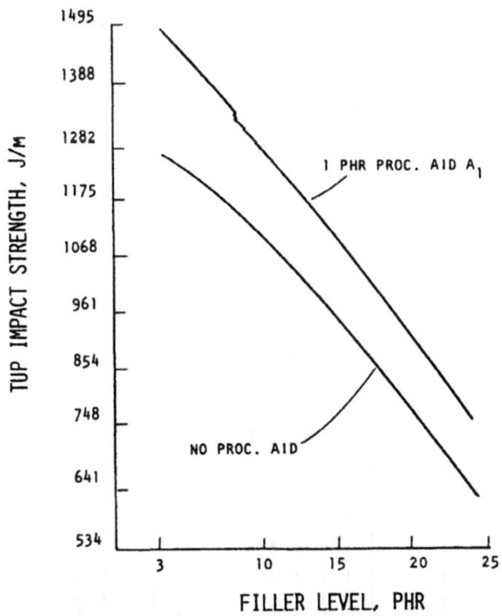

Fig. 6.15 Effect of acrylic processing aid on impact strength of filled pipe (1 in, SRD-26). Formulation: 100 PVC ($K = 69$), 0·4 tin stabiliser, 0·4 CaSt, 1·2 165° wax, $X\,CaCO_3$, 0–1 processing aid A_1.

composition, structure and molecular weight are all functional in determining how well or efficiently an acrylic processing aid will perform. The effects of these variables are shown in Fig. 6.16. All the processing aids are acrylic, but clearly they do not perform the same. At low rpm (low extrusion rate, low melt shear), all the processing aids yielded uniform, glossy, smooth extrudates. As the rpm increased, greater and greater shear stress was applied to the melts exiting the die. Compound F rapidly approached unmodified PVC for poor quality. At the highest shear rate, only one acrylic processing aid (D) survived. Processing properties are affected by processing aid concentrations (compare compounds B and C). Therefore, it is possible that the acrylic processing aids other than that in compound D could be used also at higher concentrations to give good quality extrudates. Cost then becomes a factor as does melt viscosity which tends to increase during extrusion with increasing processing aid content. This is due to earlier and earlier fusion in the extruder barrel, causing the screw to work against a melt over a longer percentage of its length.

Further illustrations of the efficiency of acrylic processing aids are presented in Table 6.4 and Fig. 6.17. Although increasing the acrylic processing aid content from 0·6 phr to 1·0 phr significantly improved the hot melt strength and the quality of the extrudates, dramatic improvements

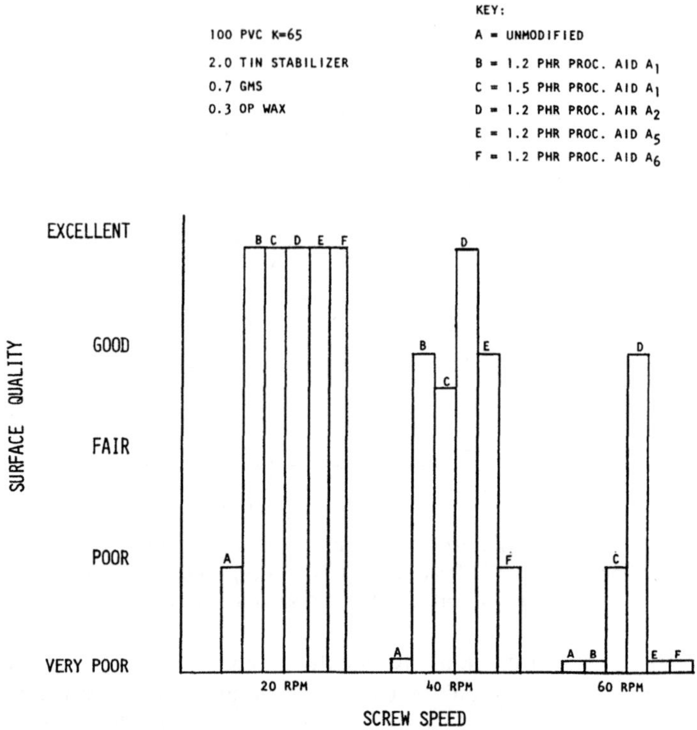

Fig. 6.16 Effects of processing aids on melt fracture v. extrusion rate.

Table 6.4 Effect of processing aid on extruded profile quality

Processing aid	*(phr)*	*Extrudate appearance*		*Extrudate hot melt strength*
		Edges	*Surface*	
None	—	Partial saw-tooth	Very rough	Fairly poor
Paraloid K125	0·6	Smooth	Slightly rough	Very good
Paraloid K120N	0·6	Smooth	Slightly rough	Very good
PVC/acrylic graft	0·6	Slightly rough	Rough	Good
Styrene–acrylic	0·6	Rough	Rough (+)	Fair
Paraloid K125	1·0	Very smooth	Smooth	Excellent
Paraloid K120N	1·0	Very smooth	Smooth	Excellent
PVC/acrylic graft	1·0	Rough	Rough (−)	Good
Styrene–acrylic	1·0	Very rough	Rough (+)	Good

Formulation: 100 PVC ($K=69$), 6·0 Acryloid KM330, 1·0 Acryloid K175, X processing aid, 2·0 tin stabiliser, 1·0 165 °F wax, 1·0 CaSt, 10·0 TiO_2.

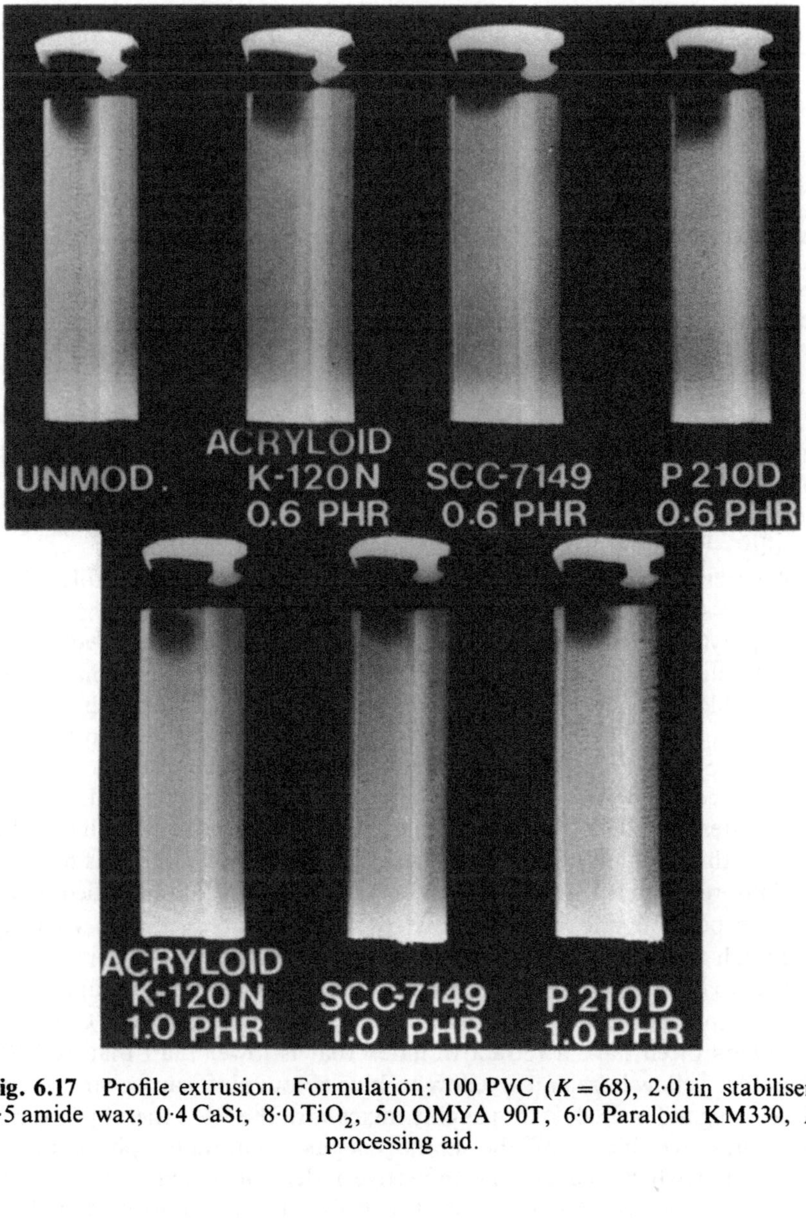

Fig. 6.17 Profile extrusion. Formulation: 100 PVC ($K = 68$), 2·0 tin stabiliser, 0·5 amide wax, 0·4 CaSt, 8·0 TiO_2, 5·0 OMYA 90T, 6·0 Paraloid KM330, X processing aid.

Table 6.5 Twin-screw (KMDL-25) pipe[a] extrusion rates

Processing aid		*Max. extruder screw speed*	*Max. puller speed setting*	*Max. ft/min extrusion before line break*
Type	*phr*			
None	—	46	180	10·2
Acrylic	1·0	46	240	14·8
Acrylic–styrene	0·6[b]	46	160	9·7
α-Methyl styrene	2·5[b]	46	240	14·6

[a] Pressure pipe, formulation: 100 PVC ($K = 69$), 0·4 methyl tin stabiliser, 0·8 calcium stearate, 1·2 165 °F paraffin wax, 0·15 low MW polyethylene, 1·5 TiO_2, 2·0 fine particle, coated, $CaCO_3$ filler, X processing aid as indicated.
[b] Levels recommended in respective product literature.

over unmodified PVC are obtained even at 0·6 phr. The non-acrylic processing aids also improve the melt strength and quality of the extrudate compared to the unmodified compound. However, neither gave good quality—even at 1·0 phr in the single-screw profile compounds.

A practical application of these differences is illustrated in Table 6.5. In this twin-screw pipe extrusion experiment, the all-acrylic, acrylic–styrene and poly(α-methyl styrene) were used at levels recommended in the manufacturer's literature. It is obvious that comparable performances are achievable, but the all-acrylic is the most efficient. Among these three processing aids, the *p*(α-methyl styrene) has several deficiencies other than the lack of efficiency. These include poor clarity which restricts its use to opaque compounds, volatility which can lead to voids and tears in extrudates, a tendency to contribute to plate-out or die-lip deposit (acrylics correct this problem) and a significant reduction in DTUFL (as much as 7 °C poorer than the unmodified or acrylic processing aid-modified PVC). On the positive side, poly(α-methyl styrene) does reduce melt viscosity, although the reduction does not result in better thermal stability.

Due to its somewhat lower refractive index, an acrylic processing aid can be used to titrate colour and clarity of clear impact modified PVC. Impact modifiers often have a refractive index that is lower than that of PVC, therefore yellow haze and a yellow-to-blue colour shift can occur. When a processing aid is used, it thoroughly mixes with the PVC thereby lowering the refractive index of the homogeneous continuous phase in the compound which reduces the refractive index mismatch in the system. These effects are illustrated in Fig. 6.18. Because impact modifiers,

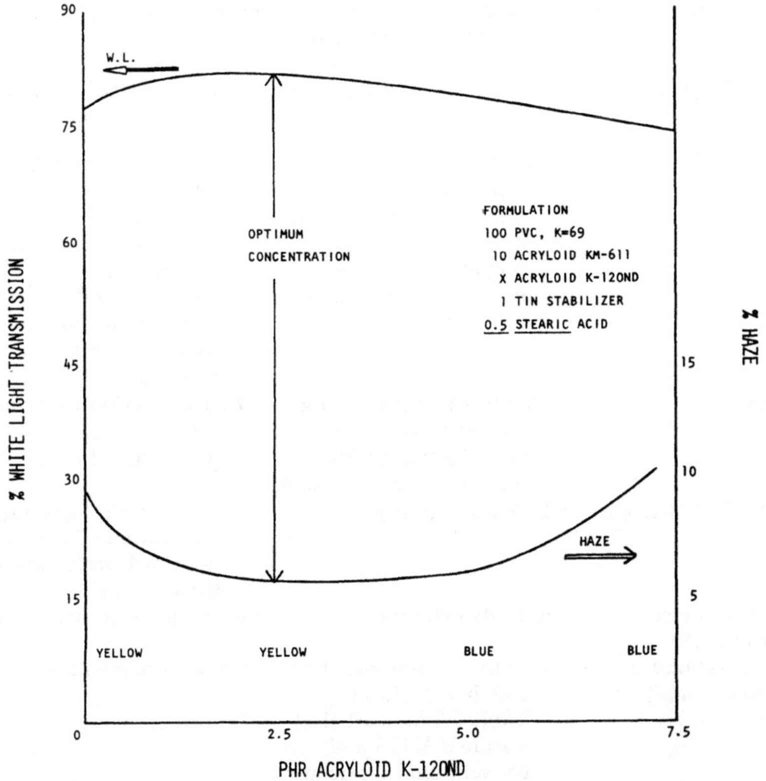

Fig. 6.18 Effect of Acryloid K120ND on colour and clarity.

lubricants and stabilisers vary from one formulation to another, the level of acrylic processing aid required to bring about the desired changes in colour and clarity will also vary.

Specially formulated processing aids have been designed to have lubricating functionality. An all-acrylic 'polymeric lubricant' (such as Paraloid® K175) has the ability to function as an external lubricant and to provide excellent release at the hot PVC melt/hot metal surface (of the processing equipment) interface. Paraloid K175 is especially useful in injection moulding where high shear rates are experienced as the melt is injected through the nozzle. Unlike conventional external lubricants that function due to incompatibility, Paraloid K175 has the advantages shown in Table 6.6.

Although processing aids are almost universally accepted as necessary

Table 6.6 PVC processing and properties: comparison of Paraloid® K175 with conventional lubricants

Property	*Paraloid K175*	*Conventional external lubricants*
Compatibility	Incompatible in melts especially at high temperatures, but as a high MW all-acrylic polymer, there is no migration at room or lower temperatures	Basically incompatible; temporary increases in compatibility as temperature increases; therefore, a loss in metal/melt release. Tends to bloom or migrate at room or lower temperature
Fusion	Due to some processing aid functionality, there is no reduction in fusion time or extent of fusion	Reduces fusion time and extent of fusion proportional to use level
Weld line strength	Enhances strength	Weakens weld lines due to retardation of fusion and poor melt/melt mixing at the weld line
Plate-out (see Fig. 6.19)	Retards plate-out	Contributes to plate-out
Heat stability during processing (see Table 6.7)	Effectively improves heat stability. Data in Table 6.8 indicate that Paraloid K175 and conventional lubricants display synergism. Normally Paraloid K175 is used in conjunction with adjusted levels of conventional external lubricants	Some improvement
Compound clarity (see Table 6.9)	Because Paraloid K175 is an all-acrylic polymer, its refractive index enhances the clarity or 'sparkle' of many PVC compounds	Detracts from clarity in proportion to use due to basic incompatibility with PVC

® Paraloid, Acryloid and Paraplex are registered trademarks of the Rohm and Haas Company. The Paraloid designation is used in Europe; Acryloid is used in the United States. Products designated as Paraloid or Acryloid with the same modifier or processing aid designation may be used interchangeably. For example, Paraloid KM323B and Acryloid KM323B are the same product.

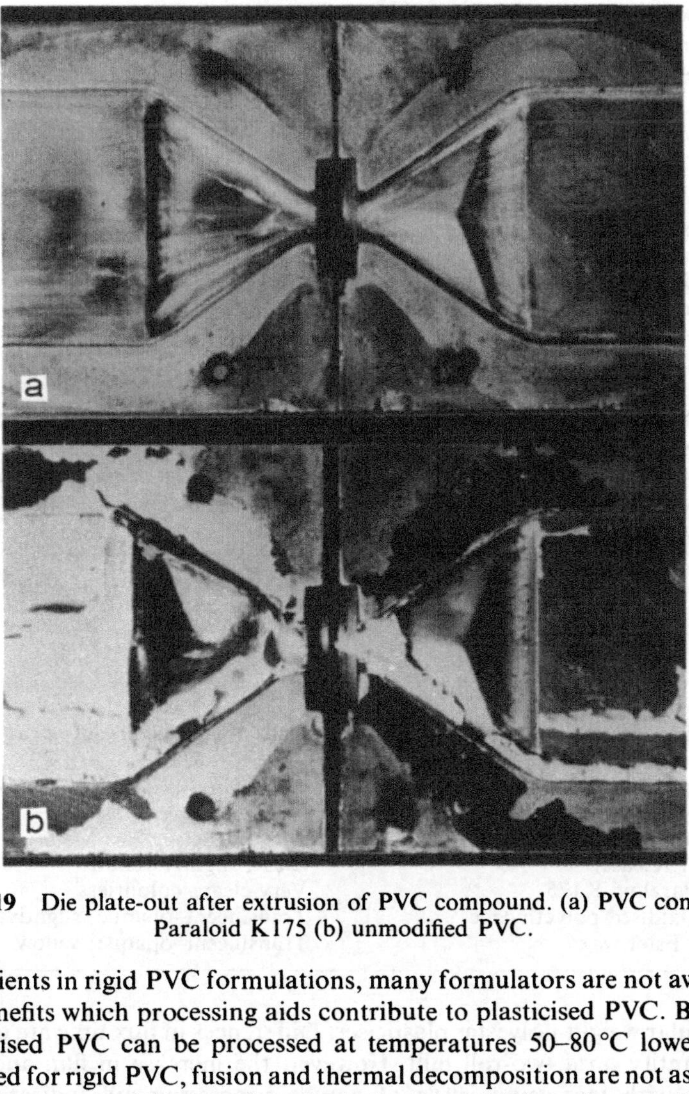

Fig. 6.19 Die plate-out after extrusion of PVC compound. (a) PVC containing Paraloid K175 (b) unmodified PVC.

ingredients in rigid PVC formulations, many formulators are not aware of the benefits which processing aids contribute to plasticised PVC. Because plasticised PVC can be processed at temperatures 50–80 °C lower than required for rigid PVC, fusion and thermal decomposition are not as severe in processing plasticised PVC. Nonetheless, the addition of processing aids to semi-rigid or highly flexible PVC improves fusion rate, hot strength, quality of the product and post-forming and physical properties of the product.

The data in Table 6.10 demonstrate the effects of acrylic processing aids on the processing properties of flexible PVC plasticised with a medium

Table 6.7 Effect of Paraloid K175 on dynamic heat stability at 385 °F

Lubricant	*Minutes to discoloration*						
	Very slightly yellow	*Slightly yellow*	*Yellow*	*Dark yellow*	*Orange*	*Brown*	*Black*
Standard lubricants	15[a]	25	25–30	30	30–35	30–35	35
3% Paraloid K175 + std. lubricants	15	25[a]	35+	35–40	35–40	40+	45
3% Paraloid K175	10	15	25	25–30	30[a]	30–35	35

[a] Stock stuck to mill at this point.

Table 6.8 Brabender plastograph, 340 °C, No. 6 roller head, 60 rpm

	Parts lubricating processing aid	*Conventional lubricant*	*Average torque at equilibrium (m-g)*	*Average temp. (°F)*	*Degradation time (min)*
A	0	1·0	1 650	386	20
B	0	0	1 500	390	17
C	3	1·0	1 650	386	40
D	3	0	1 700	390	29

Table 6.9 Clarity of lubricated PVC compounds at equivalent mill-release ratings

Lubricant system	*Internal lubricant control*
Internal lubricant control (poor release)	Clear, slightly yellow
3 phr Paraloid K175	Very clear, 'colourless'
1 phr Oxidised polyethylene	Translucent–opaque; slightly yellow
1·5 phr Ester wax	Translucent–opaque; yellow

molecular weight polyester plasticiser. Differences in flux time are difficult to quantify on a two-roll mill. However, the increase in film and sheet clarity with increasing levels of acrylic processing aid indicate more thorough fusion if not more rapid. The appearance rating was a combination of gloss, smoothness of surface and smoothness of the 'tail' of the sheet as it was removed from the rolls. In similar fashion, the processing aids improve the quality of monomeric plasticised PVC (Table 6.11). In addition, acrylic processing aids slightly improve the tensile strength and

Table 6.10 Processing improvements of polyester-plasticised PVC

PVC	100	98	95
Processing aid	—	2	5
Time to flux (min)	1·5	1·25	1·5
Rolling bank	None	Good	Fair–good
Appearance			
10 mil film	Fair–good	Good	Good +
40 mil sheet	Fair	Good	Good[a]
Clarity	Fair	Fair–good	Good

Formulation: 100*X* PVC, *X* acrylic processing aid, 50 polyester plasticiser, 1·7 Ba/Cd laurate.
[a] Free of air streaks, but rippled due to softness of compound.

Table 6.11 Effects of processing aids on DOP-plasticised PVC

PVC	100	98	95	90
Processing aid	—	2	5	10
Rolling bank	None	Fair	Fair–good	Excellent
Appearance				
10 mil film	Fair	Good	Very good	Excellent
40 mil sheet	Poor	Very good	Very good	Good
Tensile strength (psi)	2 670	2 800	2 810	2 910
100 % Modulus (psi)	1 280	1 320	1 300	1 430
Elongation (%)	380	390	395	370

Formulation: 100*X* PVC, *X* acrylic processing aid, 50 DOP, 5 Atomite, 5 Blue dispersion, 1·7 Ba/Cd laurate.

have little effect on elongation. The increase in modulus in compounds containing 10 % processing aid is an indication that the formulator should be aware of a stiffening effect when high levels of processing aid are used.

The improvement in hot strength is indicated in results from a simple test of drawing a vacuum under a hot plasticised PVC film. The extent of the 'draw' is indicated by the reduction in film thickness at the time the bubble ruptured (Table 6.12). This improvement in hot strength allows processors of calendered sheet—both clear and opaque—to increase production rates. The addition of processing aids improves a property known as 'calender tracking'. This property is somewhat paradoxical in that there is the need for adhesion of the melt to the calender rolls, but with regulated release of the film without sticking. The adhesion is necessary to eliminate entrapped air which results in voids and streaks in the calendered product. The hot strength is also vital where deep embossing is required, especially when that

Table 6.12 Hot strength—elongation at 350 °F

% Acrylic modifier	*Vacuum-formed part reduction in film thickness (%)*
0	5·4
1	22·0
2	35
5	48

Formulation: 100 PVC, 30 DOA, 10 soyabean oil epoxide, 1 Ca/Zn stabiliser, 0·2 liquid lubricant.

embossing must be retained in post-laminating procedures using heat activated adhesives. The rule-of-thumb is that embossing temperatures must be 100 °C higher than subsequent heat exposure if the embossing is to be retained. Hot strength (therefore acrylic processing aid) is critical to both the depth of embossing and the ability to retain production speed on the calendering line.

The value of the acrylic lubricating processing aid in extrusion of flexible PVC is illustrated in Table 6.13. Substituting Paraloid K175 for all of the conventional lubricants (Compound B) gave the best combination of increased output rate, quality and most-neutral die swell/die shrinkage. An advantage of eliminating conventional external lubricants whenever possible in flexible PVC is an improvement in adhesion of printed designs, etc., often applied in producing finished materials. Paraloid K175, K120N and K125 are equally effective in controlling plate-out in plasticised systems.

The hot strength imparted by the acrylic processing aids is very valuable in the production of flexible PVC foams produced from plastisol systems or post-expanded calendered or extruded sheet. In either case, lower density and finer foam texture foams can be produced over a wide temperature range. Without acrylic processing aids, the temperature range is narrow to prevent over-blowing which results in rupture of the hot cell wall and the subsequent deterioration of foam quality.[35]

6.3 IMPACT MODIFIERS

6.3.1 Impact modifiers in rigid PVC

Except for PVC pipe, where the ductility imparted by the acrylic processing aids gives adequate impact resistance to satisfy service requirements, rigid

Table 6.13 Improvements in extrusion of flexible PVC (diisodecyl phthalate-plasticised) electrical insulation compound[b]

Processing aid system:	*A* *None (control)*	*B* *Paraloid K175 (2 phr)*	*C* *DS-207/ Paraloid K175 (0·1/2 phr)*	*D* *DS-207/ Paraloid K175/ Paraloid K120ND (0·1/2/1·0 phr)*	*E* *Acrawax C/ Paraloid K175 (0·1/2 phr)*
% Increase in output rate of extrudate with a surface rating of 3[a] or better	—	>156	104	168	>158
Output rate (g/min)/ampere at:					
20 rpm	27/3·2	25·3/3·3	20·3/3·3	28·6/3·4	26·3/3·4
50 rpm	58·1/3·8	52·3/3·6	55·6/3·7	56·3/3·8	43·3/3·8
80 rpm	70/4·3	69·3/4·3	72·1/4·4	72·6/4·3	69·6/4·3
Surface rating[a]					
20 rpm	3	1	2	1	1
50 rpm	6	1	3	2	1
80 rpm	6	1	4	3	1
% Die-swell					
20 rpm	24	4	51	68	66
50 rpm	24	24	66	45	24
80 rpm	−37	17	24	66	−17

[a] Formulation: 100 PVC ($K = 69$), 30 DIDP, 7 tribasic lead sulphate, 2 antimony oxide, DS-207 (lead stearate) as indicated (0·4 in A), Acrawax C (bistearamide wax) as indicated (0·4 in A), Paraloid K175 (as indicated), Paraloid K120ND (as indicated).
[b] Surface rating: 1 = excellent (smooth, glossy, no imperfections) to 10 = bad (rough surface, melt fracture).

PVC compounds require the addition of impact modifiers to meet fabrication or end-use needs.

Several theories have been proposed to explain the function of impact modifiers in thermoplastics, especially styrenics. For a variety of reasons, none of these adequately explains the behaviour of impact modifiers in PVC. Part of the explanation may lie in the fact that impact modifiers are incorporated into styrenics during polymerisation, but the modifiers are added to PVC as a part of the formulation prior to melt compounding.

Merz *et al.*[36] suggested that the increased energy absorption represented the energy required to fracture the rubber particles as the crack passed through them. This theory was disproven when it was shown that the increase in energy absorption far exceeded that required to break the rubber, even in the unlikely event that the rubber particles were stretched to the very high elongations required for breakage.[37]

Schmitt and Keskkula[38] proposed the microcrack as an energy absorption and stress-whitening mechanism. Here the rubber particle was thought to be a stress concentrator and crack arrester. This theory requires large quantities of surface energy and does not provide an explanation for the increased elongation and cold-drawing tendencies of rubber modified plastics.[40]

Bucknall and Smith[41] propose the craze theory in which the rubber particles lower the gross craze initiation stress, but interfere with crack propagation by strengthening the craze by bearing part of the triaxial stress at the craze/rubber interface. The theory appears to be valid for HIPS where the glassy and rubbery phases are present during polymerisation. It does not appear to hold for modified PVC where the rubbery phase is dispersed in the polymer melt. With proper formulation and choice, the impact modifier will not detract from the clarity of the PVC and will not stress-whiten. Stress-whitening in impact modified PVC is rate dependent and is most often caused by birefringence due to elongation of the modifier particles under stress. Frequently, impact modified PVC sheet or film will stress-whiten preferentially in the trans-machine direction, strengthening the suggestion that elongation of the impact modifier induces the stress-whitening. However, at high rates of shear as in a tensile impact test, stress-whitening has been observed in systems that did not yield elongated modifier particles. There is some evidence that this stress-whitening is caused by microvoids resulting from loss of adhesion at the modifier/PVC interface, or may be due to voids developed within the impact modifier particles.

Newman and Strella[37,39] proposed that the increased energy absorption

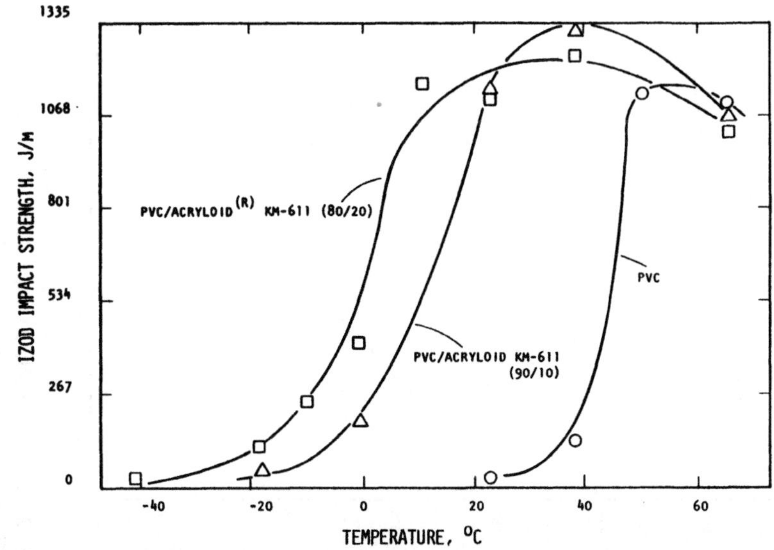

Fig. 6.20 Impact strength/temperature relationships for MBS-modified PVC.

in rubber modified plastics is caused by cold-drawing of the glassy matrix polymer. They postulated that there is an increase in free volume of the glass near the rubber particles due to the resistance to volume change of the rubber. To maintain the gross volume, the matrix must have accentuated expansion immediately around the rubber particle, making the matrix nearer its glass temperature.

Based on these older concepts, one useful explanation for the performance of impact modified PVC would appear to be a build on the Newman–Strella concept. The energy absorption is typically viscoelastic with generation of heat, which in turn lowers the glass temperature of the PVC in the immediate area. The performance results of compositional variations in impact modifiers bear out the viscoelastic concept and temperature v. impact studies confirm the effect of generated heat (Fig. 6.20).

The theories of impact modification mechanisms in PVC are undergoing revision. Unlike glassy polymers, such as polystyrene, the state of the morphology of PVC itself appears to have much to do with its impact strength. Summers *et al.*[7,8] suggest that the ultimate in PVC toughness comes from a completely fused compound. Bystedt and Lundquist[27] suggest that a melt containing a significant amount of PVC primary

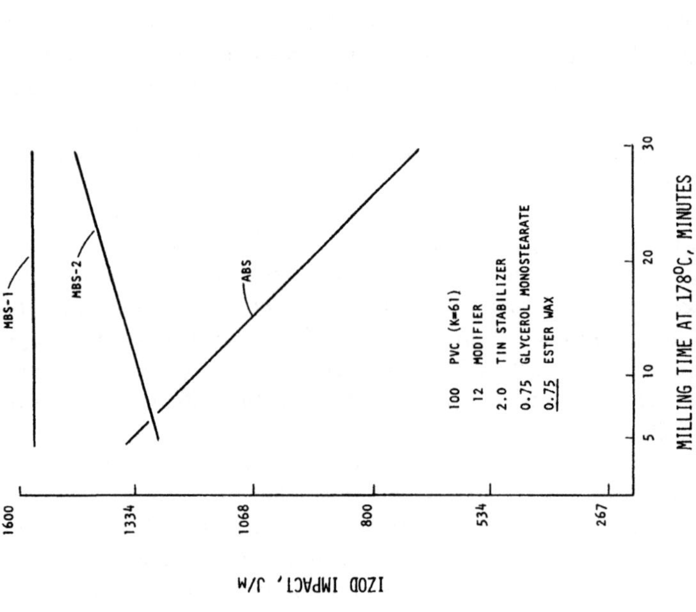

Fig. 6.21 Effect of processing time on impact strength.

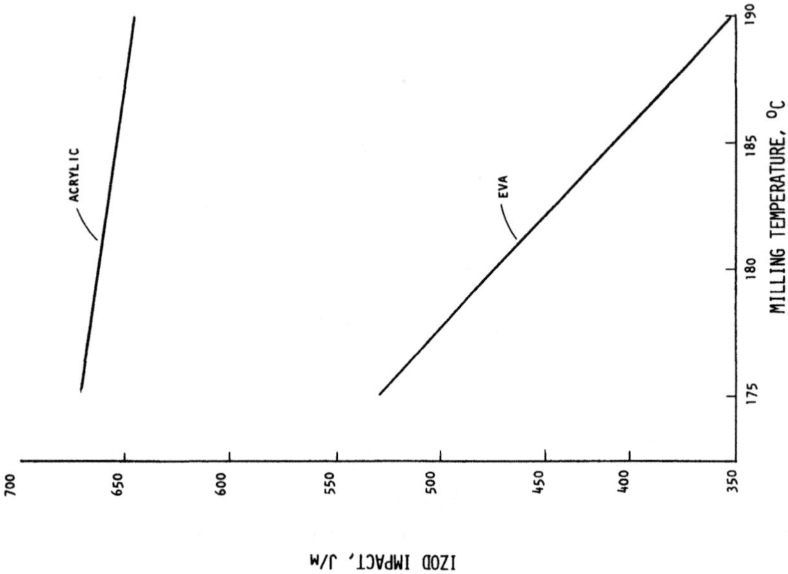

Fig. 6.22 Effect of milling temperature on impact strength of PVC modified with acrylic or EVA impact modifiers.

particles will give the best toughness (impact). Röhrl[28] uses both of these concepts to suggest that chemically different impact modifiers work by somewhat different mechanisms.

Although all impact modifiers must have a degree of compatibility to provide the necessary adhesion to PVC to be effective, the modifiers may be classified as follows:[28-30]

1. *Predefined elastomer (PDE) particle size:* 'methacrylate' butadiene styrene (MBS), methacrylate acrylonitrile butadiene styrene (MABS), all-acrylic (ACR) and modified acrylic (MACR).
2. *Not predefined elastomer (NPDE) particle size:* chlorinated polyethylene (CPE) and ethylene–vinyl acetate (EVA).
3. Acrylonitrile butadiene styrene (ABS) may be a 'transition' material.

For the PDE systems, the modifier particles are the disperse phase in the PVC continuum and maintain their particle size and shape when subjected to a wide variety of processing conditions of temperature and shear. The NPDE systems, however, depend on processing conditions to establish their integrity in the PVC matrix. Under conditions of relatively low work and temperature during processing, the NPDE modifiers form a honeycomb-like network 'cementing' the primary particles together with resulting high impact strength. When the system is 'worked' further, the primary PVC particles 'unravel' and fuse. The NPDE modifier then tends to lose its identity by 'dissolving' in the PVC, or it forms variably sized small globules. The effects of processing conditions are illustrated in Fig. 6.21 where one MBS modifier achieved and maintained excellent impact in the 'normal' cycle of fusion on a two-roll compounding mill. The second MBS modifier required additional work to complete its dispersion for optimum impact. The reduction in impact strength for the ABS modifier as the compound received increasingly more exposure to heat and shear illustrates a combination of thermal degradation of the modifier and a tendency for it to lose its identity as a dispersed phase impact modifier. The performance of EVA in Fig. 6.22 illustrates the 'phase inversion' of an NPDE impact modifier with increased processing. NPDE modifiers, then, have a relatively narrow processing range 'window' which makes it more difficult for a producer to maintain uniform product quality.

In a more elegant study of the dependence of impact strength on the time/temperature of milling, one PDE (Paraloid KM323B all-acrylic impact modifier) and two NPDE modifiers (EVA and CPE–PVC grafts) were studied in the formulation shown in Table 6.14.

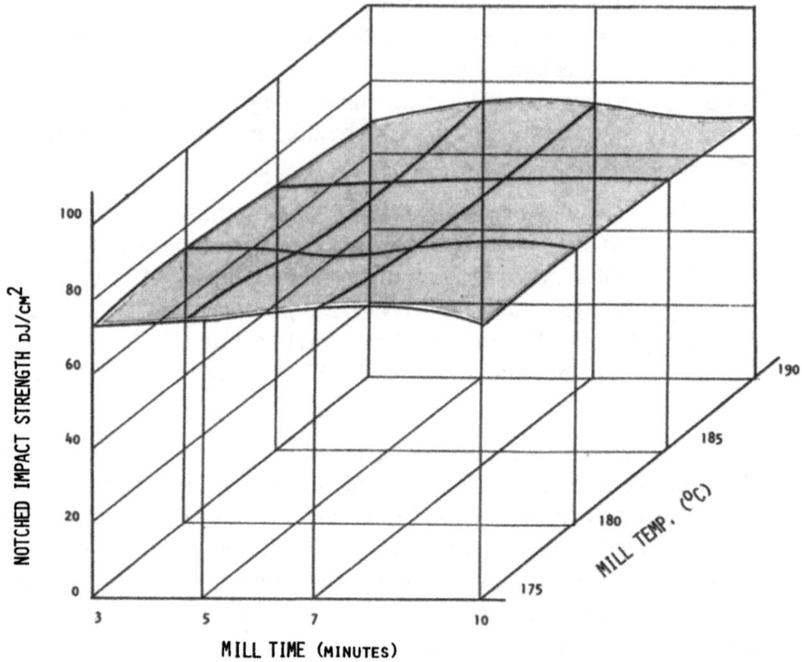

Fig. 6.23 Milling time/temperature: impact strength profile for Paraloid KM323B at 8 phr.

Table 6.14 Formulations and processing conditions for impact v. milling conditions study

	ACR modified	*EVA graft modified*	*CPE graft modified*
Impact modifier	8 (Paraloid KM323B)	60 (Vinnol K510/68)	10 (H9970)
S-PVC, $K=68$	100	40	90
Ferro® 904 (chelator)	0·5		
Paraplex® G-62	1·5		
Mark® WSX (Ba/Cd stab.)	3·0		
Loxiol® G-32	0·5		
Loxiol® G-21	0·4		
OMYA® 95T ($CaCO_3$)	4·0		
Kronos® CL220 (TiO_2)	3·0		
WAX PA520	0·05		
Paraloid® K120N	1·0		

Milling studies were carried out at each of the following temperatures—0, 175, 180, 185 and 190 °C—for each of the following times—3, 5, 7 and 10 min.

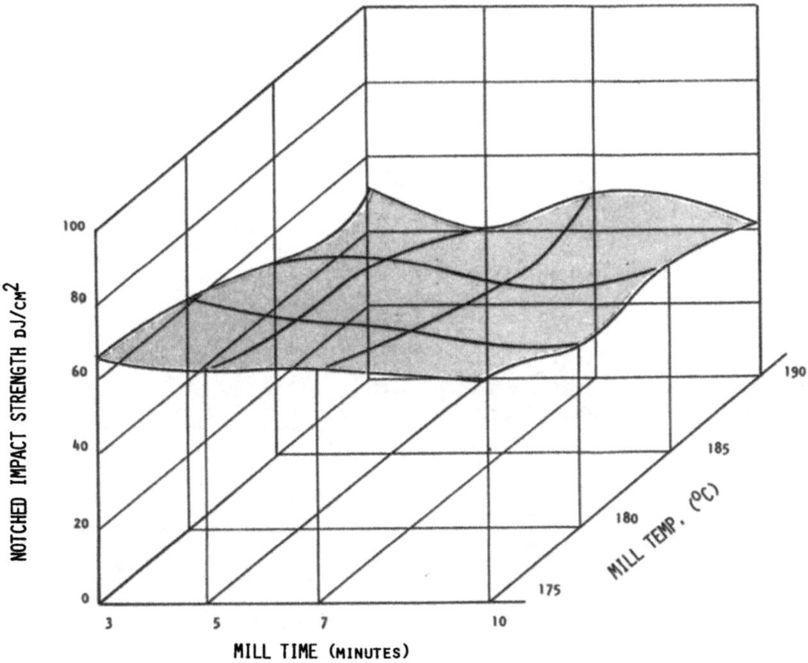

Fig. 6.24 Milling time/temperature: impact strength profile for EVA at 6%.

The three-dimensional contour plots of impact strength v. milling conditions are presented in Figs. 6.23–6.25. It is readily apparent that a PDE modifier is less dependent on the PVC morphology to develop impact strength. Another presentation of these data (Figs. 6.26 and 6.27) illustrates the consistency of impact performance for Paraloid KM323B. Uniform impact response indicates a wide processing window that is very useful in producing profiles that must be held to very close dimensional tolerances. Being able to generate consistent impact strength through the use of a PDE all-acrylic impact modifier provides the processor with many processing options to achieve an optimum balance of production rate and quality.

Proper selection of impact modifiers and other PVC compound ingredients requires a knowledge of the contributions and interactions of the various components. We will limit our discussion, mainly, to the effects of impact modifiers and their interactions with acrylic processing aids.

Properties other than impact strength are important for most applications. Using Paraloid KM-611 as a model (the magnitude of

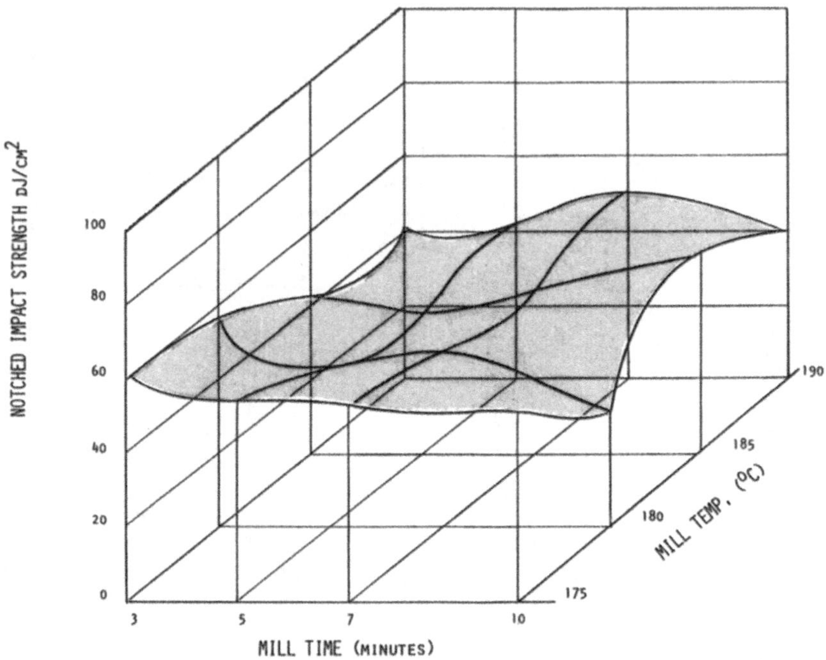

Fig. 6.25 Milling time/temperature: impact strength profile for CPE at 5%.

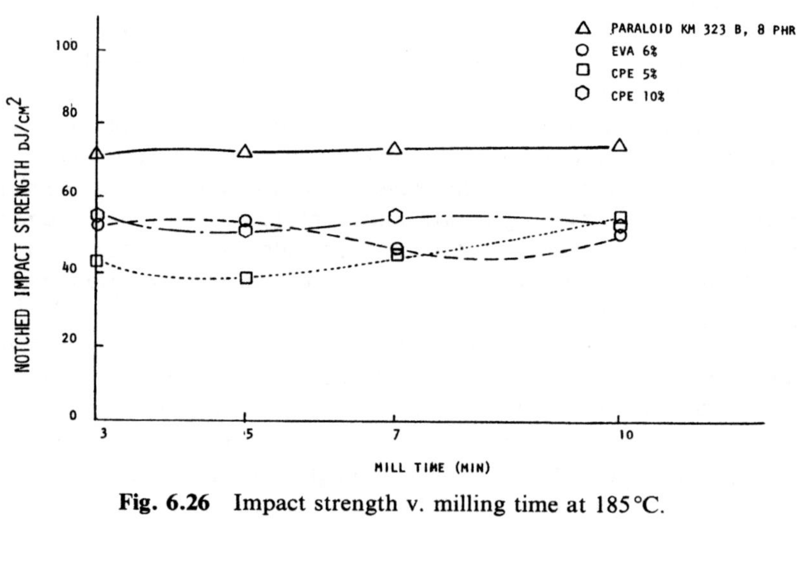

Fig. 6.26 Impact strength v. milling time at 185 °C.

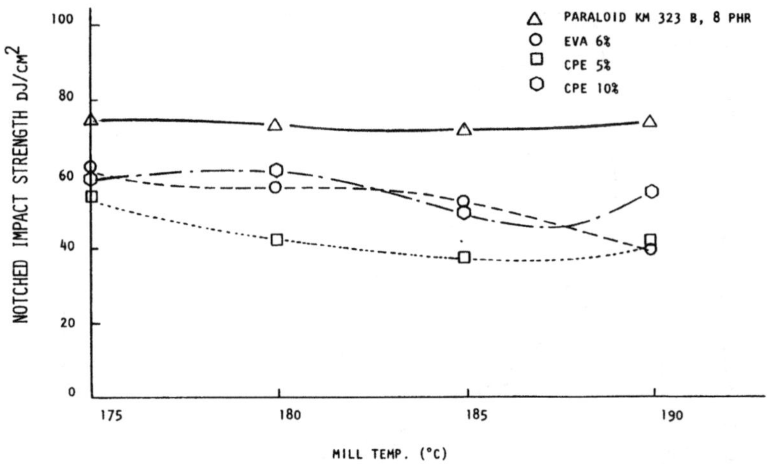

Fig. 6.27 Impact strength v. milling temperature (5 min milling time).

property responses varies with impact modifier composition, differing only in magnitude of changes) we see that impact strength improvement response is relatively abrupt and non-linear, but tensile properties tend to be linear with the concentration of impact modifier. Flexural properties tend to follow the same pattern from which it may be concluded that tensile and flexural properties are additive (Fig. 6.28). The response of impact modified PVC to changes in tensile properties with test temperature help to provide further insight into the mechanism of impact modification. Except for elongation at break (Figs. 6.29 and 6.30), unmodified PVC and impact modified PVC show essentially parallel response to test temperature. The greater elongation at break for the modified PVC indicates greater ductility, with the maximum at 30–40 °C (the same compound had a maximum impact at 20–30 °C). It may be generally concluded that the function of the impact modifier is to lower the yield stress of PVC at a given temperature.

The effects of impact modifier concentration on the rate of development and magnitude of impact strength are dependent on the molecular weight of the PVC. The curves in Fig. 6.31 show the typical S-shaped curve response, with a sharp inflection point as the PVC goes through a transition of low impact (brittle failure) to high impact (ductile failure) as the impact modifier concentration is increased. Because high molecular weight favours toughness in most polymers, PVC at higher molecular weight requires lower concentrations of impact modifier to achieve a brittle–ductile

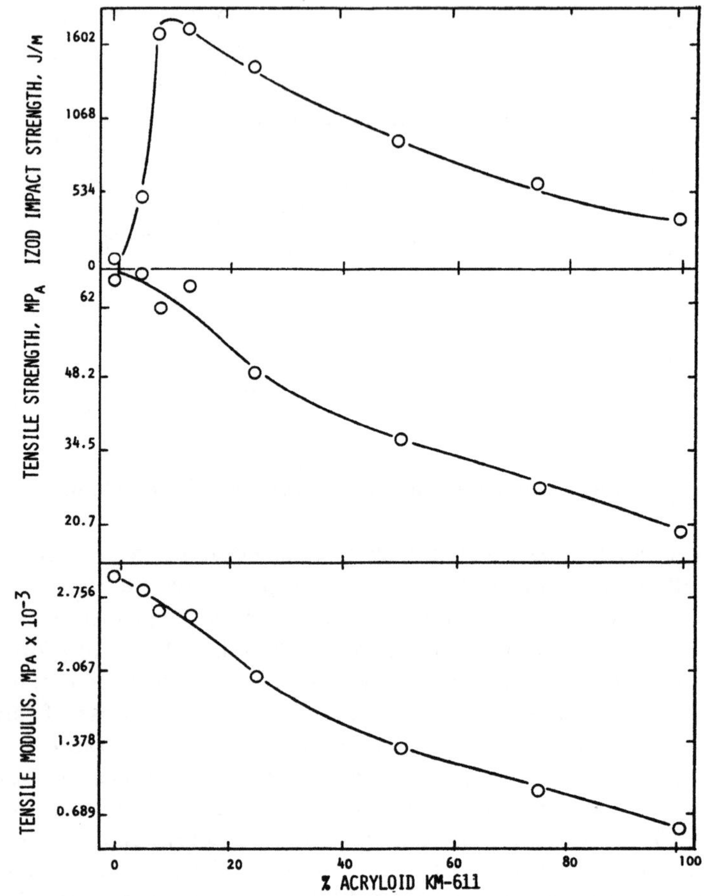

Fig. 6.28 Effect of Acryloid KM611 concentration on properties of PVC.

transition and also to achieve maximum impact. The capacity of an impact modifier to achieve brittle–ductile transition failure in PVC is referred to as its 'efficiency'. The ability of an impact modifier to achieve maximum impact strength without regard to concentration is called 'effectiveness'. It is not uncommon for one impact modifier to be more efficient than another, and yet be less effective. The drop-off in impact strength at high concentrations (Fig. 6.28) is due to less viscous energy being required for a given deformation.

The advantages for the PDE-type impact modifiers in generating good

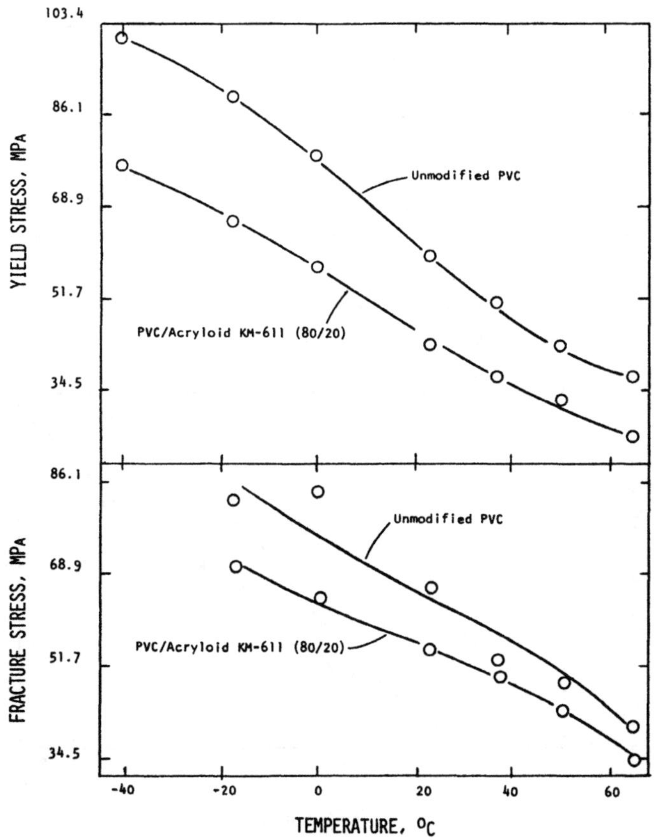

Fig. 6.29 Effect of temperature on tensile strength of PVC.

impact performance over a range of processing conditions have been discussed. By varying the structure and composition of PDE impact modifiers, further improvement in performance can be achieved.

In a study comparing PDEs (two MBS modifiers of different design and one all-acrylic modifier), transition (ABS) and NPDEs (CPE and EVA) with respect to concentration and processing temperature used to mill and mould specimens for Izod impact strength measurements, one PDE modifier (Paraloid KM680) stood out. At the 23 °C test temperature (Fig. 6.32) Paraloid KM680 consistently gave very high impact values regardless of concentration (from 4 phr to 7 phr) or processing temperature (180–190 °C). The other PDE modifier (Paraloid BTA-III-N-2) showed

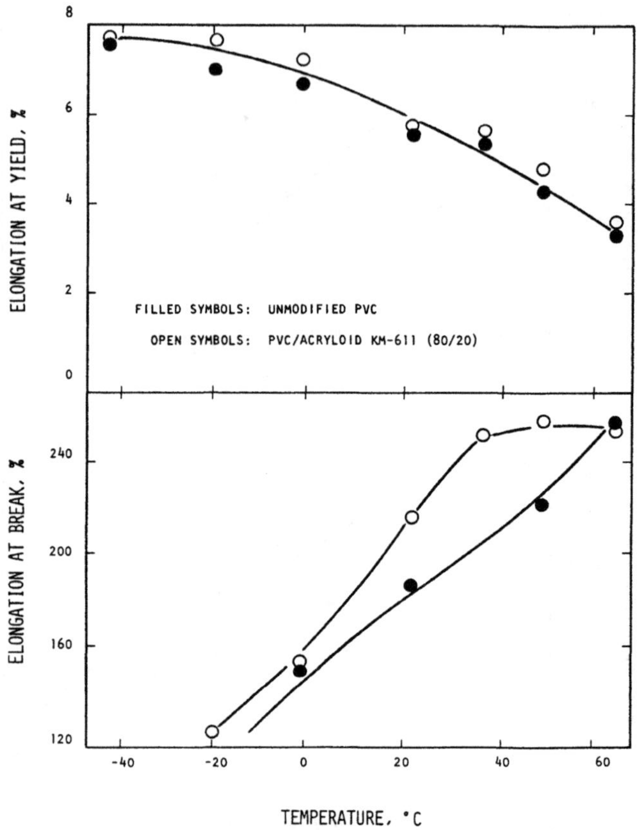

Fig. 6.30 Effect of temperature on tensile elongation of PVC.

similar behaviour, but was processing-sensitive at the lowest level (4 phr). The transition modifier (ABS), and to a lesser degree one NPDE (CPE), showed slower response to concentration and exhibited a proclivity for the lower processing melt temperature. The second NPDE (EVA) showed little response to concentration in this formulation and it also favoured the lower processing temperature.

At the 10 °C test temperature (Fig. 6.33), the same general performance patterns developed. Typical of low temperature impact testing, response to impact modifier concentration and achievement of maximum impact was less than at 23 °C. At this low temperature, Paraloid KM680 was clearly and dramatically superior in development of impact strength and in

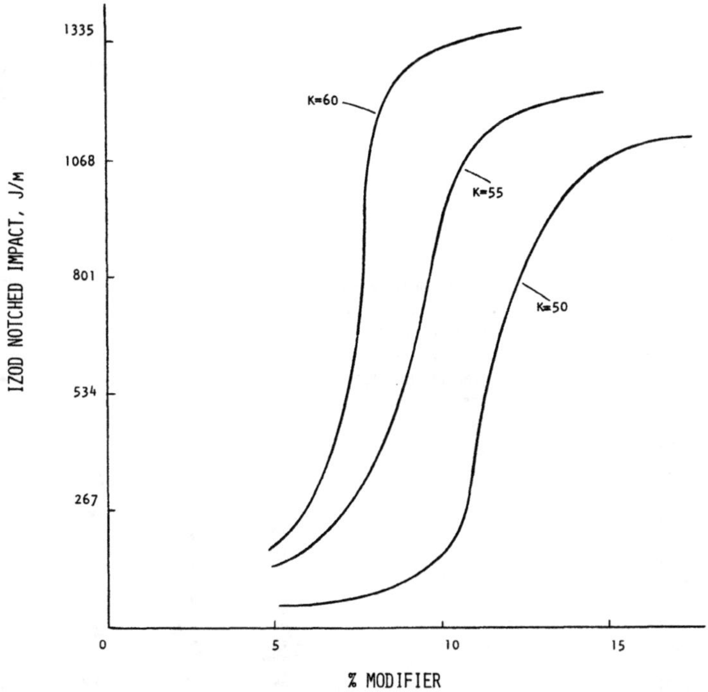

Fig. 6.31 Impact efficiency v. PVC molecular weight (*K* value).

showing less dependence on processing temperature. The second PDE modifier (Paraloid BTA-III-N-2) again was second in response to concentration, but it did show a distinct proclivity for the lower processing temperature.

From these studies it was concluded that impact strength is strongly influenced by the morphology of the PVC as a result of processing temperature.[32] At the lower temperatures, which provide some residual PVC-particle structure, all of the modifiers studied gave greater impact efficiency and strength. In all cases, the MBS (PDE) impact modifiers were the most efficient and effective. The advantages for the MBS modifiers were especially evident at the 10 °C test temperature which is a more severe measure of toughness. This study also points out that there are unique MBS impact modifiers (Paraloid KM680) that are less dependent on PVC morphology, and will perform very well and efficiently over a range of processing temperatures and test conditions. This additional processing latitude is especially important to manufacturers of merchant compound

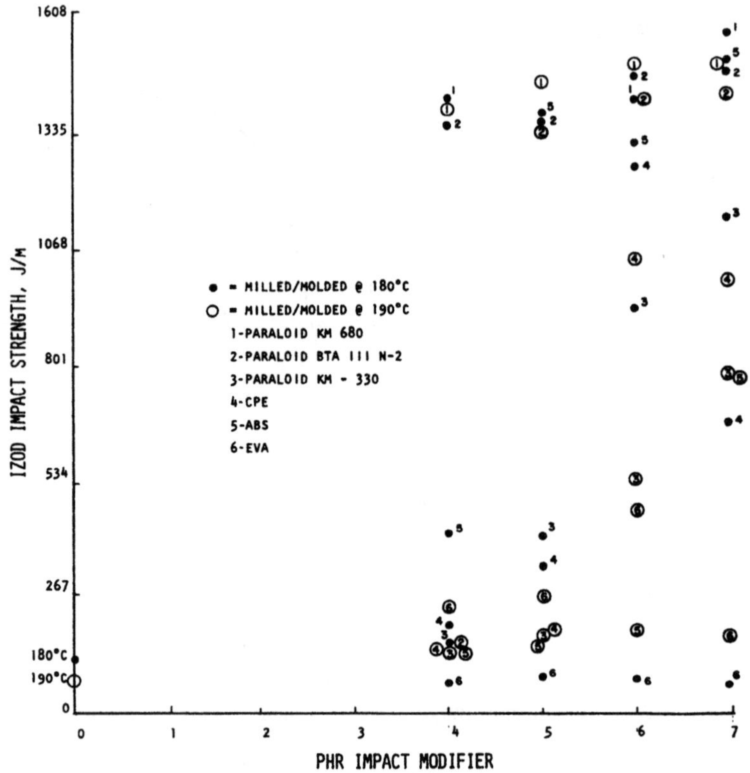

Fig. 6.32 Izod impact strength at 23 °C. Formulation: 100 PVC ($K = 62$), 0·4 Acryloid K125, 1·0 Acryloid K175, X impact modifier ($X = 4$, 5, 6 and 7 phr), 1·6 TM-181 (stabiliser), 1·2 CaSt (calcium stearate lubricant), 1·0 wax 165° (lubricant), 12·0 TiO_2 (filler–opacifier to UV light) (RCL-4).

that will subsequently be processed by a variety of converters using different equipment and conditions.

Even with highly efficient and effective impact modifiers, it is important to be aware of the interaction of impact modifier concentration and impact test temperature. The data in Fig. 6.34 illustrate typical responses, but not necessarily typical magnitudes of response, of impact modifier concentrations on impact performance over a range of test temperatures. These responses are additional expressions of the interactions shown in Fig. 6.20. At room temperature the bructile (brittle–ductile failure) transition usually occurs in the 5–10 phr range. At low temperatures,

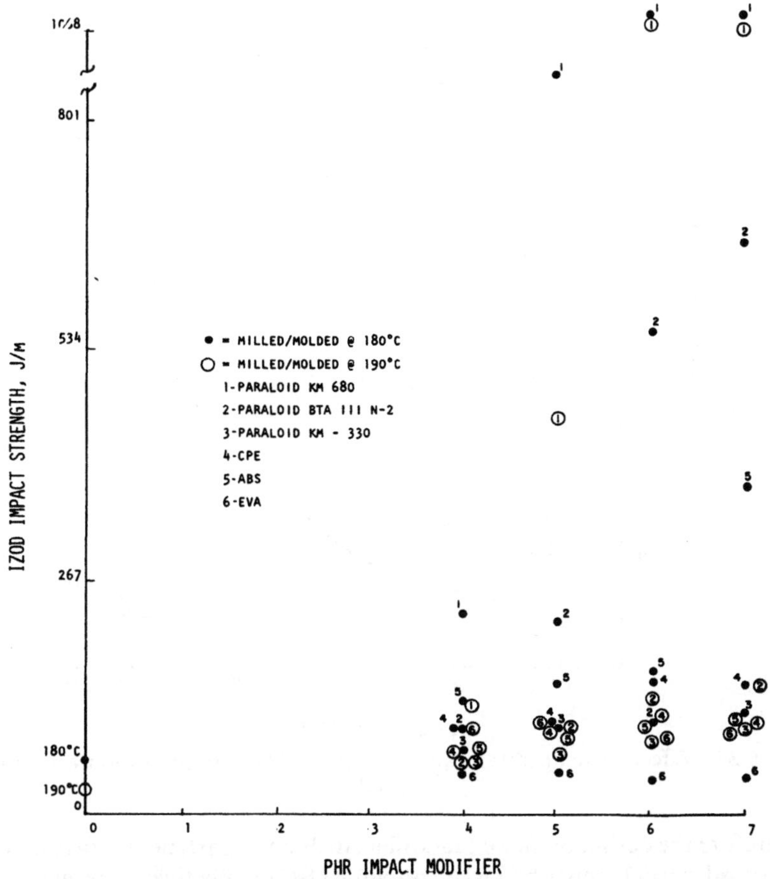

Fig. 6.33 Izod impact strength at 10 °C (formulation see Fig. 6.32).

however, the bructile transition concentrations vary widely with impact modifier composition and morphology as illustrated in Fig. 6.35. When the modifier level is sufficiently high, impact variations due to temperature are small. Where the impact modifier is at its critical concentration, the concentration–temperature response curve takes on the typical S-shaped impact curve. This behaviour suggests that a part of the impact theory includes a 'random-hit' component where the statistical chance of a modifier particle being at the right place to arrest a stress-induced crack is a function of the concentration of modifier particles. The concentration to be effective must include optimum dispersion as well as gravimetric quantity.

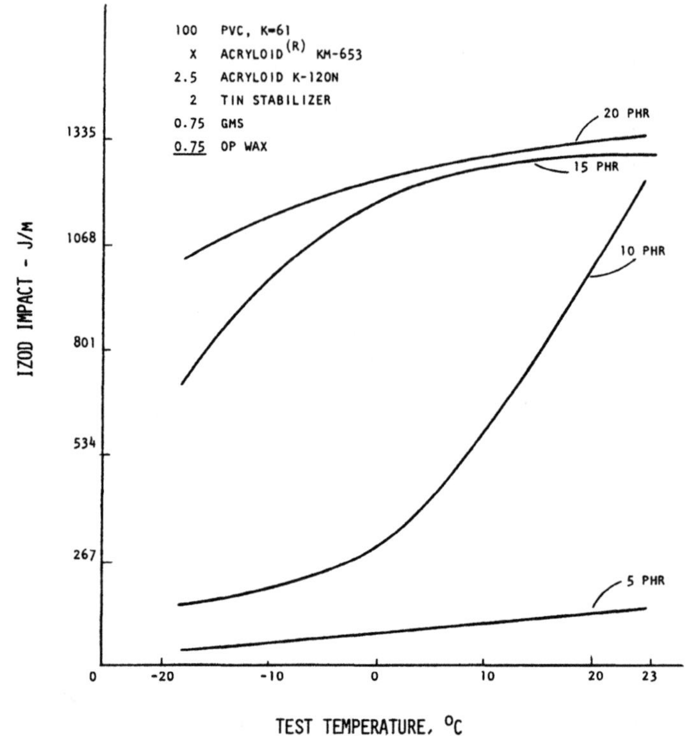

Fig. 6.34 Effect of temperature and impact modifier concentration on impact strength.

Therefore the design of impact modifiers includes a particle size factor. The dispersed particle must be large enough to be an effective crack-arresting 'shock absorber', and yet small enough to provide the necessary number of dispersed particles per unit of weight charged.

By manipulating particle size, refractive index and core–shell composition, impact modifiers can be designed to give excellent combinations of 'colour', clarity and stress-whitening resistance that are required for applications (such as packaging) where aesthetics are important. Normally, impact strength efficiency is sacrificed to obtain these critical optical properties. New MBS impact modifiers are being developed that will provide good impact efficiency along with the aesthetics. Proper choice of manufacturing process can add another performance parameter, namely low taste and odour, that is required for critical PVC food and beverage container applications.

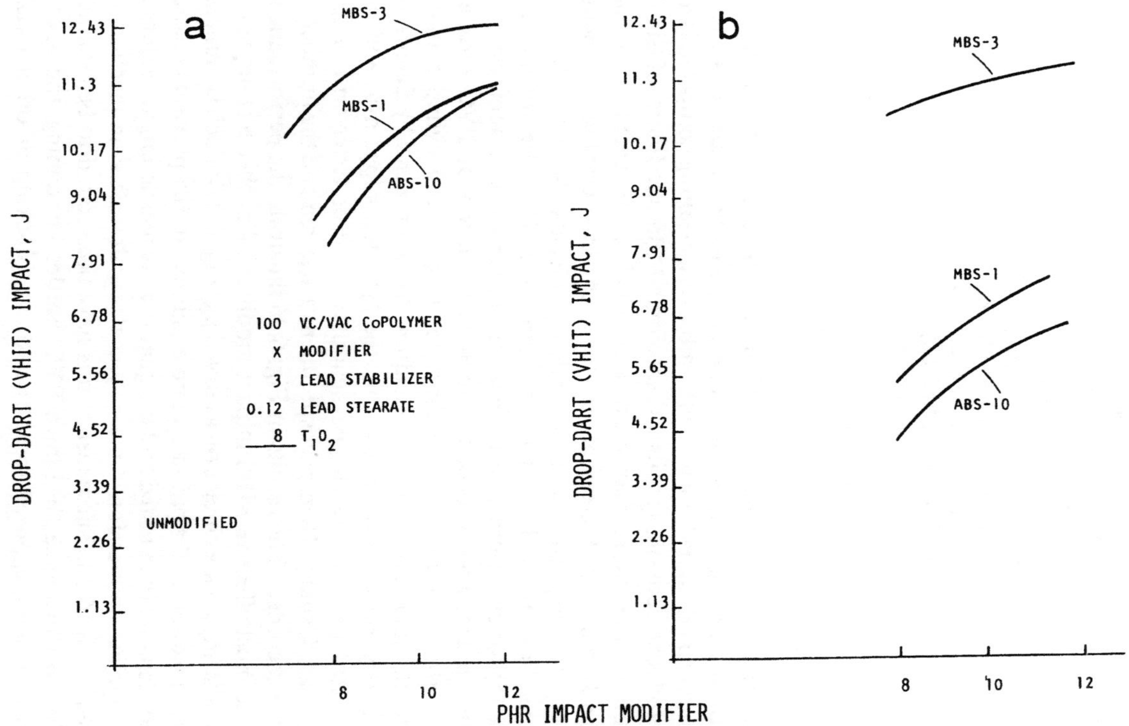

Fig. 6.35 Effect of temperature and modifier concentration on Gardner (VHIT) impact, 40 mil sheet at (a) 23 °C and (b) −18 °C.

Table 6.15 Formulation used for oil or solvent stress-crazing resistance

100·0	PVC ($K = 58$)
0·5	Acryloid K120ND
0·4	Ca behenate
0·25	Ca/Zn stabiliser
0·15	Alpha phenyl indole
0·30	Trinonyl phenyl phosphite
4.35	Paraplex G-62 (epoxy-soya bean oil)
1·35	Glycerol trihydroxy stearate
0·40	Glycerol triester of montanic acid
0·15	Paraffin wax
0·10	Ultramarine blue

Proprietary modifications of MBS impact modifiers are required to impart the increasingly important property of oil-stress-craze resistance to PVC containers designed for use with edible oils, cleaning compounds and other products containing oils or solvents. MBS modifiers of similar composition and performance in a normal environment can have significantly different degrees of resistance to stress-crazing or cracking during or after exposure to either oils or solvents. Several laboratory techniques have been devised to test typical PVC bottle formulations (Table 6.15) for stress-craze resistance during exposure to sunflower seed oil or turpentine. One procedure involves extrusion of PVC rods and wrapping them around a large diameter (e.g. 25 cm) cylinder that is subsequently immersed in the test medium. At experimentally determined time intervals, rods are removed from the cylinder, wiped dry, and then impact tested using an Izod-type pendulum impact tester with a fixed hammer load. Usually, 10 impact specimens are tested and the number (%) breaks are recorded. The results in Fig. 6.36 illustrate the performance of two of the most effective MBS impact modifiers. Another technique is to stress tensile impact specimens on a curved jig (Fig. 6.37) which is immersed in the test medium and aged in a manner similar to the previous method. The specimens are then tested and the quantitative tensile-impact values are plotted against immersion time. Typical results after immersion are presented in Fig. 6.38. These test results have been found to be reliable in predicting performance of blown-PVC bottles containing the specific medium and being exposed to stresses due to stacking or other outside stresses.

The use of PVC compounds in exterior applications—or those exposed

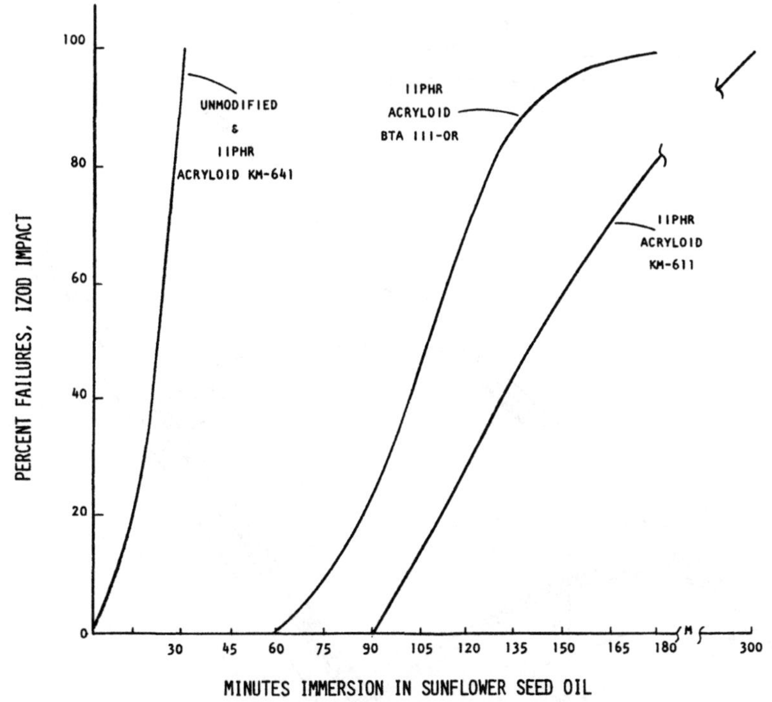

Fig. 6.36 Stress craze resistance of PVC bottle compounds in sunflower seed oil.

to the elements—has been successfully demonstrated by the long-term performance of PVC siding, window frames, rainwater gutters, etc. To be successful in these applications, the PVC must be properly stabilised against UV degradation, and the toughness of the PVC compound must be great enough to withstand abuse in fabrication, shipping and installation. In addition, the appearance (colour, gloss, dimensional uniformity) and impact strength must be maintained for extended periods of time in service.

Again, stabilisation of the PVC is critical, and is treated in depth elsewhere in this book. European experience has been that barium/cadmium (Ba/Cd) stabiliser systems and mixed Ba/Cd/Pb systems are preferable to organic (mercapto) tin compounds for cost and exterior durability. In the United States the tin stabilisers are preferred due to processing economics and, generally, commercially acceptable weathering results.

The choice of an impact modifier is much more limited in this application. The impact modifier must be stable in the presence of

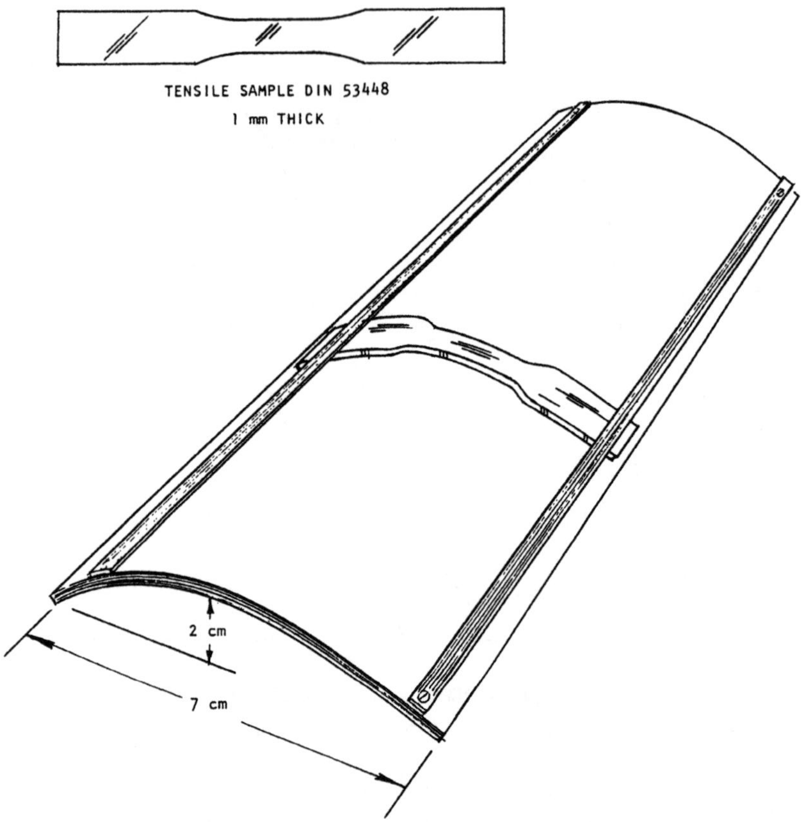

Fig. 6.37 Schematic diagram of the sample holder used for turpentine stress crazing.

ultraviolet light, thereby retaining its colour and impact functionality. Therefore, modifiers based on unsaturated (—C=C—) components (e.g. butadiene) cannot be used successfully. Modifiers that fall into the acceptable category are all-acrylic (ACR), ethylene/vinyl acetate (EVA), and chlorinated polyethylene (CPE).

The relative rate and degree of degradation on exposure to ultraviolet light is dependent on the intensity and duration of the irradiation, the angle of exposure and the formulation ingredients. For outdoor exposure sites, either high altitudes or lower latitudes provide the most accelerated exposure. Highly accelerated results are obtained by exposing PVC at 45 °C, facing south, in an area such as Phoenix, Arizona. Subtropical (e.g.

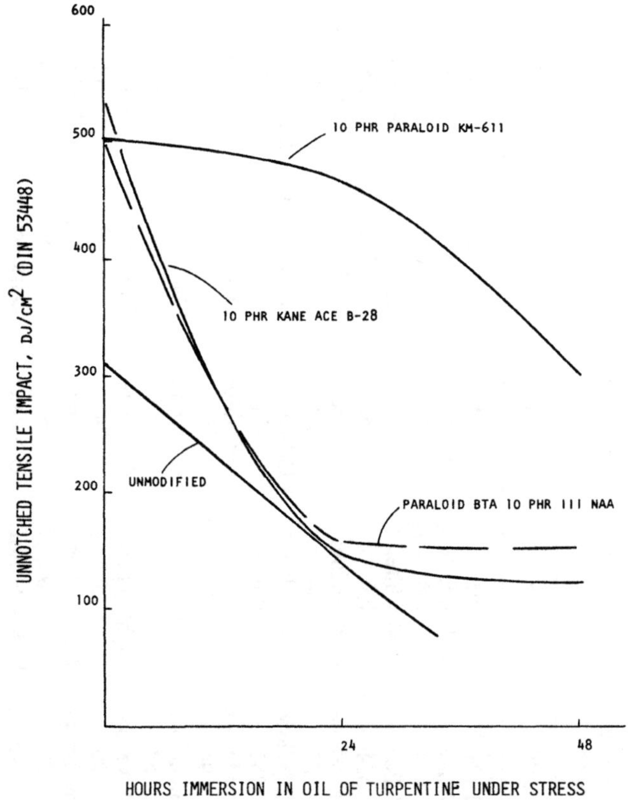

Fig. 6.38. Stress craze resistance of PVC bottle compounds in oil of turpentine.

Florida or Puerto Rico) and temperate (e.g. Philadelphia, Pennsylvania) exposure results are similar to each other and are less severe. Local conditions, such as pollutants, can also contribute to degradation which can be mistakenly included in true weatherability. The curves in Fig. 6.39 show the retention of impact strength for a typical PVC siding formulation containing various classes of impact modifier after exposure in Florida. Although the percentage decrease in impact strength over the 5-year exposure period is substantial even for the (ACR) modifier, the asymptotic shape of the ACR and CPE curves from about 3- and 4-years' exposure, respectively, indicates residual impact strength over a long exposure period. The Izod impact value of about 260 J/m for the ACR compound describes it as a 'tough' compound even after 5 years. The value of about 180 J/m for CPE suggests that it is a marginal—but definitely tougher than

Table 6.16 Weatherability of white PVC: effect of modifier type and exposure period on impact strength for two sites

Exposure period	*Arizona*			*Florida*		
	ACR	*MBS*	*CPE*	*ACR*	*MBS*	*CPE*
Original izod (J/m)	854	1 121	91	694	1 121	91
Original colour	White	White	White	White	White	White
3 months	1 068	908	100	954	641	100
	White	Yellow	Off-white	White	Dark yellow	Off-white
6 months	908	534	100	748	374	100
	Slightly yellow	Dark yellow	Dark yellow	White	Dark yellow	Light yellow
12 months	801	85	100	641	100	100
	Slightly yellow	Dark yellow	Dark yellow	Off-white	Dark yellow	Light yellow
18 months	801	251	100	801	100	100
	Slightly yellow	Yellow-brown	Yellow	Off-white	Dark yellow	Yellow
24 months	641	96	100	587	75	100
	Slightly yellow	Yellow-brown	Yellow	Off-white	Dark yellow	Yellow
36 months	352	64	100	694	64	100
	Slightly yellow	Dark yellow	Light yellow	Off-white	Dark yellow	Yellow
48 months	443	80	100	747	59	100
	Slightly yellow	Yellow-brown	Light yellow	Off-white	Dark yellow	Yellow
60 months	518	80	100	694	54	100
	Slightly yellow	Brown	Yellow-brown	Off-white	Dark yellow	Yellow

Formulation: 90 PVC ($K = 92$), 10 modifier, 4 Ba/Cd/organic stabiliser, 0·5 organic inhibitor, 2·0 Advawax 140, 4·5 TiO_2, 0·001 8 Irisol N.

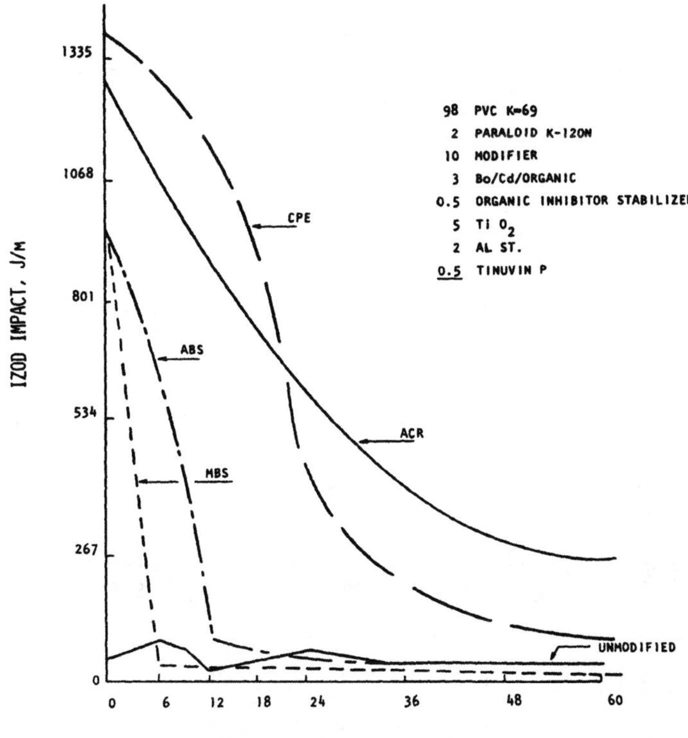

Fig. 6.39. Weatherability of white PVC: effect of modifier type.

unmodified PVC—compound after 5 years. Comparisons of results in Arizona and Florida among PVC siding-type formulations containing ACR, MBS and CPE impact modifiers are presented in Table 6.16. The retention of impact values for the ACR modified compound in this study was greater than that in Fig. 6.39, probably due to the use of a higher stabiliser level here. The importance of properly stabilising the PVC or shielding the PVC from the ultraviolet irradiation by means of high opacity pigmentation (e.g. with TiO_2) cannot be over-emphasised. Because failures due to trauma of the PVC product after installation on the building are not overwhelming, retention of colour and resistance to chalking or surface erosion become important due to aesthetics—and such changes are readily perceived by a layperson. The exterior durability of acrylic resin-based paints is legendary, and the same properties carry over to the all-acrylic impact modifiers. The all-acrylics will not prevent colour and surface

Table 6.17 Physical performance on extruded window profiles (Ba/Cd-stabilised). Extrusion production trials in a Reifenhauser twin-screw contra-rotating extruder BT2-80-16-V

Physical property	*Compound*		
	B1 (*10% CPE*)	*B2* (*12 phr Paraloid KM323B*)	*B3* (*8% EVA*)
Tensile strength at yield (kP/cm^2)	461	500	453
Shore hardness (DIN 53505)	79	81	79·5
Vicat (°C)			
(DIN 53460), 5 kg/air, 50 °C/h	78	80	77
ASTM D 1525, 5 kg/oil, 120 °C/h—0·25 mm	78	80	77
Shrinkage (%)			
(100 °C/h)	2·2	1·6	2·6
Accelerated weathering, 0·8 MW s/cm^2 (SKZ—Würzburg—Germany)			
V-V notched imp. (mJ/mm^2) (DIN 53753)			
before exposure	60·9	55·8	60·0
std deviation	4·0	3·2	4·9
after exposure	52·4	55·5	54·0
std deviation	9·9	2·9	4·9
Weathering resistance (DIN 54001) (5 = no change, 4 = slight change, 3 = noticeable change)	4/5	5	3/4

changes in PVC compounds, but among 'weatherable' impact modifiers they have demonstrated the highest resistance to those changes.

Accelerated weathering devices vary in theory and practice of operation and therefore generate different sets of data. Each of the devices is useful in providing a relative performance profile that can be used to screen out poorly performing compounds; and with pragmatic experience, some correlation can be made to predict long-term performance in an actual outdoor weathering situation. Table 6.17 contains comparisons of accelerated weathering and physical properties measured on commercial-type extruded window profiles. The superiority of the ACR modifiers (in this case, Paraloid KM323B) in weatherability and other physical properties important to profile applications is apparent.

6.3.2 Impact modifiers in plasticised PVC

Because plasticised PVC can 'get along' without impact modifiers, its utility has generally been neglected in flexible compounds. The impact modifiers can improve: (1) low temperature toughness, (2) physicals at high filler levels, and (3) drape and texture.

The use of impact modifiers allows the formulator to obtain two diametrically opposed properties, namely permanence and low temperature impact resistance (T_B) or low temperature flexibility (T_f). The data in Table 6.18 illustrate the enigma and its solution. Adding linear diester plasticisers to a polymeric plasticised PVC compound does result in significant improvement in T_f values; however, the monomeric plasticisers readily migrate into the adhesive mass of a pressure-sensitive tape to cause deterioration of adhesion. Adding an impact modifier to the polyester plasticised PVC also results in a significant improvement in T_f, but it does not significantly detract from adhesion.

Impact modifier performance varies somewhat with the type and

Table 6.18 Low temperature flexibility v. migration resistance

P(VC/VA)	100	100	75
MBS	—	—	25
Polyester plasticiser	50	35	50
DIDA	—	15	—
Shore 'C' hardness (10 s)	62	55	51
T_f (°C/°F)	+1/34	−8/19	−10/14
Peel strength (g/in) after one week, 140 °F			
against PVC	530	255	450
against steel	660	240	645

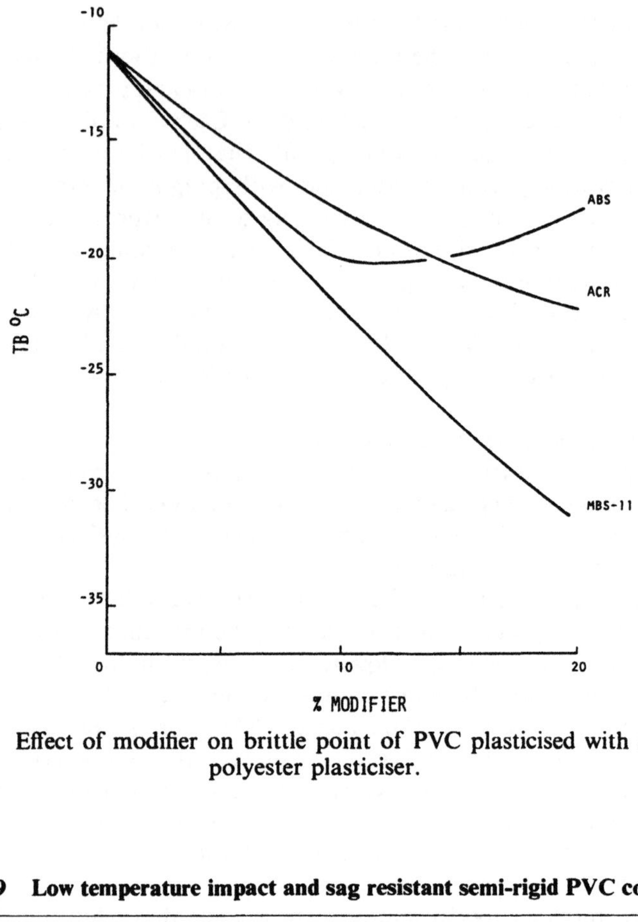

Fig. 6.40 Effect of modifier on brittle point of PVC plasticised with 50 phr of polyester plasticiser.

Table 6.19 Low temperature impact and sag resistant semi-rigid PVC compounds

% Modifier	0	20	50	50
phr XL-51[a]	54	40	25	50
T_B (°F)	−40	−67	−31	−57
Sag index[b] (× 10^{-4})	1 250	8·9	7·5	4·9
Flexural modulus (psi)				
160 °F	955	845	1 235	660
73 °F	2 255	12 090	41 830	19 545
−20 °F	60 000	96 100	97 500	67 600
% Set.	0	11	7·5	2·5

[a] Proprietary mixture of alkyl adipate plasticiser and epoxidised soybean oil plasticiser.

[b] Sag index = inches of sag/°C/s to sag 3 in.

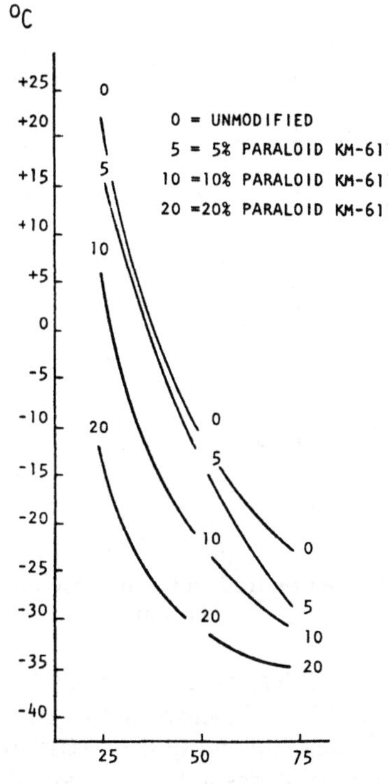

Fig. 6.41 Brittle point.

concentration of both plasticiser and modifier. Among the types of impact modifiers, the MBS modifiers are the most efficient (Fig. 6.40). The impact modifiers are most effective in reducing brittle point of low plasticiser content compounds and are thus most useful when they are needed most (Fig. 6.41).

Relatively rigid and heat sag-resistant PVC compounds having excellent low temperature impact resistance can be made by using relatively low plasticiser levels and MBS or ACR impact modifiers (Table 6.19).[43] If excellent retention of low temperature impact resistance after exposure to UV light is required, an ACR impact modifier would be used. Combinations of properties such as these suggest the use of impact-modified semi-rigids for automotive applications.

The MBS impact modifiers are the most effective in semi-rigid PVC

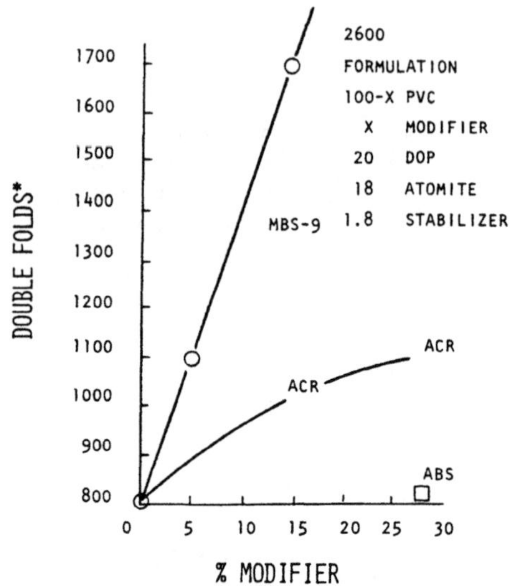

Fig. 6.42 Effect of modifier on fold endurance. *MIT fold endurance tester, 3 kg tension.

compounds containing filler, where fold endurance is a measure of toughness. The data in Fig. 6.42 show that essentially linear improvement is obtained by using increasing amounts of MBS impact modifier. Five per cent MBS, based on total resin, is dramatically more effective than 20–30 parts ABS. ACR impact modifiers are less effective than MBS, but more effective than ABS for imparting toughness in these systems. MBS impact modified plasticised PVC will permit the use of very high filler levels without serious deterioration of tensile properties (Table 6.20). At the 30 %

Table 6.20 Effects of filler level, MBS v. ABS-modified plasticised PVC

PVC	70	70	70	70	70	70
MBS	30	30	30	30	—	—
ABS	—	—	—	—	30	30
Plasticiser	65	65	65	65	60	60
Atomite	0	25	50	75	0	50
TiO_2	5	5	5	5	5	5
Ba/Cd laurate	2	2	2	2	2	2
Shore 'A' hardness (15 s)	73	76	76	76	76	77
Tensile strength (psi)	1 730	1 770	1 640	1 500	1 520	1 060
Elongation (%)	375	370	370	355	280	250

modifier level, the MBS modified PVC containing 75 phr filler is better than the unfilled ABS modified PVC for tensile strength and elongation; hardnesses are equivalent.

6.4 SPECIALITY ADDITIVES

The production of rigid (unplasticised) PVC foam structures requires a high hot melt strength, good hot metal release and good impact strength in the finished article. As indicated earlier, the acrylic processing aids provide the hot strength necessary to retain the integrity of the cell wall as the gas expands to make the foam. Paraloid K125 is especially effective in controlled expansion PVC processes like the Celuka® process. In free-blown foams, however, the integrity of the skin is maintained by good release from the die as well as good hot melt strength. For good release, Paraloid K175 is an excellent choice because it gives excellent release from hot metal surfaces without causing plate-out or interfering with physical properties or appearance. In the finished article, an impact modifier is required because part thickness that would be tough for a solid PVC is actually composed of many very thin-walled cells that are easily ruptured. Thus a combination of acrylic processing aid, lubricant and impact modifier is required for optimum performance. A multi-purpose polymer has been developed (Paraloid KM318F) that provides each of the three functions—and does the job more efficiently. Extruded profiles showing the improvement in surface integrity, reduction in density and increase in impact strength imparted by Paraloid KM318F compared to the same properties obtained through the addition of individual components are illustrated in Fig. 6.43. The practical importance of the toughness imparted by Paraloid KM318F is presented in Fig. 6.44. Although component packages containing impact modifiers are effective, the Paraloid KM318F is even more effective as evidenced by the lack of splitting even when a 10-d nail is driven through it.

Because Paraloid KM318F is all-acrylic, it has excellent weathering properties, allowing the production of foamed PVC articles for outdoor applications. Four-inch diameter foam pipe has been successfully extruded on conventional extruders with normal sizing equipment. The pipe is tough enough to pass a crush-test. Although no standards are currently available for expanded PVC, many 'non-scheduled' applications appear to be practical.

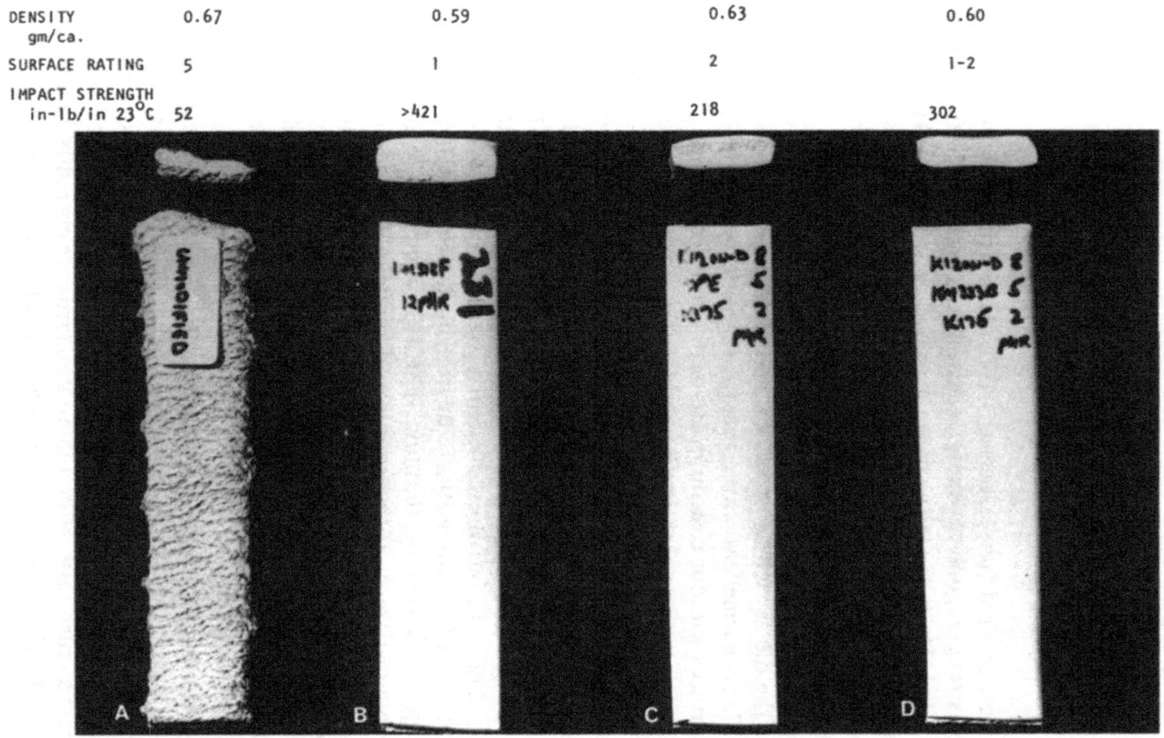

Fig. 6.43 PVC foam profiles. Density (g/cm^3), surface rating (scale: 1 = smooth/glossy to 5 = rough/pitted) and rolling dart impact strength values are given above each sample. Formulation: 100 PVC ($K = 61$), X modifier, 1·5 Advastab TM-181, 0·5 barium stearate, 0·25 stearic acid, 1·0 Celogen AZ130, 5·0 TiO_2. A, unmodified. B, KM318F 12 phr. C, K120ND 8 phr, CPE 5 phr, K175 2 phr. D, K120ND 8 phr, KM323B 5 phr, K175 2 phr.

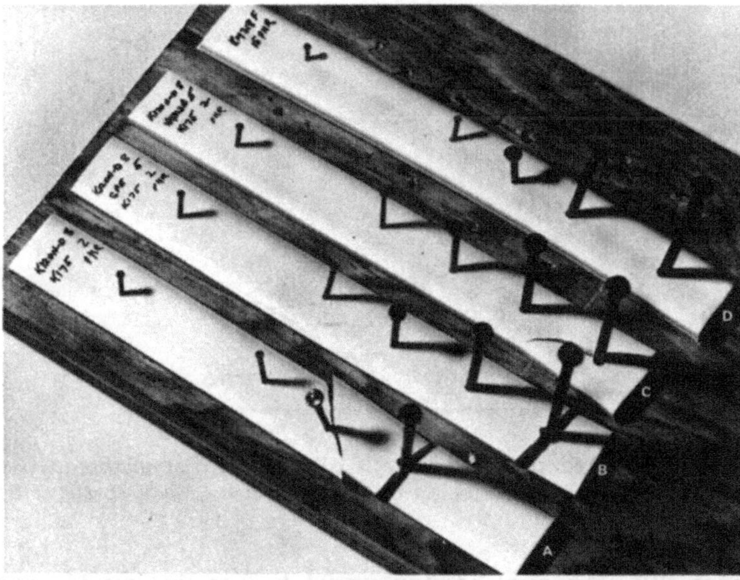

Fig. 6.44 Effect of processing aid/impact modifier systems on the impact strength of rigid PVC foam extrudates. Formulation—see Fig. 6.43. A, K120ND 8 phr, K175 2 phr. B, K120ND 8 phr, CPE 5 phr, K175 2 phr. C, K120ND, 8 phr, KM323B 5 phr, K175 2 phr. D, KM318F 15 phr.

6.5 ADDITIVES FOR ADJUSTING GLOSS (FLATTING AGENTS)[44,45]

The adjustment of surface gloss can be a matter of aesthetic or functional criteria. High-gloss surfaces generally resist staining and soiling, but mar readily and often cause blocking (sticking of one polished surface to another). Adjustment of gloss has traditionally been achieved by embossing the film and/or adding inorganic flatting agents (fillers) that protrude through the surface to scatter reflected light. The combination of embossing roll and flatting agents appears at first glance to be the ideal way to control the consistency of gloss during long production runs. However, the abrasiveness of the inorganic flatting agents makes this process self-defeating due to the polishing action of the metal embossing roll.

Due to their mechanism of light scattering by protrusion through the surface, inorganic flatting agents readily absorb soils and stains, and further act as defects which cause deterioration in notch-sensitive plastics

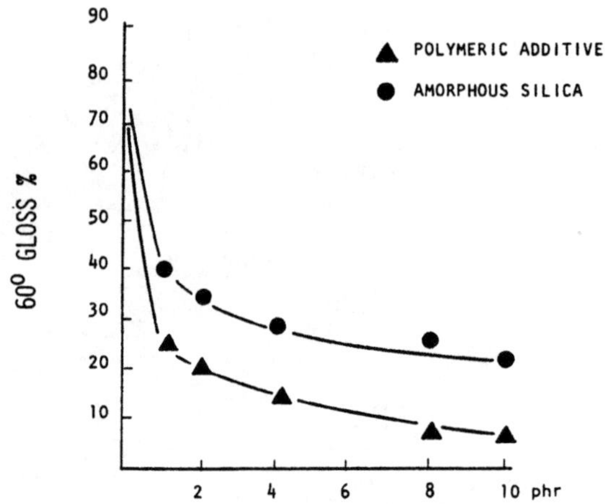

Fig. 6.45 Gloss reduction efficiency, flexible PVC. Formulation: 100 PVC ($K = 69$), 50 DOP, 2·0 tin stabiliser, 0·25 stearic acid, 2 black paste, X flatting agent.

such as PVC. An ideal additive would be one that would impart a 'mechanically embossed' surface without detracting from physical properties or soil resistance. This additive is an acrylic polymer, Paraloid KF710.

The gloss-reducing properties of the polymeric additive are greater than those of an amorphous silica popularly used as a flatting agent (Fig. 6.45). Performances are similar in plasticised or rigid PVC.

Because the polymeric additive is all-acrylic, the surface of the PVC compound has an acrylic-rich 'layer' which presents a homogeneous surface unlike the 'broken' surface due to inorganic flatting agents. Therefore, the impact strength of rigid PVC (modified with 12 phr MBS impact modifier) is not seriously impaired even when 10 phr of Paraloid KF710 is used. In contrast, 1 phr of amorphous silica causes a catastrophic loss in impact, rendering the compound brittle (Fig. 6.46). Similar effects are seen in low temperature impact tests (brittle point) on plasticised PVC.

Retention of embossing is also an important consideration. Flatted sheet is frequently thermoformed, and the gloss increases as the thermoplastic sheet relaxes due to plastic memory. With the acrylic additive, gloss appears to increase as the sheet is heated, but when the sheet cools the controlled-pattern surface shrinkage induced by the acrylic polymer once again takes place and the original dullness is restored (Fig. 6.47). Mechanically embossed or inorganically flatted sheets 'gloss-up' rapidly

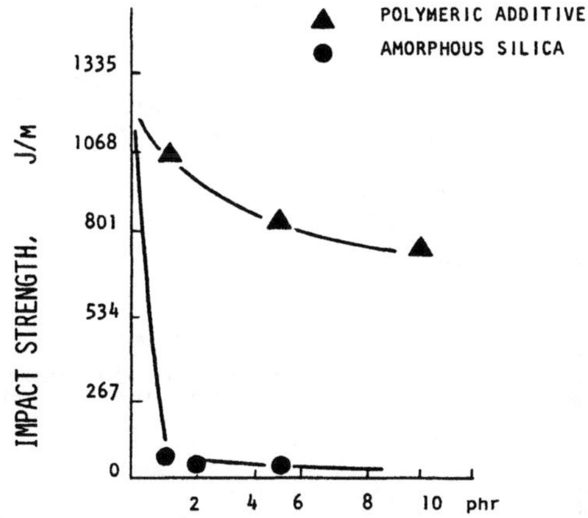

Fig. 6.46 Effect of flatting agents on impact retention (formulation A). Formulation: 100 PVC ($K = 69$), 2·0 tin stabiliser, 12 MBS I.N., 3·0 ACR P.A., 1·5 lubes, X flatting agent.

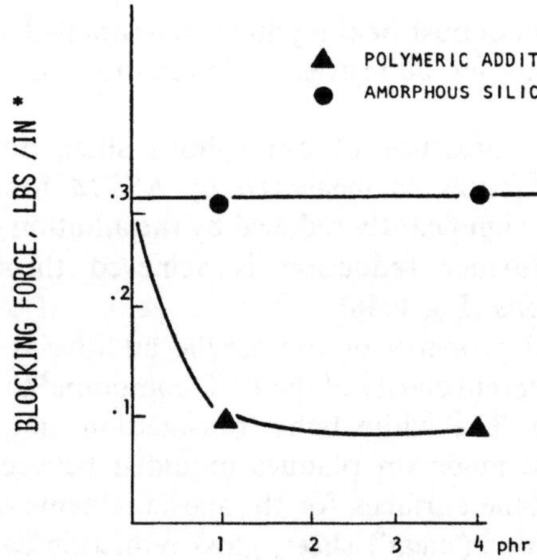

Fig. 6.47 Anti-blocking efficiency, 170 °F, formulation C. Formulation: 100 PVC ($K = 69$), 50 DOP, 2·0 tin stabiliser, 0·25 stearic acid, X flatting agent. * ASTMD-1893-67.

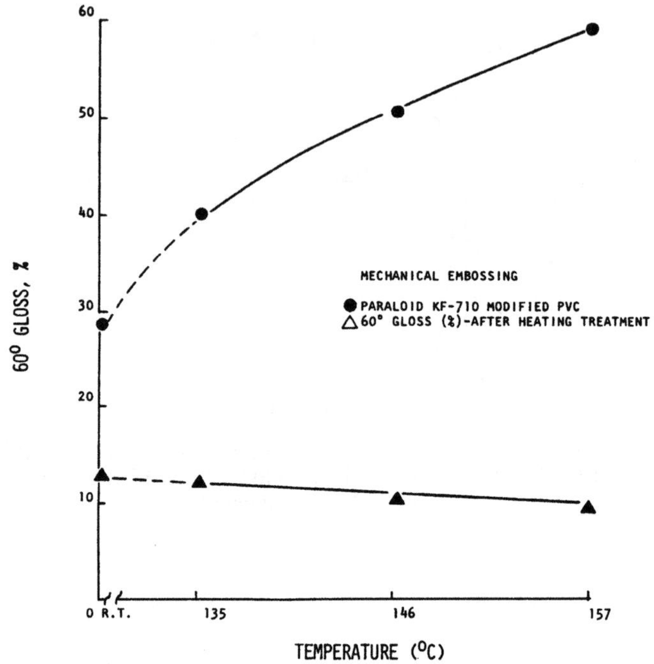

Fig. 6.48 Effect of elevated temperatures on gloss.

and limit the amount of post-heating that can be applied to thermoform the sheet, to use thermoset adhesives in lamination or to provide other treatments.

Surprisingly, the presence of amorphous silica does not assist in reducing blocking forces as measured by ASTM Method D-1893-67. Blocking forces are significantly reduced by the addition of 1 phr Paraloid KF710, but no further reduction is achieved through the use of higher concentrations (Fig. 6.48).

Another unusual property of the acrylic additive is that it does not detract from the inherent clarity of the PVC compound to which it has been added (Table 6.21). The white light transmission and percentage haze measurements were made on plaques moulded between chrome-plated cauls to provide plane surfaces for the measurements. In calendered or extruded unpigmented ('clear') sheet, gloss reduction can make even the acrylic modified sheet appear to be hazy. However, when the sheet is in intimate contact with a substrate, excellent 'contact-clarity' is observed. With silica, as indicated by 98–100% haze, the films are effectively

Table 6.21 Effect of Acryloid KF710 on clarity

Gloss control agent	Rigid		Plasticised	
	% WL^a	% $Haze^a$	% WL^b	% $Haze^b$
None	79·0	9·9	89·4	5·9
Paraloid KF710	83·0	6·3	90·0	15·7
Silica	56·2	100·0	83·5	98·0

[a] Formulation: 85 PVC ($K = 62$), 3 K120N, 12 KM641, 2·25 Advastab TM-180, 0·2 Aldo MS, 0 OP wax, 2 gloss control agent.
[b] Formulation: 100 PVC ($K = 69$), 50 plasticiser, 2 Advastab TM-180, 0·25 stearic acid, 0 or 2 gloss control agent.

pigmented–opaque. With the acrylic additive, high-fidelity second-surface printing is feasible, as is production of decorative laminates and non-glare coverings for documents, etc.

ACKNOWLEDGEMENTS

Special acknowledgement is due to I. S. Rabinovic (Rohm and Haas Co., Valbonne, France) and Robert P. Petrich (Rohm and Haas Co., Bristol, Pennsylvania, USA) for their major contributions to this chapter.

REFERENCES

1. Fleischer, D., Brandrup, J. and Heinzmann, O. *Künststoffe*, **67** (1977), 312.
2. Terselius, B. and Ranby, B. *Pure Appl. Chem.*, **53** (1981), 421.
3. Berens, A. R. and Folt, V. L. *Trans. Soc. Rheol.*, **11** (1967), 95.
4. Collins, E. A. and Krier, C. A. *Trans. Soc. Rheol.*, **11** (1967), 225.
5. Hori, Y. *Japan Plast.*, **3** (1969), 48.
6. Shinagawa, Y. *Plast. Ind. News*, May (1973), 65.
7. Faulkner, P. G. *J. Macromol. Sci. Phys.*, **B11** (1975), 961.
8. Summers, J. W., Isner, J. D. and Rabinovitch, E. B. *36th SPE ANTEC*, 1978, p. 757.
9. Gouinlock, E. V. *J. Polym. Sci., Polym. Phys.*, **13** (1975), 961.
10. Juijn, J. A., Gisolf, J. H. and DeJong, W. A. *Colloid Polym. Sci.*, **251** (1973), 456.
11. Illers, K. H. *Macromol. Chem.*, **127** (1969), 1.
12. Gonze, A. *Plastica*, **24** (1971), 49.

13. Lambertz, P. *Plast. Mod. Elast.*, **26** (1974), 82.
14. Berens, A. and Folt, V. *Polym. Eng. Sci.*, **8** (1968), 5.
15. Menges, G. and Berndsten, N. *Künststoffe*, **66** (1976), 735.
16. Menges, G. *et al. Künststoffe*, **69** (1979), 562.
17. Rabinovitch, E. and Summers, J. *J. Vinyl Technol.*, **2**(3) (1980), 165.
18. US Patent 2 646 417 (1953) B. F. Goodrich Co.
19. US Patent 2 791 600 (1957) B. F. Goodrich Co.
20. US Patent 3 373 229 (1965) ICI plc.
21. US Patent 3 764 638 (1973) Stauffer Chem. Co.
22. US Patent 3 833 686 (1974) Rohm and Haas Co.
23. British Patent 1 378 434 (1974) Kanegafuchi.
24. US Patent 3 673 283 (1972) Japanese Geon.
25. US Patent 3 874 740 (1974) Tenneco.
26. Wilson, A. and Raimondi, V. *Reprints of PRI Conf. on Proc. of PVC*, London, April 1978.
27. Bystedt, J. and Lundquist, J. *J. Vinyl Technol.*, **2** (1980), 209.
28. Röhrl, E. *Künststoffe*, **70** (1980), 41.
29. Menges, G. *et al. Plast. Rubb. Process.* (December 1979), 156.
30. Haaf, K. *et al. Angew Makromol. Chem.*, **58** (29) (1977), 95–119 (NR853).
31. Faulkner, P. G. *J. Macromol. Sci. Phys.*, **B-20** (1981), 319.
32. Rabinovitch, E. B. *J. Vinyl Technol.*, **4** (1982), 62.
33. Krzewki, R. J. and Collins, E. A. *J. Vinyl Technol.*, **3** (1981), 116.
34. Brown, R. *J. Vinyl Technol.*, **3** (1981), 125 *et seq.*
35. Lutz, J. T., Jr. *Plast. Des. Process.* (September 1971).
36. Merz, E. H., Cleaver, G. C. and Baer, M. *J. Polym. Sci.*, **22** (1956), 325.
37. Newman, S. and Strella, S. *J. Appl. Polym. Sci.*, **9** (1965), 2297.
38. Schmitt, J. A. and Keskkula, H. *J. Appl. Polym. Sci.*, **3** (1960), 132.
39. Strella, S. *J. Polym. Sci.*, **4** (1966), 527.
40. Berry, J. P. *Fracture Processes in Polymeric Solids*, B. Rosen (Ed.), Wiley, New York, 1964, pp. 157–195.
41. Bucknall, C. B. and Smith, R. R. *Polymer*, **6** (1965), 437.
42. Rohm and Haas Co. *Case Study No. 1 and No. 2—Paraloid Modifiers in Pipes, Bulletins PL 210/M/1078 and PL 211/M/1078.*
43. Lutz, J. T., Jr. *Plast. Eng.* (September 1974), 40.
44. Rabinovic, I. S. *A Polymeric Flatting Agent for Control of Surface Gloss in PVC*, ACS, Atlantic City, 1974.
45. Rabinovic, I. S. *32nd SPE ANTEC*, San Francisco, California, 1974, 522 *et seq.*

INDEX

GPSR Compliance

The European Union's (EU) General Product Safety Regulation (GPSR) is a set of rules that requires consumer products to be safe and our obligations to ensure this.

If you have any concerns about our products, you can contact us on ProductSafety@springernature.com

In case Publisher is established outside the EU, the EU authorized representative is:

Springer Nature Customer Service Center GmbH
Europaplatz 3
69115 Heidelberg, Germany

Batch number: 10607212

Printed by Printforce, the Netherlands